GEOLOGY
AND
ENGINEERING

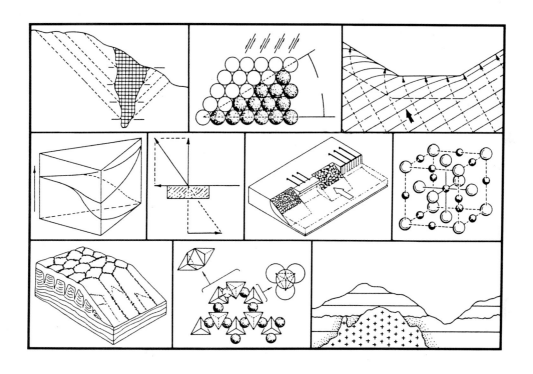

GEOLOGY AND ENGINEERING

William H. Dennen
University of Kentucky

Bruce R. Moore
University of Kentucky

wcb

Wm. C. Brown Publishers
Dubuque, Iowa

BOOK TEAM

Edward G. Jaffe *Executive Editor*
Lynne M. Meyers *Associate Developmental Editor*
Natalie Gould *Production Editor*
Mark E. Christianson *Designer*
Jeanne Rhomberg *Design Layout Assistant*
Carla D. Arnold *Permissions Editor*
Mary M. Heller *Picture Editor*

Part Opening Photos by W. H. Dennen

Printed in the United States of America
10 9 8 7 6 5 4 3 2 1

wcb group

Wm. C. Brown *Chairman of the Board*
Mark C. Falb *President and Chief Executive Officer*

wcb

WM. C. BROWN PUBLISHERS, COLLEGE DIVISION

Lawrence E. Cremer *President*
James L. Romig *Vice-President, Product Development*
David A. Corona *Vice-President, Production and Design*
E. F. Jogerst *Vice-President, Cost Analyst*
Bob McLaughlin *National Sales Manager*
Catherine M. Faduska *Director of Marketing Services*
Craig S. Marty *Director of Marketing Research*
Eugenia M. Collins *Production Editorial Manager*
Marilyn A. Phelps *Manager of Design*

The cover images are illustrations from the interior of this book. Top row (left to right)—figure 18.2, figure 11.2, figure 7.15. Middle row (left to right)—figure 4.6, figure 11.3, figure 7.23, figure 1.3d. Bottom row (left to right)—figure 6.22, figure 1.2, figure 3.4b.

For our grandchildren, who must live in the world you make for them

CONTENTS

PART I
Geologic Considerations

1
Minerals 3

2
Rocks 21

3

Geologic Time, Weathering, and the Rock Cycle   41

4

Soils   61

5

The Hydrologic Cycle and Sediment Transport   79

6

Groundwater and Frost   99

7

Oceans and Shorelines   125

8
Mechanical Properties of Rocks
151

9
Geologic Structures and the Geometrical Description of Deformed Rocks 171

PART II
Engineering Considerations

10
Laboratory and Field Description of Soil and Sediment 195

11
Mechanics of Unconsolidated Materials 211

17
Industrial Rocks and Minerals 347

18
Mining 363

TABLES

PREFACE

Engineering geology involves the integration of two distinct bodies of knowledge—the facts and principles of physical geology and certain aspects of civil engineering practise and design. Unfortunately, the number of definitions of engineering geology is nearly as great as the number of practitioners, but the following distillation of views expressed by participants in a recent workshop is representative:

> Engineering Geology is an interdisciplinary field in which pertinent studies in geology and other geoscience areas are applied toward the solution of problems involved in engineering works and resource uses.[1]

Engineering has been described as the art of providing safe, economical, and workable solutions to real problems; and it is taken for granted that a competent engineer can design and construct adequate works to meet any requirement. The skill of the engineer in this art, however, is in large part predicated upon an understanding of site conditions;

1. N. Sitar, et al. 1983. *Goals for Basic Research in Engineering Geology.* Dep't. Civil Engineering, Univ. California, Berkeley.

and failures, although rare, can usually be traced to incomplete knowledge of the problem faced. Many engineering problems and their solutions are rooted in the nature and variability of natural materials and in the ways these materials respond to natural forces. Geology, which addresses these materials and phenomena, makes important contributions to this understanding.

This book is intended for use in a one-semester introductory course for civil and mining engineering majors. It is assumed that these students will have no background in geology, so the pertinent aspects of this discipline are presented in order that the nature and setting of geology-related engineering problems is made clear. On the other hand, engineering students will take intensive course work in such fields as geotechnics, hydrology and hydraulics, rock mechanics and foundations, so although these topics are introduced they are not belabored.

It is felt that the approach to the subject matter is at least as important as the material itself, and the attempt is made to bring out underlying principles and underscore the interrelated nature and properties of geologic materials and their engineering applications. The goal is not to make an engineer into a professional geologist but to provide an insight into the geologist's fields of competence. Particularly, it is essential that the engineer have sufficient grasp of geologic materials, processes, and hazards to know when further geologic information is needed for the design or implementation of a project.

A traditional organization of subject matter providing for a logical arrangement of topics has been developed in both engineering and geology. These traditional approaches are necessarily different, and in our view, neither provides the best means of interconnection between the disciplines. In consequence, we have taken liberties in the organization of this text in order to highlight related material. Specifically, the first section of the text emphasizes geologic materials and processes and the second section covers topics of particular concern in civil and mining engineering practise.

Special features of the text include the liberal use of line diagrams integrated with the written material, illustrative photographs, and the inclusion of short case histories to highlight important points. A glossary is provided for ready explanation of the many technical terms that must be employed. Although mathematical expressions are necessary to describe many important relationships, the mathematics used has been held to a precalculus level.

Acknowledgments

We appreciated the comments and suggestions of the following reviewers:

Robert A. Matthews   *University of California, Davis*

Lyle V. A. Sendlein   *University of Kentucky*

Ian Hutcheon   *University of Calgary*

Robert G. Font   *Conoco*

PART I
Geologic Considerations

Engineering projects have an intimate connection with the earth, so the engineer must deal directly with soils, rocks, and water and be aware of the geologic processes that produce and modify them. Geologic materials may have to be built upon, excavated, tunneled, or drained, or may become the engineer's aggregate, foundation, or land fill. Geologic processes make, modify, and move these materials while shaping the earth's surface and can cause engineering problems or hazards to be overcome. An understanding of the complex interplay of processes and materials both on and within the earth is thus essential in order to grasp the setting in which engineering activities are carried on.

Since the engineer must deal with many different kinds of rocks and soils as well as surface and subsurface water under widely different climatic and topographic conditions, it is essential that he or she have an understanding of the basic properties of these materials in order to anticipate their engineering aspects wherever they may be encountered. The properties of rocks and soils derive from the kind of mineral particles that compose them and the manner in which they are aggregated; hard grains make hard and brittle rocks, connected pore space allows water flow within a rock, soluble materials may be expected to dissolve, and plastic substances will flow under sufficient load.

Minerals are the primary building blocks of rocks, which in turn are the source of a wide range of natural resources including building materials (stone, aggregate, cement, lime, gypsum), pigments, refractories, fillers, abrasives, and many others. Rocks also include the solid fossil fuels (lignite, coal) formed by the consolidation of vegetable debris and are often the host for both solid and fluid materials of value. The contained fluids include petroleum, natural gas, and groundwater; the solids may include the ores of metals and nonmetals extracted and produced by the mining industry. The winning of these raw materials involves close coordination of geologic activities and engineering practice.

Human activities on the earth are restricted to its outer crustal portion—the deepest penetration is about 9 000m by boring and 3 000m by mining—but the surface materials have a more or less close geologic connection to materials and processes of the earth's interior. The internal structure of the earth is not usually of direct concern in engineering practice, however, and will only be mentioned briefly in this text.

The present knowledge of the interior of the earth has been largely derived from studies of volcanic materials extruded from within the earth and from the interpretation of earthquake waves passing through the earth. The earth has apparently solidified from molten or quasi-molten material and has a general layered structure. The crust is composed of the thicker continental rocks, most frequently encountered in engineering, and of the thinner ocean floor materials. The continental rocks are dominantly granitic material and sedimentary rock while the oceanic crust is typically heavier basaltic rock. These rocks will be discussed in more detail later. The mantle underlying the crust occupies 80% of the earth's volume and is made up of magnesium-and iron-rich silicate minerals. Earthquake data indicate the core to be mostly liquid with a solid central zone. The core material is probably iron with small amounts of either sulfur or silicon.

Because the earth is geologically complex, not only in its interior but also on its surface, wide variability in the kinds and properties of the rocks and soils of engineering concern must be anticipated. It is the aim of the authors to develop basic concepts that allow these differences to be understood from first principles and to integrate the pertinent knowledge of geologic materials, properties, processes, and techniques with the needs and practices of civil engineering.

A note on scale is pertinent before beginning serious study. The geologist typically works with rock relationships over large areas from the basic data presented in quadrangle maps covering 15 minutes, 30 minutes, or 4 degrees of latitude and longitude; state maps; or regional maps. The engineer, on the other hand, is usually concerned with geologic phenomena on a more local scale suitable to the engineering work. Because of this, the geologic literature will often serve as a framework into which the engineering geology is placed, but will seldom replace detailed on-site studies.

1

Minerals

INTRODUCTION

Almost all solid geologic materials are made up of aggregates of **minerals,** which are naturally occurring, inorganic, solid substances having a definite crystal structure, specific physical properties, and a composition that is fixed or varies within fixed limits. There are about 2 000 known mineral *species,* but only about 100 minerals are common and perhaps 20 truly abundant; fewer than 10 make up over 99% of all rocks. These numbers are perhaps surprisingly small in view of the potential permutations of the 90-odd chemical **elements,** but are readily explained by the fact that most elements are rare in the earth's **crust.** Permissible arrangements of atoms are severely limited by geometric and atomic bonding considerations, and physicochemical conditions greatly limit the possible compounds that may be formed in nature.

The average crustal composition of the earth is given in table 1.1 where it may be seen that oxygen is absolutely dominant in terms of both weight and volume and silicon and aluminum are the only electropositive elements present at greater than 10 weight %. More than 99 weight % of the earth's outer

Table 1.1 Average Crustal Composition of the Earth

Element	Chemical Symbol	Weight %	Subtotal	Volume %	Subtotal
Oxygen	O	46.60		93.77	
Silicon	Si	27.72	74.32	.86	94.63
Aluminum	Al	8.13	82.45	.47	95.10
Iron	Fe	5.00	87.45	.43	95.53
Calcium	Ca	3.63	91.08	1.03	96.56
Sodium	Na	2.83	93.91	1.32	97.88
Potassium	K	2.59	96.50	1.83	99.71
Magnesium	Mg	2.09	98.59	.29	100.00

solid portion is made up of only eight chemical elements. From this it may be predicted that the more common minerals will be oxygen-bearing compounds of silicon, aluminum, iron, magnesium, calcium, sodium, and potassium. Geologic processes, later described, can affect significant local changes in the composition of the crust and produce **aggregates** of minerals called *rocks* whose chemical makeup is markedly different from the average for the crust.

MINERAL STRUCTURES AND PROPERTIES

The common minerals are built up of infinitely repeating electropositive and electronegative components. These may be singly charged atoms (ions) or charged atomic groups (radicals), which combine in a regular and symmetrical manner. Every mineral is composed of atoms of definite kind and proportion as shown by its formula, and these atoms are arranged in a definite way in space. Since every mineral has both a definite chemical composition and a crystalline structure, these aspects of the substance are characteristic of each particular species and both the nature and the arrangement of the component atoms lead to distinctive chemical, physical, and geometrical properties useful for mineral identification and utilization. The properties are related to the arrangement and the types of packing of the atoms and to the strengths of the chemical bonds between the individual atoms.

The possible regular arrangements of points in three dimensions are sharply limited by geometrical considerations. Regular arrays of imaginary points will lie on the corners of identical imaginary parallelepipeds or cells (fig. 1.1). In turn, these cells stack to fill all space and their corners define a coordinate system or lattice to which the atoms of the structure are related in a regular way. The term *structure* properly refers to the actual matter and its arrangement in a mineral. The term *lattice,* which refers to a regular array of imaginary points to which atom positions are geometrically related, is often and improperly used in the context of structure.

There are only six basic cells that may be employed to generate a space-filling lattice: triclinic, monoclinic, orthorhombic, tetragonal, hexagonal, isometric (table 1.2).

PHYSICAL PROPERTIES

The physical properties of minerals are important to both their identification and use. Those of greatest importance to an engineer are hardness, manner of breaking, specific gravity, and stability (reactivity with common gases, liquids, or temperature changes). Properties of lesser importance in engineering applications include color; thermal, electrical, and magnetic properties; habit; density; and transparency.

Figure 1.1 Orientation and labeling of crystallographic axes

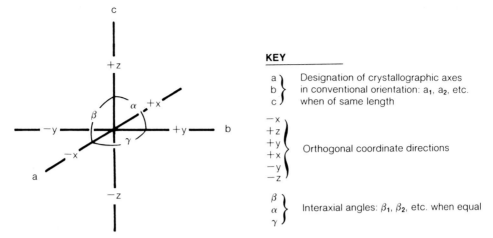

KEY

$\left. \begin{array}{l} a \\ b \\ c \end{array} \right\}$ Designation of crystallographic axes in conventional orientation: a_1, a_2, etc. when of same length

$\left. \begin{array}{l} -x \\ +z \\ +y \\ +x \\ -y \\ -z \end{array} \right\}$ Orthogonal coordinate directions

$\left. \begin{array}{l} \beta \\ \alpha \\ \gamma \end{array} \right\}$ Interaxial angles: β_1, β_2, etc. when equal

Table 1.2 The Crystal Systems

Name	Description	Coordinate System	Cell Types
Triclinic	$a \neq b \neq c$ $\alpha \neq \beta \neq \gamma$		
Monoclinic	$a \neq b \neq c$ $\beta = \gamma = 90°$		
Orthorhombic	$a \neq b \neq c$ $\alpha = \beta = \gamma = 90°$		
Tetragonal	$a_1 = a_2 \neq c$ $\alpha = \beta = \gamma = 90°$		
Hexagonal	$a_1 = a_2 = a_3 \neq c$ $\beta_1 = \beta_2 = \beta_3 = 90°$ $\gamma = 120°$		
Isometric	$a_1 = a_2 = a_3$ $\alpha = \beta = \gamma = 90°$		

Diagrams of cell types redrawn from M. J. Buerger, *Elementary Crystallography*, John Wiley & Sons, N.Y. © 1956 M. J. Buerger. Used with permission of the author.

Table 1.3   Mohs Scale of Hardness

1 Talc	5½ Window glass
2 Gypsum	6 Orthoclase
2½ Fingernail	6½ Steel file
3 Calcium	7 Quartz
3 Copper coin	7½ Procelain streak plate
4 Fluorite	8 Topaz
5 Apatite	9 Corundum
5¼ Knife blade	10 Diamond

Hardness is usually taken as the resistance of a mineral to abrasion and is measured by scratch tests on materials of known relative hardness. The Mohs mineral hardness scale is given in table 1.3,[1] with some common materials included. Each mineral is capable of scratching the one below it on the scale.

In practice, it is common to assign unknown minerals an upper or lower hardness limit as compared with some common material (i.e., fluorite will scratch a copper coin and be scratched by a knife blade). Care should be taken in testing that a true scratch is made since surface coatings, cracks, or mechanical powdering can produce misleading results. Generally, the hard mineral can be felt to dig into the softer in a short, say 1 mm, scratch. Further, the materials should be cross-checked with each other since they may mutually scratch, or not scratch, if of the same hardness.

Minerals break in characteristic ways. Some **fracture** to yield irregular or smoothly scalloped surfaces (conchoidal fracture, e.g. quartz). The internal structure of many minerals, however, is such that parallel surfaces of lower cohesion occur at a periodicity of a few atomic diameters and provide a

1. Friederich Mohs, *Grundriss der Mineralogie,* ca. 1825.

locus for planar rupture called **cleavage.** None, one, two, three, or more directions of such ready breakage may be present depending on the mineral species; and the presence, perfection, and orientation of cleavage are extremely useful determinative criteria (table 1.4).

The **specific gravity,** *SG* or *G,* of a mineral or other substance is the ratio of its weight to that of an equal volume of water, which is usually determined by use of a specially designed balance that allows the mineral to be weighed in air and when suspended in water (see also chapter 10):

$$G = \frac{\text{weight in air}}{\text{weight in air} - \text{weight in water}}.$$

An alternative method, particularly useful for small amounts of granular or porous materials, employs a special bottle called a *pycnometer.* This is a small bottle fitted with a ground glass stopper through which a capillary hole passes. The bottle is filled with water to the top of the capillary, but with no excess, and weighed, *B;* a weighed amount of sample, *S,* is next introduced; the capillary aperture wiped off; and the filled bottle plus sample weighed. The difference between this weight, *W,* and the sum of the other two weights, *B* and *S,* is the weight of the displaced water, thus

$$G = \frac{S}{B + S - W}.$$

Mineral *colors* are the combined effect of a mineral's constituents and their arrangement, chemical impurities, and various crystal imperfections. As a rule, minerals whose colors are normally dark will not suffer color changes because of impurities or imperfections. On the other hand, impurities and imperfections in a mineral may interact strongly with light and markedly modify the colors of normally light-colored minerals. The amount of impurity need not be large, indeed the amount of a coloring agent such as ferric iron needed to produce strong coloration may be a fraction of a percent.

Table 1.4 Shapes of Typical Cleavage Fragments

Number of Cleavage Directions	Characteristic Fragment	Example
0		Quartz
1		Muscovite
2		Augite
		Orthoclase
		Hornblende
3		Halite
		Anhydrite
		Calcite
4		Fluorite
6		Sphalerite

From W. H. Dennen, *Principles of Mineralogy.* © 1960 The Ronald Press Co., N.Y. Reprinted by permission of John Wiley & Sons, Inc.

Table 1.5 Mineral Habits

Terms Used to Describe Single Crystals		Terms Used to Describe Crystal Groups and Mineral Aggregates	

Capillary, filiform, acicular—hairlike, threadlike, or needlelike crystals.

Bladed—crystals in elongate, flattened blades.

Tabular, lamellar—booklike in shape.

Foliated, micaceous—easily separated into sheets or leaves, micalike.

Plumose—featherlike arrangement of fine scales.

Stout or stubby—usually applied to pyramidally terminated crystals whose *c* axis is short compared with its other axes.

Blocky—brick-shaped.

Columnar—columnlike crystals.

Geometrical terms—various geometrical terms are used as applicable, (e.g., *cubic, tetrahedral, octahedral, prismatic, dodecahedral, scalenohedral,* etc.)

Columnar—an aggregate of columnlike individuals.

Bladed—an aggregate of bladed individuals.

Fibrous—an aggregate of capillary or filiform individuals.

Dendritic—treelike or mosslike form.

Granular—an aggregate of mineral grains.

Massive—a compact aggregate without distinctive form.

Divergent, radiated, stellated—individuals arranged in fan-shaped groups or rosettes.

Colloform (botryoidal, reniform, mammillary, globular)—radiating individuals forming spherical or hemispherical groups. The various terms have been used to designate the extent and radius of the hemispherical surfaces developed. Colloform includes all other terms.

Reticulated—slender crystals arranged in a latticelike array.

Pisolitic, oolitic—composed of rounded masses respectively the size of peas or BB shot.

Banded—bands or layers of different color and/or texture.

Concentric—onion-like banding.

From W. H. Dennen, *Principles of Mineralogy.* © 1960 The Ronald Press Co., N.Y. Reprinted by permission of John Wiley & Sons, Inc.

Table 1.6 Tests for Short-term Stability

Test Reaction	Mineral or Element Indicated
Effervesces in cold, dilute, hydrochloric acid	Carbonate : calcite
Effervesces in hot, dilute, hydrochloric acid	Carbonate : dolomite
Gelatinizes in hot hydrochloric acid	Silicate : olivine
Magnetic without heating	Magnetite (strong), pyrrhotite or ilmentie (weak)
Magnetic after heating	Iron present
Evolves water in heated tube (strongly heat sample in test tube; water condenses in upper portion)	Water

Color should always be determined by the examination of a fresh surface since surface alterations and tarnishes can obscure the normal color. *Streak* is the color of a powdered mineral and has a much more constant color than does the bulk material. For example, a series of colored glasses, when powdered, all show a very similar white streak. Streak tests are usually made by rubbing the mineral on a piece of unglazed porcelain (a streak plate) and examining the resultant powder against the white background of the plate. If no streak plate is available, the color of the mineral powder may be conveniently observed against any white background.

Habit, as the term implies, is the commonly assumed external form, or nature of aggregation, of mineral grains. Habit is, of course, not a constant mineral attribute, but a particular habit is so often associated with a particular mineral species that it is one of the more powerful means of mineral identification. Nevertheless most practicing mineralogists tend to make tentative identifications based on habit and color—two inconstant mineral properties! Terms used in describing habits together with some sketches of the habits is given in table 1.5.

The *stability* of minerals refers to their ability to remain unchanged in the presence of solvents or temperature changes. Most common minerals are rather stable substances, but a few react readily under conditions that are often met in engineering situations. Of particular concern is the expansion of **clay minerals** and opaline silica in the presence of water; the oxidation of sulfide minerals such as pyrite; the transformation of calcite to gypsum on exposure to a sulfurous atmosphere; and the ready solubility of such minerals as gypsum or halite and other sulfates and halides.

For engineering purposes, short-term stability is easily tested by exposing the mineral to the anticipated conditions for an appropriate time. The reactions of minerals to various chemical reagents or to heat, with or without the presence of fluxing agents, are the basis for determinative testing beyond the scope of this text. A few of the simpler and more definitive tests are listed, however, in table 1.6, and many others may be found in any standard mineralogy text.

A key for the identification of common minerals is included at the end of this chapter in table 1.10.

MINERAL CLASSIFICATION

Minerals are grouped into classes on the basis of their electronegative component (table 1.7). Because of their common formation in the presence of oxygen, the most abundant minerals contain either oxygen alone or oxygen-bearing radicals. A few mineral classes (halides, sulfides, sulfosalts) have electronegative components other than those involving oxygen. Finally, a limited number of native elements such as gold, copper graphite, and sulfur exist.

Table 1.7 Mineral Classification

Mineral Class	Electronegative Component	Shape of Radical
Involving oxygen		
Oxides	O^{2-}	
Hydroxides	$(OH)^{-}$	
Carbonates	$(CO_3)^{2-}$	Equilateral triangle
Sulfates	$(SO_4)^{2-}$	Tetrahedron
Phosphates	$(PO_4)^{3-}$	Tetrahedron
Silicates	$(Si_xO_y)^{n-}$	Tetrahedra, often linked
Not involving oxygen		
Halides	Cl^{-}, F, etc.	
Sulfides and sulfosalts	S^{2-}, As^{2-}, Sb^{2-}	

The most common minerals are *silicates,* wherein a positively charged ion (cation) alternates with a negatively charged silicon–oxygen or silicon–aluminum–oxygen radical. The possible arrangements are many and the resultant silicate minerals are varied in appearance and properties. All, however, tend to be hard, mechanically resistant, chemically inert, and of low specific gravity.

The basic unit in silicates is the *silicon–oxygen tetrahedron* of ions (SiO_4): a silicon ion nestled in the central hole of four close-packed oxygen ions. Each oxygen ion contributes one of its two negative units of charge toward neutralization of the $4+$ charge of the silicon ion, and the resultant charge on the tetrahedral group is thus $4-$, or $Si^{4+} + 4\,O^{2-} = (SiO_4)^{4-}$. This radical alternates with cations in some regular way to yield an electrically neutral mineral structure; for example (fig. 1.2), the mineral olivine, $Mg_2^{2+} + (SiO_4)^{4-} = Mg_2(SiO_4)$.

Other silicates are based upon more complex silicate radicals that arise when oxygen ions are shared between adjacent tetrahedra. Sharing of a single oxygen between two SiO_4 tetrahedra leads to paired tetrahedral radicals containing 2 silicon ions and $8-1$ oxygen ions, $(Si_2O_7)^{6-}$; sharing of two oxygens per tetrahedron leads to an infinite chain, $(SiO_3)^{2-}$; sharing three a sheet, $(Si_2O_5)^{2-}$; and sharing of all of the tetrahedral oxygens a framework, $(SiO_2)^0$. Some basic silicate structural skeleta are illustrated in table 1.8 together with mineral examples.

Figure 1.2 Alternation of Mg^{2+} (spheres) and $(SiO_4)^{4-}$ tetrahedra in the structure of olivine, $Mg_2(SiO_4)$

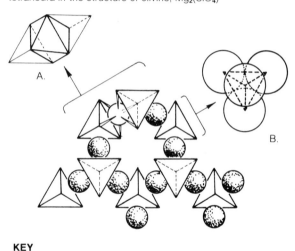

A.

B.

KEY

A. Octahedral array of oxygen ions around Mg^{2+}
B. Tetrahedral arrangement of oxygen ion around Si^{4+} site in the central hole

Table 1.8 Silicate Skeleta

Structural Group	Diagrammatic Representation	Number of Shared Oxygens per Silicon	Si-O Repeat Unit	Si:O	Example
Tetrahedral subsaturate	+0	0	$(SiO_4)O_2^{(4 + 2x)-}$	$1:(4 + x)$	Kyanite, $Al_2(SiO_4)O$
Tetrahedral unit		0	$(SiO_4)^{4-}$	1:4	Forsterite, $Mg_2(SiO_4)$ (an olivine)
Tetrahedral pair		1	$(Si_2O_7)^{6-}$	1:3½	Akermanite, $Ca_2Mg(Si_2O_7)$
Tetrahedral single chain		2	$(Si_2O_6)^{4-}$	1:3	Enstatite, $Mg(SiO_3)$ or $Mg_2(Si_2O_6)$ (a pyroxene)
Tetrahedral double chain		2½	$(Si_4O_{11})^{6-}$	1:2¾	Tremolite, $Ca_2Mg_5(Si_8O_{22})(OH)_2$ (an amphibole)
Tetrahedral sheet		3	$(Si_2O_5)^{2-}$	1:2½	Kaolinite, $Al_2(Si_2O_5)(OH)_4$ (a clay mineral)
Tetrahedral framework		4	$(SiO_2)^{0}$	1:2	Quartz, SiO_2

From W. H. Dennen, *Principles of Mineralogy.* © 1960 The Ronald Press Co., N.Y. Reprinted by permission of John Wiley & Sons, Inc.

Figure 1.3 Examples of mineral structures

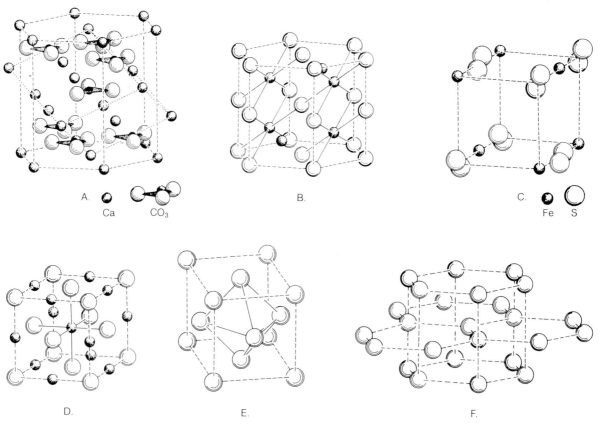

A.   Ca CO_3

B.

C.   Fe   S

D.

E.

F.

KEY

A. Calcite
B. Hematite
C. Pyrite
D. Halite
E. Copper
F. Graphite

Because of the requirement of alternating charges, which is necessary to maintain electrical neutrality of the compound, each ion is surrounded by (coordinates) nearest neighbors of opposite sign, and these neighbors are typically disposed on the corners of a regular polyhedron centered on the co-ordinating ion. A mineral structure may thus be visualized as either a rather tightly packed array of spherical ions or as the filling of space by repeating coordination polyhedra.

Common, nonsilicate, oxygen-bearing mineral classes include the *carbonates* in which cations such as calcium, Ca^{2+}, magnesium, Mg^{2+}, or iron, Fe^{2+}, alternate with the carbonate radical, $(CO_3)^{2-}$, (fig. 1.3a). These minerals are typically soluble, soft, and of low specific gravity.

A final important class of abundant-oxygen-bearing minerals are those in which either oxygen ions, O^{2-}, hydroxyl radicals, $(OH)^-$, or both alternate with various cations. An example of such a mineral is hematite, Fe_2O_3, (fig. 1.3b). These *oxides, hydroxides,* and *hydrated oxides* are often dark colored, mechanically and chemically resistant, and heavy.

Figure 1.4 Two-component diagram for the olivine series. The dot identifies an olivine ($Fo_{75}Fa_{25}$).

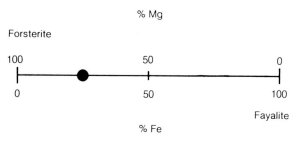

An example of a common oxygen-free mineral whose properties of metallic luster, high specific gravity, and electrical conductivity are typical of its class is the *sulfide* pyrite, FeS_2. In this mineral, Fe and S_2 pairs alternate in a cubic array (fig. 1.3c) in a geometrical analog to the structure of the halide halite, NaCl, (fig. 1.3d).

A few *native elements* are found as minerals although they are relatively rare. These minerals may be either metallic substances such as gold, silver, or copper (fig. 1.3e) or nonmetallic minerals such as sulfur or graphite (fig. 1.3f).

Solid Solution

When the ionic properties of size, electrical charge or valence, and nature of the chemical bonding of two ions are closely similar, the ions may substitute for one another in a mineral structure in much the same way that some yellow bricks could be used in the making of a dominantly red-brick fireplace. This phenomenon, called **solid solution,** is rather common in many minerals. In silicates, for example, solid solution occurs in both the silicate radical (e.g. Al for Si) and the cationic "glue" that binds silicate units, chains, and sheets into three-dimensional crystals. In minerals such as the previously mentioned olivine, iron typically replaces more or less magnesium; and the general formula for an olivine, (Mg,Fe_2) (SiO_4), indicates this. The continuous chemical variation possible in the olivine series of minerals may be represented by a line (fig. 1.4) whose ends are the pure Mg and Fe end members—respectively the minerals forsterite, $Mg_2(SiO_4)$, and fayalite, $Fe_2(SiO_4)$—and along which all intermediate compounds may be shown. In figure 1.4 an ol-

Figure 1.5 Three-component diagram for the olivine series

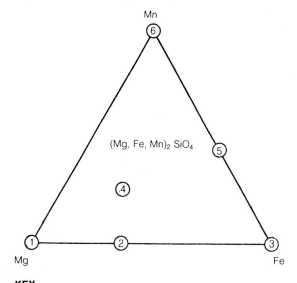

KEY

Point	Mineral	Formula	Proportions
1	Forsterite (Fo)	Mg_2SiO_4	Fo_{100}
2	Olivine	$(Mg,Fe)_2SiO_4$	$Fo_{63}Fa_{37}$
3	Fayalite (Fa)	Fe_2SiO_4	Fa_{100}
4	Hortonolite	$(Mg, Fe, Mn)_2SiO_4$	$Fo_{50}Fa_{25}Te_{25}$
5	Knebelite	$(Fe,Mn)_2SiO_4$	$Fa_{56}Te_{44}$
6	Tephroite (Te)	Mn_2SiO_4	Te_{100}

Mg Magnesium end member
Fe Iron end member
Mn Manganese end member

ivine whose composition is 75 atom % magnesium and 25 atom % iron is identified. Since this olivine contains 75% of the forsterite (Fo) and 25% of the fayalite (Fa) end members, it may be described as $Fo_{75}Fa_{25}$ or simply Fo_{75}.

Not only may Fe and Mg be interchanged in olivines, but other geochemically similar elements such as manganese, Mn, may enter the structure in their places. To represent the compositions of this three-component system, a triangular or **ternary diagram** is often used (fig. 1.5). Each angle of such a diagram represents an end member, each side the solid solution series between end-member pairs, and the interior all possible end-member combinations.

Natrolite

Calcite, dogtooth spar

Calcite

Electron photograph of halite

Angles represent 100% of a constituent and the opposite side 0%. Since the sum of mutually interchanging ions is 100%, any composition may be shown on the diagram.

Aluminum, with a charge of 3+, often substitutes in silicates for silicon whose charge is 4+ and thus changes the charge on the silicate radical. For example, in a framework structure the substitution for one out of every four silicon ions by an aluminum ion changes the charge on the radical from zero to 1−; $Al^{3+} + 4(SiO_2)^0 = (AlSiO_3O_8)^- + Si^{4+}$. If a monovalent cation such as potassium, K^+, or sodium, Na^+, is utilized for charge balance in an appropriate structural site, the resulting compound might be feldspar, the most common of all mineral groups. A ternary plot for the feldspar family is given in figure 1.6 in which it may be seen that no significant solid solution exists between potassium feldspar and anorthite, that the potassium feldspar–albite series is discontinuous, and that the albite–anorthite series is one of complete solid solution. The diagram includes a field in which solid solution is possible at elevated temperature to make the point that the degree of solid solution in this and in other systems is a function of temperature.

It should be noted parenthetically that the principles of the ternary diagram developed in figure 1.6 to describe solid solution in minerals are also useful in illustrating interrelationships in any three-component system and will often be used in other contexts; for example, the classification of soil and sediment based on grain size.

Geometrical Relationships

In addition to the chemical variability of minerals just described, there also exist various kinds of geometrical relationships among minerals. If, for example, two mineral species are based on the same structural arrangements of their constituents (*isostructures*), those physical properties that are geometrically based will be the same whereas those related to the nature of the atoms may be different. As an example, the geometrical disposition of sodium and chlorine atoms in halite, NaCl, and the lead and sulfur atoms in galena, PbS, are identical,

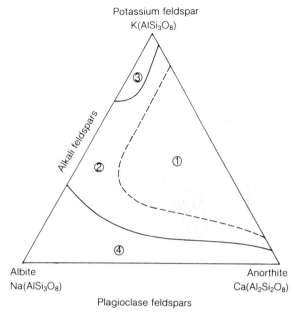

Figure 1.6 Stability fields of feldspars

KEY

1. No feldspars
2. Crystals stable at high temperature only
3. Potash (K) feldspar
4. Plagioclase feldspar

and both minerals thus crystallize and break in cubes although halite is nonmetallic, transparent, and has a low specific gravity whereas the opposite is true for galena.

Structural similarity rather than identity also leads to similarity in physical properties, thus all minerals having a structure based on stacks of parallel sheets tend to separate into flakes on intersheet planes. The silicates muscovite, $KAl_2 (AlSi_3O_{10})(OH)_2$; talc, $Mg_3(Si_4O_{10})(OH)_2$; and kaolinite, $Al_2(Si_2O_5)(OH)_4$; the sulfide molybdenite, MoS_2; and the native element graphite, C, are all based on sheetlike structures of one kind or another and display this property.

Another structural variation is the ability in some instances for the same chemical constituents to assume different structural arrangements. In such **polymorphs** the mineral composition is unchanged,

Table 1.9 Clay Mineral Classification

Single-Sheet Clays	Double-Sheet Clays
Kaolinite Group (kandites)	*Montmorillonite Group* (smectites)
Kaolinite	Montmorillonite series
Anauxite	Beidellite
Nacrite	Hectorite
Dickite	Nontronite
Halloysite–endellite series	Sauconite
	Saponite
	Illite Group
	Members not well established—includes clays that are similar but not identical with the white micas
	Palygorskite Group
	Attapulgite series
	Vermiculite Group

Modified from W. H. Dennen, *Principles of Mineralogy.* © 1960 The Ronald Press Co., N.Y. Reprinted by permission of John Wiley & Sons, Inc.

but the geometry and properties are often quite different. Some examples of polymorphs follow:

C	graphite, diamond
$Ca(CO_3)$	calcite, aragonite
SiO_2	quartz, tridymite, cristobalite
$K(AlSi_3O_8)$	orthoclase, microcline
$Al_2(Si_2O_5)(OH)_4$	kaolinite, dickite, nacrite, halloysite
FeS_2	pyrite, marcasite

CLAY MINERALS

Clay is one of the more important groups of minerals from an engineering point of view. The term **clay** is variously used to describe either a group of minerals, a particle size (< 4 μm), or a set of properties. *Clay minerals* are important due to their common occurrence and peculiar properties in association with water, which may lead to hazardous conditions if they make up the bulk of material at an engineering site. On the other hand, clays are widely used as sealants in dams or landfills, as lubricants in drilling, and are an important industrial resource. Their role in these applications will be further discussed later.

Clay minerals are built up from basically simple units, but because these units may be arranged in a number of ways and have interleaved water and adsorbed (physically attached) ions, a very complex mineralogy results. The more important clay mineral species are listed in table 1.9.

The structural units of importance are tetrahedral sheets of silicon and oxygen ions and octahedral sheets of magnesium or aluminum, hydroxyl, and oxygen ions. Mineralogical complexity arises because the combination of these tetrahedral (T) and octahedral (O) layers may repeat in either a T–O T–O T–O or a T–O–T T–O–T sequence generating "openfaced" or "regular" sandwiches (fig. 1.7). The layers may alternate in a regular (ordered) or irregular (disordered) manner in the orientation of the displacement of the octahedral layer, and magnesium and aluminum octahedral layers may alternate in a patterned or unpatterned way. These arrangements of layered sheets are directly responsible for the many species of clay minerals.

Typically, clay mineral crystals are very tiny (no more than a few micrometers in diameter), and hence a clay mineral aggregate will have a very large specific surface causing surface-related properties to dominate over bulk properties. Measured surface

Figure 1.7   Arrangement of sheets in clay minerals

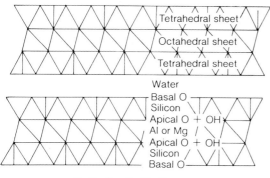

T - O - T   T - O - T

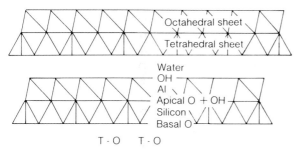

T - O   T - O

KEY

T   Tetrahedral sheet
O   Octahedral sheet

is of the order of 10 to 100 m² per gram. One reason for limited crystal size may be found in the slight misfit of the octahedral and tetrahedral layers resulting in strain that limits the crystal growth. Electrically charged surfaces exist on these very small mineral grains because the electrical charges of the ions in the structure are not completely neutralized by the presence of their nearest ionic neighbors of opposite sign. Further, the surface charges may be of opposite sign on the faces and edges of the tabular crystals. For large crystals, the ever-present surface effects are outweighed by the bulk of the crystal, but for tiny crystals such as clays, surface charges control many of the mechanical properties of an aggregate.

Water is always present in natural clay aggregates and serves to neutralize the negative surface charge by **adsorption** (attachment) to the particle surface. Regardless of the mechanism, which is controversial, aggregates of clay particles and adsorbed water (the clay–water system) display marked *plasticity*. This property may be ascribed to a bonding between clay particles caused by the union of layers of oriented water molecules; optimum plasticity is developed when all of the requirements for rigidly fixed water are met and additional interstitial water can act as a lubricant. This plasticity can be either an advantage or a disadvantage depending on the application, and properties of the clay–water system are so important that this brief introduction will be amplified in chapter 11.

Clays are also capable of ionic substitutions within the mineral structure leading to variations in the surface charge and thus to differing thickness of the rigid water envelope and the strength and distribution of attractive forces for adsorbed ions. Maximum *swelling* (water uptake) and probably maximum dispersion of particles occur when substitution gives the mineral structure a net residual negative charge. The kind and amount of adsorbed ions can markedly change the mechanical properties of clay aggregates because of their effect on the water-holding capacity of clay minerals. Most natural clays have adsorbed calcium ions and a relatively low capacity, but when the adsorbed ions are hydrogen, sodium, or lithium, the water-holding capacity soars (fig. 1.8, p. 18).

SUMMARY

The composition of the earth, the nature of inorganic chemical bonds, and the geometrical constraints on the arrangement of atoms in crystals serve to limit the total of mineral species to a few thousand and the rock-forming minerals to only a few tens. Table 1.10 provides a simplified resumé of the more abundant rock-forming minerals and the rocks that they commonly make. These rocks will be discussed in the next chapter.

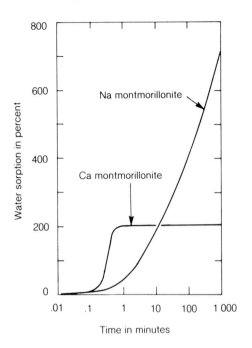

Figure 1.8 Water-holding capacity of different clay minerals

Na montmorillonite

Ca montmorillonite

Water sorption in percent

Time in minutes

Minerals are divided into classes on the basis of their negatively charged component (i.e., into oxides, silicates, carbonates, etc.) and further classified according to their atomic arrangement or structure and the cations that are present. The composition, bonding, and structure of minerals give them intrinsic properties, which may be used for their identification and are particularly important in the exploration for and extraction of mineral wealth (table 1.11). Rocks and sediments are built of mineral particles, and the bulk of rock and sediment properties arise from the nature of these particles coupled with the manner in which they are arranged and held together.

Sediments are unconsolidated aggregates of grains, and their bulk mechanical properties are strongly dependent upon their granularity. If composed of particles above about 4μm in diameter, the mass acts like sand, but if smaller than about 4μm like modeling clay or soft mud depending upon the water content. Since clay minerals have particle sizes below 4μm and may be very abundant, the mechanical response of a material such as a soil is strongly conditioned by its clay mineral content. For the coarser fraction, however, size and stability and not mineral kind are important.

Table 1.10 Rock-forming Minerals

Minerals	Rock Types		
	Igneous	Sedimentary	Metamorphic
Feldspar (potash feldspar and plagioclase)	X	x	X
Quartz	x	X	X
Olivine	x		
Pyroxene (eg., augite)	x		x
Amphibole (eg., hornblende)	x		x
Mica	x		x
Calcite		X	x
Dolomite		X	x
Iron oxide (eg., hematite)	x	x	x
Clay (eg., kaolinite)		X	
Gypsum and *Anhydrite*		x	
Halite (rock salt)		x	
Apatite (phosphorite)		x	

X = common and often dominant
x = common or locally abundant

Table 1.11 Key for the Identification of Common Minerals

Minerals with Metallic Luster

Name	Color	Luster	Streak	Hardness	Cleavage	Form	Special properties
Limonite, $Fe_2O_3 \cdot nH_2O$	brown-black, yellow-brown	submetallic to earthy	brown to yellow-brown	1–4	none	massive, intergrown, rounded masses	yields water, magnetic when heated
Hematite, Fe_2O_3	dark red, steel gray	metallic to earthy	dark red	1–6	occasionally like fine mica	micaceous to massive	magnetic when heated
Magnetite, Fe_3O_4	steel gray to black	metallic	black	5½–6½	indistinct	granular to massive	attracted by magnet, heavy
Pyrite, FeS_2	brass-yellow	metallic	greenish black	6–6½	none	cubes or crystals with 5-sided faces	commonly has striations on crystal faces

Minerals with a Hardness of 1–3
(softer than a fingernail)

Name	Color	Luster	Streak	Hardness	Cleavage	Form	Special properties
Kaolinite, $Al_2(Si_2O_5)(OH)_4$ (a clay mineral)	white, brown, gray	earthy-dull	white	1	none visible	massive	earthy odor, plastic when wet
Gypsum, $CaSO_4 \cdot 2H_2O$	colorless, white, pink, gray	glassy, pearly or silky	white	2	3 good in crystalline varieties	platy crystals, massive, or fibrous	
Muscovite, complex silicate	colorless	pearly	white	2½–3	1 perfect	sheets, flakes	flexible, elastic sheets
Biotite, a complex silicate	dark brown to black	glassy	light gray, brownish	2½–3	1 perfect	sheets, flakes	flexible, elastic sheets
Halite, NaCl	gray, colorless, blue	glassy, waxy	white	2	3 perfect at right angles	granular, cubic crystals, cubic cleavage fragments	salty taste
Limonite, $Fe_2O_3 \cdot nH_2O$	yellow-brown	earthy to submetallic	yellow-brown	1–4	none	acicular to earthy, spongy masses	appearance of iron rust

Table 1.11 Key for the Identification of Common Minerals (continued)

Minerals with a Hardness of 3, 4, or 5 (harder than a fingernail but softer than a nail)							
Calcite, $CaCO_3$	colorless, any color	glassy	white	3	3 perfect, not at right angles	massive rhombs, "dogtooth" crystals	effervesces in weak, cold acids
Dolomite, $(CaMg)(CO_3)_2$	white, pink, colorless, any color	glassy	white	3½–4	3 perfect, not at right angles	massive rhombs	effervesces in warm, weak acids

Minerals with a Hardness Greater than 5 (harder than a nail)							
Hornblende, a complex silicate	black or greenish black	glassy	grayish-green	5–6	2 very good, not at right angles	massive, elongate crystals	"Pyribole"
Augite, a complex silicate	greenish black or black	dull, glassy	grayish or brownish-green	5–6	2 fair at right angles	massive, stubby crystals	"Pyribole"
K-Feldspar, $K(AlSi_3O_8)$	white, gray, pink, rarely green, brown	glassy to pearly	white	6	2 very good at right angles	cleavage, fragments, massive	rock former
Quartz, SiO_2	colorless, white, any color	glassy	white	7	none	massive, crystals, six-sided prisms	rock former
Chalcedony, SiO_2	white, red, black, agate banded	waxy	white	7	none	massive, layered encrustations	nodules in limestone
Olivine, $(FeMg)_2SiO_4$	green	glassy	white	6½–7	none	granular, massive	sugary appearance
Plagioclase, a complex silicate	white to gray	glassy	white	6	2 at nearly right angles	cleavage fragments, massive	rock former

ADDITIONAL READING

Casagrande, A. 1932. Research on the Atterberg limits of soils. *Public Roads,* v.13. Washington, D.C.: Bureau of Public Roads.

Deer, W. A., Howie, R. A., and Zussman, J. 1966. *An introduction to the rock-forming minerals.* New York: John Wiley & Sons.

Dennen, W. H. 1960. *Principles of mineralogy.* New York: Ronald Press.

Gillot, J. E. 1968. *Clay in engineering geology.* New York: American Elsevier.

Grim, R. E. 1962. *Applied clay mineralogy.* New York: McGraw-Hill.

Hurlbut, C. S., Jr., and Klein, C., Jr. 1977. *Dana's manual of mineralogy.* 19th ed. New York: John Wiley & Sons.

Mason, B., and Berry, L. G. 1968. *Elements of mineralogy.* San Francisco: W. H. Freeman.

Sudo, T., Shimoda, S., Yotsumoto, H., and Aita, S. 1981. *Electron micrographs of clay minerals.* New York: Elsevier Science.

2

Rocks

INTRODUCTION

Rocks are solid aggregates of mineral or other particles held together by the interlocking of grains or an intergranular cement. Rocks constitute the outer portion, or *lithosphere,* of the earth and are formed and transformed in the mantle, crust, or at the earth's surface. The rock-forming processes in the deeper subsurface go on in a region of high temperature and pressure. They involve both the recrystallization of mineral aggregates in the solid state, **metamorphism,** and the production of silicate melts called **magma.** Magma bodies, being liquid, tend to rise and intrude the surrounding country rock where they cool and crystallize into deep-seated **igneous** (fire formed) **rocks,** which are variable in shape and often of very large size. Sometimes the magma is extruded on the surface as **lava** or as other products of volcanic action.

Rock-forming and decay processes on the surface of the earth involve the physical and chemical transformation of previously formed rocks to materials more or less stable in the markedly different surface environment. Here lower pressure and temperature together with an abundance of such active chemical agents as water and free oxygen tend to disaggregate, transform, and dissolve rocks in the processes termed **weathering.** The products of weathering may remain in place as a residual blanket called *soil,* or regolith, or be transported and deposited as *sediments.* These sediments, in turn, may form into solid *sedimentary rocks* by compaction or the introduction of cementing materials precipitating from permeating **groundwater.** Substances dissolved in water can be transformed into solid sedimentary rocks—*evaporites*—by the evaporation of surface waters.

Rock units in the field are recognized as being aggregates of fairly constant constitution bounded by more or less regular and sharply defined surfaces, which serve to separate the unit from adjacent rock bodies. The diverse and individually variable properties of rocks, ranging from color to mechanical strength and porosity to economic value, are ultimately based upon four simple features: (1) the nature of the constituent solids, (2) their arrangement in space, (3) the mechanics of interparticle bonding, and (4) the grain size of the particles.

DESCRIPTION AND CLASSIFICATION OF ROCKS

Geologically, rocks are classified into three major divisions in accordance with their origins.

1. *Igneous rocks* are formed by the cooling and crystallization of molten material called magma.
2. *Metamorphic rocks* arise by recrystallization of previously formed rocks at elevated temperature and in the solid state.
3. *Sedimentary rocks* are formed by aggregation and cohesion of mineral or other grains at or near the earth's surface.

IGNEOUS ROCKS

Igneous rocks form from siliceous melts, which are called magma and originate within the earth's crust and **mantle.** The molten magma crystallizes at depths within the earth or pours out as lava to crystallize

Figure 2.1 Igneous rock bodies

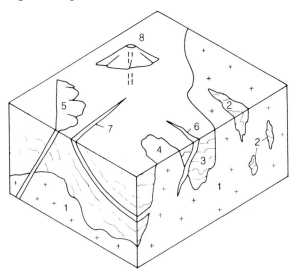

KEY

1. Batholith
2. Xenoliths or inclusions
3. Roof pendant
4. Stock
5. Dike and lava flow
6. Dike
7. Sill
8. Volcano over pipe

upon the surface. The crystallization generates a mass of mineral grains that intimately interlock to produce an impermeable, mechanically strong, isotropic mass whose shape depends upon the manner of its emplacement. Igneous rocks are divided into three general groups depending on the depth at which emplacement occurs: (1) **plutonic rock,** or deep-seated, intrusive rock; (2) **hypabyssal rock,** or shallow, intrusive rock; and (3) **volcanic rock** extruded at the surface. In figure 2.1 a block of crustal rock is intruded by magma in a variety of intrusive and extrusive bodies. As the large magma body works its way upward, a large body of igneous rock may crystallize as (*1*) a **batholith.** Such a rock body may be a few tens to several hundred kilometers in diameter and cool slowly over perhaps 10 million years at some 5 to 50 kilometers deep within the crust. Blocks of country rock (*2*) called **xenoliths** have been incorporated into the magma chamber and partially dissolved. A lobe of downward-projecting roof rock (*3*) is called a roof *pendant.* A relatively

Dike, Pilbara region, western Australia

small upward projection (*4*) of the magma will crystallize as a **stock.** A narrow sheetlike intrusion (*5*), which shoots to the surface as a **dike,** may produce a lava flow at the surface. Another dike (*6*) should be contrasted with the **sill** (*7*) since dikes cut across existing strata and sills are parallel to them. The final structure (*8*) is a central-vent volcano over a volcanic pipe. Stocks, dikes, and sills are typical of shallow, igneous, intrusive rock bodies and volcanic piles and sheetlike lava flows are typical extrusive bodies.

Magma always contains dissolved water plus other volatile constituents held in solution by the great pressures at depth. These constituents boil out of the magma when it is extruded as lava, sometimes turning the lava into a froth and explosively ejecting fragments and liquid globules into the air. Porous, or vesicular, lava has a Swiss-cheese texture formed by gas expanding in a liquid matrix. Explosively erupted **pyroclastic** material includes volcanic ash, cinders, and large fragments. Because the composition, temperature, and confining pressure may vary in time, it is possible to produce a layered pile of alternating dense and porous flow rocks.

Small body of intrusive rock

Ropy lava, Craters of the Moon, Idaho

Classification of Igneous Rocks

The general classification of igneous rocks is based on mineral composition, grain size, texture, and silica content (table 2.1). The texture of an igneous rock is related to the rate of cooling of magma. Fast-cooling volcanic rocks yield the finer crystals or glass. Slow cooling of deep-seated magma results in coarser textures. The main textures used in the classification are **pegmatitic, phaneritic, aphanitic, porphyritic, glassy, vesicular, amygdaloidal, and pyroclastic.** High silica rocks are typically lighter in color and become darker and heavier with decreasing silica content.

Crystallization of Magma

Magma is a mutual solution of many chemical constituents at 600 to 1 200°C and 2 to 12 kilobars (1 kilobar = 986.92 atmospheres), so the details of the crystallization of magma may be anticipated to be complex. In general, the freezing point, or temperature of crystallization, of a particular substance is lowered by the presence of other substances in the solution as, for example, the freezing point of water is lowered in a mixture of alcohol and water (fig. 2.2). A mixture whose temperature and composition is represented by point *1* will, when cooled, move to point *2* on the **liquidus** line separating the liquid (water + alcohol) field from the solid + liquid (ice + water + alcohol) field. At this point, a tiny amount of ice crystallizes, the amount of water is fractionally reduced, and the composition of the remaining liquid is slightly enriched in alcohol. Continued cooling, crystallization, and consequent enrichment of the remaining liquid moves the liquid composition down the liquidus line, and at any particular temperature, there will be an equilibrium mixture of solid and liquid phases present. (This system has been used both in antifreeze and in the making of applejack.)

When two or more solids form simultaneously from the same liquid or melt, phase diagrams are

Table 2.1 Classification of Igneous Rocks

		Mineral Composition			
	Determining minerals	Orthoclase Quartz	Plagioclase	Plagioclase Pyroxene Olivine	Pyroxene Olivine
	Accessory minerals	Biotite Muscovite Amphibole	Amphibole Biotite	Magnetite	
Texture and Grain Size	*Pegmatitic* (very coarse-grained—crystals > 1 cm)	Pegmatite			
	Phaneritic (medium-grained—crystals 0.25–2.0 cm)	Granite	Diorite	Gabbro	Peridotite Pyroxenite Dunite
	Aphanitic (fine-grained, visibly crystalline)	Felsite (Rhyolite)	Trap (Andesite) (Basalt)		
	Porphyritic (larger crystals set in finer-grained ground mass)	—Porphyry or Porphyritic—			
	Glassy (glasslike)	Obsidian			
	Vesicular (Swiss-cheese texture)	Pumice	Scoria		
	Amygdaloidal (vesicles filled with later minerals)	Amygdaloid			
	Pyroclastic (fine to coarse volcanic fragments)	Volcanic tuff Volcanic breccia			
	Silica content	Salic (over 65%)	Intermediate (55%–65%)	Mafic (45%–55%)	Ultramafic (less than 45%)
	Color	◄—— Increasingly light		Increasingly dark ——►	

See also figure 2.5 for the chemical and mineralogic composition of igneous rocks

instructive in showing the sequence of crystal formation and the change in liquid composition that take place. For example in figure 2.3, a liquid of temperature and composition represented by point *1* and cooled to point *2* begins to crystallize with the formation of solid phase *B* while the liquid composition moves down the liquidus line as in the previous example. At point *3*, however, the liquid composition is simultaneously on two liquidus lines and both solids *A* and *B* simultaneously crystallize to the exhaustion of the liquid without further lowering of temperature. This point of simultaneous crystallization of two solid phases is called a **eutectic point.** Crystallization of the system diopside–anorthite ($CaMgSi_2O_8 - CaAl_2Si_2O_8$) is a geologic example.

Photomicrograph of aphanitic rock

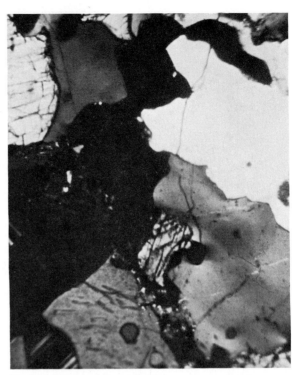

Photomicrograph of phaneritic rock

Figure 2.2 Alcohol–water freezing relations

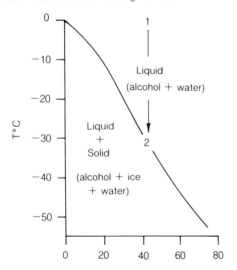

KEY

1. Initial temperature and composition
2. Temperature and composition at the freezing point

Figure 2.3 Eutectic crystallization

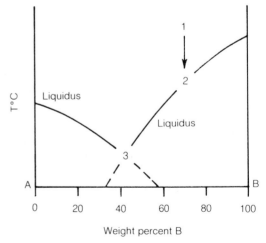

Weight percent B

KEY

1. Initial temperature and composition
2. Crystallization begins
3. Eutectic point; crystallization at fixed temperature and composition
A. 100% composition A
B. 100% composition B

Figure 2.4 Plagioclase phase diagram

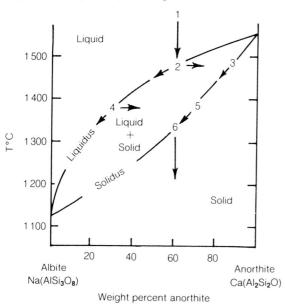

Some solid phases can readjust their composition after crystallization through a continuing reaction between the solid and liquid phases. Plagioclase feldspars, a dominant rock-forming mineral group, are particularly representative of this complex kind of compositional adjustment or reaction series (fig. 2.4). Cooling brings a melt of particular composition to the liquidus where crystallization occurs as previously described. The composition of the crystal formed, however, is not on a bounding ordinate but on the *solidus* phase boundary between the solid and the solid + liquid fields, point *3*. Crystallization causes the liquid composition to move down the liquidus with consequent enrichment in the albite component, say to point *4*, while the composition of the solid phase simultaneously moves along the solidus to point *5*,

which is at the same temperature as point *4*. Crystallization finally ceases at point *6* when the composition of the crystals forming is the same as that of the initial melt, point *1*. It should be noted that if physical separation of liquid and solids was accomplished by any means during the crystallization history, the liquid would be sodium and silica rich compared with the calcium-aluminum-rich mineral phase.

The sequence of mineral formation from a magma is epitomized by the reaction series of Bowen (1928), which serves both as a model of the physical chemistry of magma crystallization and as the basis for classification of igneous rocks. The series (fig. 2.5) is divided into two limbs representing the solid (mineral) phases that are reacting (sequentially from top to bottom of the diagram) with the magma to form the next lower phases. Mineral phases crystallize simultaneously down the continuous (plagioclase) limb and the discontinuous (ferromagnesian) limb, thus the mineral composition at any particular stage is approximately given by horizontally equivalent minerals. For example, early formed olivine and calcic plagioclase react with the melt as temperature falls to form pyroxene and calcium-sodium plagioclase. The composition of the earliest formed rock is olivine and calcic plagioclase; at somewhat lower temperature, olivine plus pyroxene and slightly sodic plagioclase; and cooler still, pyroxene and more sodic plagioclase; etc.

The presence of the remaining magma is disregarded in this sequence of mineral formation except as a source of nutrient material for the crystals. If no physical separation of liquid and crystal occurs, the process continues until crystallization is complete and the chemical composition of the rock is that of the initial magma. Some separation of solid and liquid phases may reasonably be anticipated, however, in most magmatic bodies. The mechanisms are (1) *settling* of early formed crystals, which depletes the upper portions of the magma in their constituents; (2) **filter-pressing,** whereby liquid is squeezed out of the chamber through porous walls; or (3) *armoring* of early formed minerals by a covering of a later solid phase, which effectively removes the core from further reaction. Contamination

Figure 2.5 Bowen reaction series

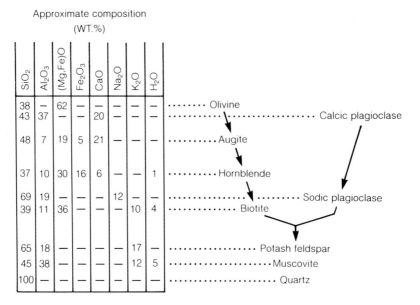

Approximate composition
(WT.%)

SiO$_2$	Al$_2$O$_3$	(Mg,Fe)O	Fe$_2$O$_3$	CaO	Na$_2$O	K$_2$O	H$_2$O	
38	—	62	—	—	—	—	—	Olivine
43	37	—	—	20	—	—	—	Calcic plagioclase
48	7	19	5	21	—	—	—	Augite
37	10	30	16	6	—	—	1	Hornblende
69	19	—	—	—	12	—	—	Sodic plagioclase
39	11	36	—	—	—	10	4	Biotite
65	18	—	—	—	—	17	—	Potash feldspar
45	38	—	—	—	—	12	5	Muscovite
100	—	—	—	—	—	—	—	Quartz

of the magma by reaction with its walls or inclusions of country rock, loss of volatiles, or rapid drops in temperature on extrusion may also alter the course of crystallization. Even in the face of a nonuniform cooling history, however, the mineral assemblages tend toward relatively few common groupings.

Following these principles, igneous rocks may be classified in accordance with their mineralogic makeup and their grain size (table 2.1). It should be noted that this classification is a compartmentalization of continuously variable features.

The generation of magma is incompletely understood but is a local phenomenon within the lower crust or upper mantle of the earth (i.e., within a few hundred kilometers of the surface). Local accumulations of radiogenic heat and downwarping or downthrusting of low-melting-point rocks to hot depths are generally called upon. Once formed, magma as a liquid is less dense than its solid surroundings and tends to move upward and intrude its

roof. Very large masses may make upward progress by shouldering aside their walls, by permeating the wall or country rocks, or by incorporating and assimilating blocks of the country rock (called inclusions, or xenoliths) in a process called **stoping.** Smaller intrusions near the surface exploit planar zones of weakness in the country rocks to insert tabular masses of magma, to cause doming of the cover, or to penetrate to the surface up pipelike vents. (Some typical shapes of igneous rock bodies were shown previously in fig. 2.1.)

Most igneous rocks make excellent engineering material due to their compact crystalline texture and constitute top-quality material for aggregates, road ballast, and foundation material. The hardness of some of these rocks can make tunneling and excavation expensive, but the result is usually a freestanding structure of great strength without reinforcement.

Table 2.2 Sedimentary Rocks Classified by Their Dominant Mineral

Rock Name	Principal Constituent
Limestone	Calcite
Dolostone	Dolomite
Rock salt	Halite
Gypsum	Gypsum
Chert	Chalcedonic quartz
Coal	Coalified plants

Table 2.3 Detrital Sedimentary Rocks

Particle Size Range	Rock Name	Constituents
> 2 mm	Conglomerate	Coarse rounded rock or mineral particles in a finer matrix
.062 6 − 2 mm	Sandstone	Quartz, feldspar, mica
.004 − .062 5 mm	Siltstone	Quartz, mica, clay
< .004† mm	Shale	Clay, quartz, organic matter

†This value is taken as .002 mm in some classifications.

SEDIMENTARY ROCKS

Sedimentary rocks are formed by the **aggregation** of almost any solid particles found on the earth's surface and may consist of mineral grains, fossil remains, rock fragments, or materials chemically precipitated such as salt, gypsum, iron oxides, and many others. Many subdivisions and classifications of sedimentary rocks are in use, but the most common groups are clastic and nonclastic categories. **Clastic rock** is made of solid **detrital** (derived from other rocks) particles. **Nonclastic rock** is usually of chemical or biochemical origin and forms by precipitation in a body of water. There are many overlaps between the two groups, and the general classification based upon their dominant mineral composition is useful (table 2.2).

Evaporites constitute a special group of sedimentary rocks formed by evaporation of saline waters in partially or wholly enclosed water bodies such as lagoons or salt lakes in arid regions. Rhythmic bedding is often present with the bedding surfaces marked by accumulations of wind-blown dust. Typical are deposits of calcium sulfate (gypsum, $CaSo_4 \cdot 2H_2O$, and anhydrite, $CaSO_4$) and sodium chloride (halite, $NaCl$).

Solid mineral and rock fragments released by weathering (see chapter 3) may be moved from the weathering site by downslope migration or by the action of running water, wind, or ice movement. In transport, these detrital particles will tend to be sorted according to their *size, shape,* and *density* with more or less efficiency depending on the transporting agent and the length of transport. The dominant feature of the particles in this process is their *size* (since most are of similar shape and density). It is thus common to describe detrital sedimentary rocks as in table 2.3.

Transportation of sedimentary debris is typically by fits and starts to some location such as the ocean or an enclosed basin where its residence time is sufficient to allow for **lithification** by cementation or other processes.

Detrital sediments and sedimentary rocks exhibit a spectrum of physical properties related to their degree of lithification, which ranges from loose material through lightly lithified, often highly porous aggregates to dense and impermeable masses.

Interbedded shale and sandstone

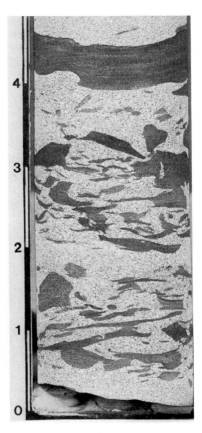

Shale-pebble conglomerate

Physical Properties of Sediments

The *size* of particles in sediments is very important in their classification and in their engineering properties. Several measures of size are in common use. The Wentworth scale is a logarithmic scale in which each grade limit is twice as large as the next smaller grade limit. U.S. standard sieve mesh sizes are commonly used in engineering for particles separated by sieving, and granular material will often be designated as a percentage plus (+) or minus (−) a certain mesh size. A modern system is the ∅ scale (phi scale), which is a logarithmic transformation of the Wentworth scale (table 2.4). Although the coarser sizes may be measured directly with calipers, and the grades of granule through coarse silt are measured by sieve size, the very fine sizes (clay and fine silt) are determined by their settling rate in water.

The composition of constituent particles is also important in the classification and physical properties of sediments and sedimentary rocks. The average abundance of the more common detrital minerals in some sedimentary rocks is given in table 2.5.

Grain morphology, or particle shape, is an important property from an engineering point of view since particles may be required to have a certain roundness or angularity or shape for a particular application or to give particular properties to a material such as concrete or blacktop in which they are incorporated. The degree of roundness of particles is usually simply stated on a rather subjective scale ranging from angular to spherical (fig. 2.6).

The shape of sediment particles may be measured as the relative lengths of the longest, intermediate, and shortest dimension and thus be referred to a triaxial ellipsoid. This is generally not done, however, because of measurement difficulties. A simpler and generally more useful criterion is the **aspect ratio** of grains—the ratio of longest to shortest

Table 2.4 Grain-size Scales for Sediments

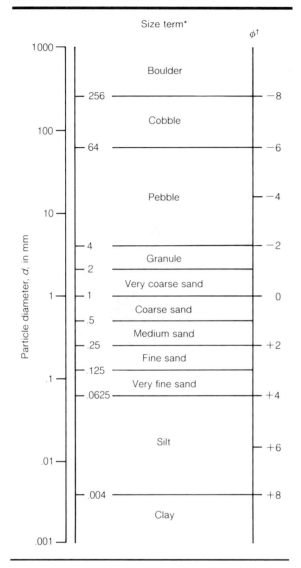

Size term*

ϕ†

Particle diameter, d, in mm		Size term*	ϕ†
1000			
	256	Boulder	−8
100		Cobble	
	64		−6
10		Pebble	−4
	4		−2
	2	Granule	
1	1	Very coarse sand	0
	.5	Coarse sand	
	.25	Medium sand	+2
.1	.125	Fine sand	
	.0625	Very fine sand	+4
.01		Silt	+6
	.004		+8
.001		Clay	

*Following C. K. Wentworth, A scale of grade and class terms for clastic sediments. *Journal of Geology*, 1922.

†$\phi = -\log_2 d$

diameters. For most purposes, the tedious measurement of grain dimensions is unnecessary, and their shape may be adequately described by geometric terms (table 2.6).

The surfaces of sediment grains may be fresh, polished, striated, pitted, or coated, and all of these

Figure 2.6 Roundness of particles

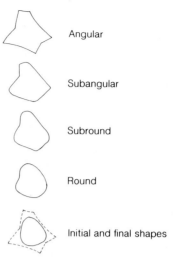

Angular

Subangular

Subround

Round

Initial and final shapes

features can give clues as to the history of the aggregate. For example, wind-transported grains may show numerous impact scars that give the grain a frosted appearance; fresh surfaces on angular grains suggest recent release from the host rock; and coatings that form after deposition are a clue to the chemistry of the interstitial fluids.

The arrangement of the particles in a rock is called the **fabric** and may be homogeneous, implying uniformity of constituents and properties, or nonhomogeneous. The fabric of a sediment is important in determining mechanical **anisotropy** and in such properties as permeability, which is important in the passage of fluids such as groundwater and petroleum or natural gas. Fabric terms for sediments are not well established, but any description should include the degree of sorting and orientation of the constituent grains and the spacing and interrelationships of bedding and other structures.

The degree of **sorting** describes the grain sizes present in sediments and in the sedimentary rocks that may be formed from them. Sorting is good when most of the particles are of similar size, and poor when a wide range of grain sizes is present. The degree of sorting suggests the process and the distance of sediment transport to the deposition site, the type of depositional environment, and it may control the

Table 2.5 Mineral Composition of Detrital Sedimentary Rocks

Mineral	Shale	Sandstone	Limestone
Quartz	32%	70%	4%
Feldspar	18	8	2
Kaolinite and sericite	28	8	1
Calcite and dolomite	8	11	93
Iron oxides	5	1	—
Gypsum	1	0.1	0.1
Chlorite	6	1	—
Carbonaceous material	1	—	—
Miscellaneous	1	1	0.2

After C. K. Leith and W. J. Mead, *Metamorphic Geology.* H. Holt and Company, N.Y., 1915.

engineering uses of a deposit. It should be noted that the geologic terminology focuses on the degree of similarity of grain size, a *well-sorted* sediment being composed of similarly sized fragments. In contrast, engineering usage focuses on the range of grain sizes present, a *well-graded* sediment being one in which the amounts of successively smaller grains are just sufficient to fill the pore space between the larger sizes.

Sorting is accomplished by the selective transport and deposition of the constituent fragments. In streams, some particles that are oversize, overweight, or of the wrong shape will be unmoved whereas others above some hydraulic threshold will be transported either by suspension, rolling, sliding, or jumping (**saltation**). Particles of equivalent hydraulic properties will tend to move together and hence a sorting of both size and nature of particles is achieved.

Primary Sedimentary Structures

A characteristic feature of sedimentary rocks is their layering, or **bedding.** Each layer, or **stratum,** marks a particular set of geological conditions that were either constant or changed regularly during its deposition. Successive beds mark later incidents and the past environmental conditions are recorded in the set of layered beds, or *stratigraphic* sequence. Periods of erosion or nondeposition may also be present in a sedimentary sequence. The individual

Table 2.6 Shapes of Sediment Particles

Term	Shape
Cylindrical	
Discoidal	
Spherical	
Tabular	
Ellipsoidal	
Equant	
Irregular	

beds may vary in thickness from a fraction of a millimeter to tens of meters and are bounded above and below by *bedding planes,* which are mechanically weaker surfaces within the stack along which the rock tends to split. The spacings of bedding planes is related in a general way to the amount of clayey, or **argillaceous,** material present in the rock. Finely bedded fissile shales contain more than 70% argillaceous matter, thinly bedded rocks generally between 40% and 70%, heavily bedded rocks 20% to

Figure 2.7 Typical sedimentary structures

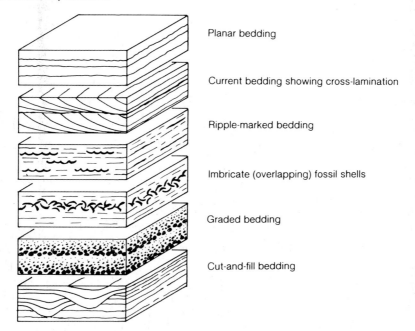

Planar bedding

Current bedding showing cross-lamination

Ripple-marked bedding

Imbricate (overlapping) fossil shells

Graded bedding

Cut-and-fill bedding

40%, and massive limestones and shales less than 20%.

Bedding need not be laminar nor beds be composed of material of identical size. Examples of typical sedimentary structures (fig. 2.7) include the following:

1. *Planar, or parallel, bedding* is the simplest and is initially horizontal due to the smooth laminar flow of the fluid carrying the sediment. The individual bedding planes usually mark a change in grain size of the sediment.
2. *Current bedding* is produced by sediment deposited from a current moving from left to right.
3. *Ripple-marked bedding* may be either symmetrical or asymmetrical, the former being produced by small oscillations of water on the bottom and the latter from essentially the same source plus a unidirectional current.

4. *Graded bedding* evidences a decrease of grain size upward due to decreasing current velocity.
5. *Cut-and-fill bedding* is produced by deposition followed by erosion by currents producing channels that are then filled by later sediment deposition.

Sedimentary rocks may also contain fossils, which are evidence of the former existence of life during the deposition period. Fossils may be shells, animal bones or teeth, tracks, plants, or a wide range of other remains and are used to establish the age of the sedimentary rock.

Lithification of Sediment

Lithification is the conversion of a newly deposited mass of unconsolidated sediment into a coherent and solid rock and usually involves the processes of recrystallization, **compaction,** or **cementation.** The principal mechanisms for these changes are consequences of the fact that unconsolidated sediment deposited in water contains a high percentage of water and the weight of more sediment deposited on top of it will progressively compact it and squeeze the interstitial water from the pores. This usually results in low-temperature chemical reactions, which may transform the grains or introduce material into the remaining pore spaces to cement the rock. Figure 2.8 indicates some of the effects of grain size, sorting, and cementation on porosity.

The processes of lithification, although varied, follow one or both of two themes: interlocking and cementation. Interlocking may occur either in consequence of direct crystallization and recrystallization as in some carbonate rocks or salt beds, by grain enlargement as in quartz sandstones, or by the squeezing together of grains by compaction as in coal or **compaction shale.** In the latter instance, a particularly unstable material for engineering purposes is formed since a shale composed of interlocked clay particles, although apparently hard, has little resistance to deforming forces and will flow when stressed. Also, this material quickly takes up water when exposed and slakes readily to a mud. Fortunately, this latter phenomenon allows simple testing by placing a piece of the shale in water and observing after several hours if it has retained its form or slaked.

If the component particles are in the size range of clay, the water molecules that are attached to particle surfaces form intergrain bonds. Since, however, the water content of a clay aggregate can be readily changed, the strength of such bonds is highly variable and fine-grained clay aggregates and shales that are so bonded must be viewed as dangerous materials.

The randomly oriented and edge-bonded clay plates in a freshly deposited clay aggregate are lightly attached and the interstitial water may be easily expressed by compression. Compressed aggregates of parallel plates are also held together by water bonds and are easily deformed plastically since the plates may slide over one another. Further dessication generates a brittle aggregate but one of low strength, and rehydration returns the solid aggregate to a dispersed state. Charged ions dissolved in the water may increase the strength of bonding and

Figure 2.8 Porosity in granular sediments and sedimentary rock

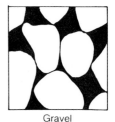

Sand   Gravel

Different sizes, but each well sorted and of about equal porosity

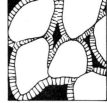

Poorly sorted; low porosity   Porosity decreased by cement

deposit solid cements on dessication. A rock so formed is much stronger than one held together by water bonds only and will not disaggregate when wetted.

The effectiveness of water bonding decreases rapidly with increased particle size and above the clay range becomes negligible. The coarser-grained sediments may be thought of as aggregates of grains of various sizes and shapes with properties analogous to loose sand. Lithification by cementation of such materials usually proceeds by the deposition of solid films on the grain surfaces from materials dissolved in the pervasive groundwater. As these grain coatings thicken they bridge between adjacent grains and form a bond whose strength is that of the cementing mineral matter.

In time, these coatings thicken to reduce the pore space and eventually all openings are filled. Sometimes cementing materials are not deposited unless the sediments are exposed at the surface, so a freshly quarried rock may be relatively soft and workable but harden on exposure, and an uncemented sand encountered in a drill hole may be equivalent to a case-hardened sandstone seen in outcrop.

CASE HISTORY 2.1 The Cannelton Lock and Dam

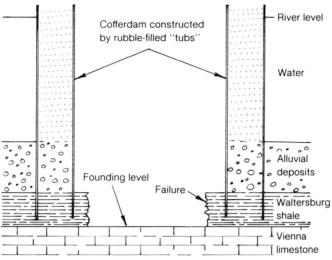

Figure 1 Cross section at cofferdam

The Cannelton Lock and Dam, an integral part of the Ohio River navigation system, suffered a cofferdam failure during its construction in 1968. Original plans, calling for the use of a caisson in the excavation for foundations of the tainter gates, were changed and a cofferdam was employed. This cofferdam was constructed of linked cylindrical tubs made of Z piling and filled with rubble.

The piling was driven through a thick layer of alluvial deposits and nearly through the Waltersburg Shale, which was about 12 meters thick. Excavation within the cofferdam removed the gravels and shale to expose the Vienna Limestone, which was to serve as the founding rock. When excavation was complete, the excavation was walled with the shale in which the pilings were footed. The shale failed and the bottoms of several of the tubs slid into the excavation; fortunately, no workers were present. Salvage operations and internal bracing of the cofferdam allowed completion of the work.

This failure can be traced to either disregard or misinterpretation of well-known geologic and engineering information. Preconstruction information included the following:

1. Good quality subsurface data that included drilling records and samples were available.
2. Studies related to the earlier proposal for caisson work anticipated shale creep under the cutting shoe of the caisson.
3. Tests of driving properties showed that the piling could be driven nearly to the founding level through the shale.
4. A change in plan had been made from founding on shale to founding on the underlying Vienna Limestone.
5. There were numerous references in the geologic literature to the type of clay shale represented by the Waltersburg Shale in western Kentucky.
6. The reputation of the Waltersburg Shale among well drillers who call it the "Big Mud" having found that holes may be easily lost by caving in this rock.

The descriptions and available samples of the Waltersburg Shale should have led inevitably to its characterization as an overconsolidated compaction shale whose low strength was an invitation to failure.

The more commonly occurring cements are *carbonate minerals,* which are acid soluble, mechanically weak, and usually light in color; *iron oxides,* which have a strong red-brown color; and *silica,* which is acid insoluble, usually white, and mechanically strong. Sometimes a silica cement will be deposited in optical continuity with detrital quartz grains to form an interlocking mosaic of grains. A simple test for the relative strength of a cement is whether or not individual grains may be rubbed off the surface of a specimen.

Carbonates are rather easily recrystallized, dissolved, and precipitated under the conditions that exist at or near the earth's surface, so rocks containing calcite or dolomite grains or cement may be markedly modified. Their easy recrystallization may lead to crystalline limestone distinguished with difficulty from metamorphic marble only by the preservation of such features as color or bedding. Easy solubility of carbonate grains and cements generates "secondary porosity" if the composition of the groundwater changes; as does the transformation of calcite to dolomite, which is accompanied by a large volume reduction. Extremes of solution activity may be marked by the development of **karst topography** with sinkholes and caverns.

The mechanical strength of lithified sedimentary aggregates is proportional to grain strength for materials in compression and to cement strength for materials in tension.

METAMORPHIC ROCKS

Metamorphic rocks are generated by the recrystallization of preexisting rocks, either igneous or sedimentary. Since the crystallization of most minerals requires elevated temperature and pressure, the lower limits of metamorphism are taken as about 150°C and the upper limits fixed by the production of a rock melt. The formation of a melt depends upon such parameters as rock composition, water content, and pressure as well as temperature. Generally speaking, a melt will form in the range of 700–800°C and 1–6 kb. Strictly defined, metamorphism (change of form) refers only to the rearrangement of constituent atoms into mineral assemblages stable at the particular conditions, i.e., they are isochemical and the bulk composition of the rock remains unchanged. However, at elevated temperatures small amounts of active fluids may carry material into or out of the system thus changing its overall composition. Such chemical changes are termed *metasomatic* (changed in body or substance).

Since metamorphic rocks may usually be considered as equilibrium assemblages of minerals, the particular rock formed will depend upon the starting composition and the temperature and pressure attained. Considering initial composition only, some metamorphic equivalents of common rocks follow:

Coarse-grained igneous rock = Orthogneiss
Fine-grained (volcanic) igneous rock =
    Metavolcanic rock
Sandstone (quartz rich) = Quartzite
    (distinguished from silica-cemented
    sandstones by interlocking quartz grains
    with sutured boundaries)
Limestone = Marble (distinguished from
    limestone recrystallized at low temperature
    by uniform grain, light color, presence of
    metacrysts)
Shale = Argillite, no cleavage. Slate, good
    cleavage. Phyllite, cleavage and sheen.
    Schist, cleavage and visible grains

For rocks of appropriate composition, such as shales or impure sandstones, the temperature or metamorphic grade reached under metamorphism may be roughly ascertained by the presence of particular indicator minerals in the rock. With increasing temperature the simplified sequence is

chlorite	biotite	garnet	staurolite	kyanite	sillimanite
150°C					700°C

Figure 2.9 Metamorphic facies

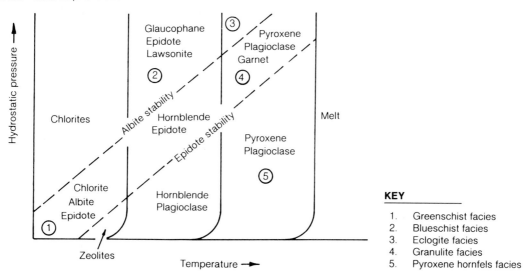

KEY

1. Greenschist facies
2. Blueschist facies
3. Eclogite facies
4. Granulite facies
5. Pyroxene hornfels facies

Since this is a reaction series, the lower-temperature mineral will persist into the higher-grade rocks until reaction is complete and the boundary, or *isograd,* between rocks of different metamorphic grade is marked by the first appearance of the higher-temperature species.

When pressure as well as temperature is considered, metamorphic rocks may be grouped into equilibrium assemblages, or *facies* for particular pressure-temperature ranges. These facies are named after common assemblages (fig. 2.9).

The fabric of metamorphic rocks, as for igneous rocks, is the result of crystal growth, which leads to the interlocking of grains, negligible porosity, and high mechanical strength. In many metamorphic rocks, however, the presence of directed pressure during their formation has resulted in dimensional orientation of the grains and hence to mechanical anisotropy of the mass. Many metamorphic rocks are characterized by the organization of their constituent minerals into parallel bands or *lenticles;* i.e., they are foliated. The mechanical anisotropy of metamorphic rocks, which can be inferred from the presence of **foliation,** is of particular importance in many engineering activities. It should always be considered in the design of foundations and the layout of surface excavations and tunnels.

Since metamorphic rocks may be generated from a variety of starting materials and recrystallized under a wide range of temperature and pressure conditions, their makeup and geologic classification are complex. For engineering purposes, however, they are crystalline rocks similar in properties to those of deep-seated igneous origin with the added complexities of a wider compositional range and sometimes well-developed foliation, and their classification may thus be simplified (table 2.7).

ROCKS AS ENGINEERING MATERIALS

The engineer is more concerned with the properties of rocks than their genesis or classification. Is the rock a good foundation material? Can it be ripped rather than blasted? Will it stand in a road cut? Is the rock permeable? These are typical questions of

Table 2.7 Classification of Metamorphic Rocks

Metamorphic Rock	Fabric	Key Features
	Homogeneous fabric	
Quartzite	Visible grained	Quartz dominant
Marble		Calcite dominant
Dolomitic marble		Dolomite dominant
Argillite	Subvisibly grained	Massive, shaley-looking
	Foliated fabric	
Gneiss	Visibly grained	Blocky minerals dominant (e.g., quartz, feldspar)
		Bands or pods of different composition
Schist		Platey, prismatic, or bladed minerals prominent (e.g., mica, hornblende)
Slate	Subvisibly grained	Splits readily on closely spaced ruptures (cleaves)

engineering importance. Because rocks vary, no text can substitute for experience in answering such questions and experience at one site may, unfortunately, not serve at another location. Careful observation and meaningful testing are thus of utmost importance. One rule: different rocks have different properties, and situations where two or more rocks are present call for particular care.

Scale and geometry are particularly important to the engineer when assessing the properties of rocks at a site. The properties of a rock in hand specimen or outcrop may be of little value in an assessment of road cut stability, soundness of foundation, ease of excavation, or quality of aggregate. Obviously, any site examination should take into account the local properties of each rock type present, but unless the job is small and confined to a single rock unit, the orientation and location of significant discontinuities within the rock mass must also be considered. The type and degree of weathering can markedly affect the mechanical properties of the rock. The presence of more or less regularly spaced cracks or joints is to be expected in any rock mass and markedly affects such features as its physical continuity, water-transporting ability and breakage on blasting. For sites that contain more than one rock type it must be remembered that each unit is a chemical–physical entity, and this must be taken into account for any excavation or building that crosses their boundary. Further, the boundary itself may impose special problems due to fracturing or water channelization.

Other aspects of importance in assessing rocks at a construction site are the various effects superimposed on the mass by later geologic phenomena. The distribution of water in pores and cracks, the nature and extent of rock decay (weathering) or deposition of postrock material (mineralization), erosion, and the results of rock deformation as seen in folded, contorted, cracked, and ruptured (faulted) units all contribute to the understanding needed for intelligent planning and design.

For engineering purposes, the geologic view of rocks and their formation may be modified to emphasize those features of concern for construction materials, excavation, or foundation. From this point of view the important considerations are *dimensions, continuity, mechanical properties,* and to some extent, *chemical properties.* Dimensionally, uniform rock bodies may be simply considered as being larger or smaller than the area of a given engineering project and as having a thickness greater or less than the project's vertical extent or effect. However, the engineering site should always be examined in terms of its location within the regional geologic framework. Continuity involves the number and spacing of actual or imminent surfaces of separation within the body such as joints, shear surfaces, or bedding planes. Mechanical properties are primarily those of bearing-strength and breakability (compressive strength, cohesion) and are related to both the rock fabric and continuity factors. Chemical properties of importance are the relative solubility and chemical stability under the conditions imposed. Many other properties such as bulk density, porosity, permeability, and even color may have importance in particular circumstances.

Since not all of the names and properties of rocks that concern the geologist are of interest to the engineer, it is appropriate to include here a general rock classification scheme (after Duncan, 1969), which incorporates the principal parameters of importance to the engineer. Basically, the scheme recognizes rock fabric together with composition and color as important aspects of any rock and subdivides these into readily recognizable, coded categories:

Texture Code

I. **Crystalline.** Rock consists entirely of interlocking grains. No particles freed when scratched.

II. **Crystalline—indurated** Isolated grains embedded in indurated matrix. No particles freed when scratched.

III. **Indurated** Interlocking grains not visible. No particles freed when scratched.

IV. **Compact** No grains visible. Particles freed when scratched.

V. **Cemented** Grains visible. Particles freed when scratched.

Structure Code

h. **Homogeneous** No visible linear or planar arrangement of grains.

l. **Lineated** Grains show a preferred linear orientation.

i. **Intact foliated** Planar arrangement of grains or color visible. No visible fractures.

j. **Fractured—foliated** Planar structure such as bedding or cleavage visible.

Composition Code

N. **Noncalcareous**
P. **Partly calcareous**
C. **Calcareous**

Color Code

1. **Light colored**
2. **Dark colored**

Grain Size and Sorting Code

a. **Coarse-grained** Particles > 2mm in diameter or longest dimension (corresponds to sand-gravel boundary).

b. **Medium-grained** Particles between 0.1 and 2mm. Visible to the naked eye.

c. **Fine-grained** Particles < 0.1mm. Not visible to the naked eye.

x. **Equigranular** Grains of approximately equal size throughout.

y. **Inequigranular** Grains showing a range of sizes present.

SUMMARY

The aggregates of minerals called rocks can be formed by crystallization from hot, liquid magma (igneous rocks), by lithification of particulate debris or precipitation of dissolved substances on the earth's surface and in bodies of water (sedimentary rocks), or by recrystallization of previously formed rocks in the absence of a melt at elevated temperature and pressure (metamorphic rocks). Each of these classes of rock and their various subdivisions are characterized by particular mineral assemblages, textures, and structures that provide information as to their origin and history.

Rocks are changed over geologic time from one to another kind by such geologic processes as weathering, lithification, and metamorphism. The interplay of process and product can be epitomized by the rock cycle and illustrated by a plate tectonic model (see chapter 3).

Rocks and their properties are of fundamental importance to engineering practise since they are the material to be tunneled, excavated, used as fill and aggregate, or serve as foundations. Some rocks are intrinsically valuable as building materials and ores while others serve as containers of groundwater supplies or reservoirs of oil and gas.

ADDITIONAL READING

Greensmith, J. T. 1978. *Petrology of the sedimentary rocks.* Winchester, Mass.: Allen & Unwin.

Ham, W. E., ed. 1982. Classification of carbonate rocks. American Association of Petroleum Geologists *Memoir 14.* Tulsa, Okla.: The American Association of Petroleum Geologists.

Holmes, A. 1965. *Principles of physical geology.* 2d ed. New York: Ronald Press.

Mason, R. 1978. *Petrology of the metamorphic rocks.* Winchester, Mass.: Allen & Unwin.

Press, F., and Siever, R. 1982. *Earth.* 3d ed. San Francisco: W. H. Freeman.

Simpson, B. 1983. *Rocks and minerals.* rev. ed. Elmsford, N.Y.: Pergamon Press.

Spock, L. E. 1953. *Guide to the study of rocks.* New York: Harper & Row.

Tennissen, A. C. 1974. *Nature of earth materials.* Englewood Cliffs, N.J.: Prentice-Hall.

Tucker, M. E. 1982. *The field description of sedimentary rocks.* New York: John Wiley & Sons.

Underwood, L. B. 1967. Classification and identification of shales. *Proceedings of the American Society of Civil Engineers,* v. 93, paper 5560. J. Soil Mechanics and Foundations Division no. SM6. New York: American Society of Civil Engineers.

Watson, J. 1979. *Rocks and minerals.* Winchester, Mass.: Allen & Unwin.

3

Geologic Time, Weathering, and the Rock Cycle

INTRODUCTION

The basic mineral and rock materials of the earth were introduced in the preceding chapters. We live, however, on the earth's surface where the solid rocks are being decomposed by various physical and chemical processes. This decomposition typically proceeds rather slowly as do a majority of geologic processes. It seems pertinent, therefore, to digress briefly into the concepts of geologic time before the interactions that make and remake rocks are summarized.

GEOLOGIC TIME

One of the principal contributions of geology to science is the recognition of the fact that the earth has had a long and complex history. The enormity of geologic time is a difficult concept to accept and understand; the *age of the earth* approximates 4.6 billion (4 600 million) years, and in this awesome span of time the abundant remains of plant and animal life date back only about 600 million years and human ancestors a mere 3 million (table 3.1). On

Table 3.1 Milestones in the Evolution of Life on Earth

Time in Millions of Years before Present	Animal Life on Earth
3	First humans
70	First important mammals
	Extinction of dinosaurs
235	First dinosaurs
300	First reptiles
350	First land vertebrates
400	First fishes abundant
600	First abundant fossils
3 000	First one-celled organisms
4 600	Origin of the earth

this scale, if the distance from New York to Los Angeles (4 200 km) represents the age of the earth, then the length of time man has been on earth is only 2.7 km, well within the reach of downtown New York. An even more graphic model assigns 1 million years of earth history to each sheet of toilet tissue; at 1 000 sheets per roll the earth's age is a bit over four and a half rolls. The detailed portion of earth history, which has largely been worked out by global correlations using the fossil remains in sedimentary rocks, can be written on six-tenths of a roll and human history and prehistory on the last 3 sheets.

With such a sweep of time, geological processes can proceed at very slow rates yet be completed within a relatively short geologic time interval. A mountain range might take only a few million years to form or a batholith 10 million years to cool; a thickness of 5 or more kilometers of sediment might be laid down in a depositional basin in perhaps 100 million years; the Atlantic Ocean opened during a period of about 200 million years at a rate of a few centimeters per year—about as fast as fingernails grow. Of course, all geologic processes are not so slow; volcanic eruptions, earthquakes, and landslides represent nearly instantaneous processes at the other end of the time spectrum.

Two principal types of geologic time are recognized, namely, absolute time and relative time. *Absolute time* ascribes the actual age of a rock or geologic event in years, whereas *relative time* gives the age of one rock or event relative to another to establish the sequence of geologic events.

Absolute Age

The absolute time or age of occurrence of a geologic event—such as the intrusion of igneous rock magma, rock deformation, erosion, or deposition of sediments—with respect to the present is placed in a time frame by means of an absolute measure of elapsed time based upon the *decay of radioactive elements*. The general principle is that radioactive elements incorporated in primary minerals at the time of rock formation decay steadily with time. By measuring the amount of a radioelement present together with its daughter products at any subsequent time and knowing the rate of radioactive decay, it is possible to calculate the time over which decay has taken place. This is the age of the rock—assuming that no gains or losses of parent or daughter elements have occurred as a result of secondary processes.

The particular radioelements used for geologic age determination must have a long **half-life**—that is, take a long time for the initial number of atoms to be reduced by one-half through decay—in order that measurable concentrations remain. Figure 3.1 shows exponential *decay curves* for four hypothetical radioactive elements whose half-lives are taken as 0.1, 0.5, 1.0, and 2.0×10^9 years together with *growth curves* of their daughter products. All else being equal, best results should be obtained when the half-life of the element used is of the order of the age to be measured.

The term *radioelements* used earlier actually refers to the radioactive **isotopes** of the elements. Every atom of an element possesses the same number

Figure 3.1 Decay and growth curves of hypothetical radioelements and their daughter products

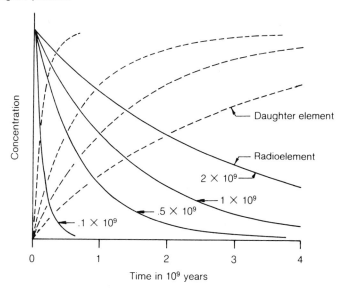

Concentration

Daughter element

Radioelement

2×10^9

1×10^9

$.5 \times 10^9$

$.1 \times 10^9$

0   1   2   3   4

Time in 10^9 years

Table 3.2 Radioisotopes Commonly Employed in Age Determination

Element	Isotope	Half-life, Years	Daughter Product
Thorium	^{232}Th	14.1×10^9	^{208}Pb
Rubidium	^{87}Rb	4.7×10^9	^{87}Sr
Uranium	^{238}U	4.5×10^9	^{206}Pb
Potassium	^{40}K	1.3×10^9	^{40}Ar
Uranium	^{235}U	0.713×10^9	^{207}Pb
Carbon	^{14}C	5.71×10^3	^{14}N

of protons in its nucleus, but the number of neutrons may vary in atoms of the same kind resulting in different atomic weights or isotopes. Uranium, for example, commonly occurs in two isotopes, ^{235}U and ^{238}U; the former has 235 neutrons and protons in its nucleus while the latter has 238. When an atom of ^{238}U disintegrates radioactively it loses 32 neutrons and protons and becomes a stable isotope of lead, ^{206}Pb.

The half-life of some isotopes is measured in fractions of a second and these are obviously of no use in geological age determinations. Others, such

as ^{238}U have very long half-lives, in this case 4.5 billion years, which approximates the age of the earth. Hence ^{238}U and similarly long-lived radioisotopes are useful in determining the ages of very old rocks (table 3.2). Because of its relatively short half-life, ^{14}C is not useful for dating most rocks but can be used for dating fossils, artifacts, bones, or plants not older than about 30 000 years.

Absolute dating methods are commonly applied to igneous and metamorphic rocks rather than to sedimentary rocks, which are often made up of

Figure 3.2 Geologic time scale

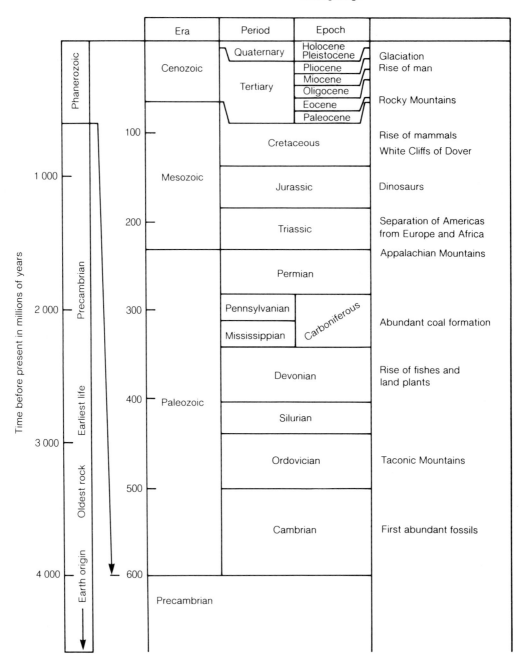

earlier-formed minerals. Sedimentary rocks are more usually dated by relative age methods discussed later. Absolute age determinations are subject to the following limitations:

1. The rock must contain a mineral with a suitable radioactive isotope.
2. The isotope must have a suitable half-life relative to the age of the rock.
3. Ages can be distorted by later events (i.e., the radioactive clock can be reset by recrystallization or parent and daughter products added or lost by alteration).
4. The method is relatively expensive.

Geologists have established a time scale (fig. 3.2) to divide geologic time into recognizable units. The three great Phanerozoic **eras** are named for the nature of the fossil remains that are found in the rocks, representing times of ancient (paleo-), middle (meso-) and recent (ceno-) life (-zoic). These eras are further subdivided into **periods,** generally named for places where the representative rocks were first recognized (e.g., *Devonian* for Devon in England). These, in turn, are divided into **epochs,** for example, *Eocene* (dawn life). Finer divisions are commonly used in local geologic descriptions and terms such as *group,* **formation,** *member,* or *bed* are employed.

Relative Age

The relative geologic age for sedimentary rocks may often be determined by comparing the fossils they contain with the evolutionary sequence established by specialists in the developmentary aspects of organisms. During the past 600 million years the hard parts of plants and animals (e.g., spores, shells, teeth, and bones) have often been preserved in the sediments after the death of the animal. The nature of these fossils as individuals or in assemblages is characteristic of their evolutionary stage, and this stage defines a particular time period of the past. Sedimentary rocks are obviously of the same age as the fossils they contain except in the unusual circumstance of the fossil being an eroded fragment from an earlier rock. Thus fossils provide one means of determining relative geologic age. This leads to a fundamental law used in the interpretation of the relative age of a sedimentary sequence known as the law of **superposition,** which states, "In an undisturbed sedimentary sequence, the oldest bed is at the bottom and the youngest at the top."

This principle of superposition can be extended to evaluate the following examples of other types of relationships among sedimentary, igneous, and metamorphic rocks and their associated structures, intrusions, and erosion surfaces:

1. Younger beds are deposited on top of older beds.
2. Inclusions are older than their host.
3. Intrusions are younger than the country rock.
4. Erosion is after rock formation.
5. Erosion typically accompanies and follows uplift.
6. Deformed rocks are older than the deformation.

The key to the determination of the relative age of rock units will be found at their geologic **contacts.** The main types of geologic contacts (fig. 3.3) include the following:

Conformity   the bedding contact surfaces are parallel and may be either horizontal or inclined. This kind of contact in sedimentary rocks indicates that sedimentation proceeded continuously layer upon layer without a break (fig. 3.3a).

Disconformity   an irregular bedding contact indicating a break (*x*) in sedimentation due to nondeposition or erosion or both between conformable beds above and below (fig. 3.3b).

Unconformity   contact represented by an erosion surface (*y*) between dipping beds below and horizontal beds above (fig. 3.3c).

Nonconformity   contact between igneous rocks and others (fig. 3.3d).

Figure 3.3 Geologic contacts

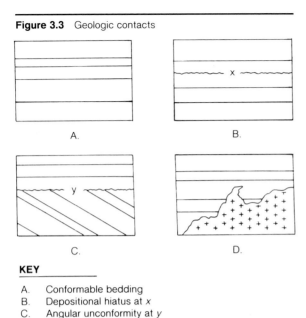

A.

B.

C.

D.

KEY

A. Conformable bedding
B. Depositional hiatus at *x*
C. Angular unconformity at *y*
D. Nonconformity

When relative age can be determined from contact relationships and an absolute age is available for some of the formations it is possible to approximate the absolute age of the undated rocks. Some examples are given in figure 3.4. In cross section *A* a lower inclined series is overlain by a horizontal series. Both series are conformable within themselves but are separated by an angular unconformity. The sequence of events from oldest (*a*) to youngest (*i*) is the deposition of the conformable sequence *a* through *d* followed by the tilting of the sequence, either accompanied or followed by erosion, producing the unconformable contact *e*. This was followed by the deposition of the conformable horizontal sequence *f* through *h* and finally by the present erosion surface *i*. Based upon fossil content, if series *a–d* is Ordovician and series *f–h* is Silurian (see fig. 3.2), then the erosional event that produced the unconformable contact took place between 400 and 500 million years ago.

In cross section *B* the conformable sequence *a* and *b* is intruded by plutonic igneous rock *c,* and there must have been considerably greater thickness of overlying rock at that time for the pluton to form and metamorphose *a* and *b*. This overlying rock was removed by erosion to produce the erosion surface *d*. The erosional event was followed by the deposition of the sedimentary rock sequence *e, f,* and *g,* conformable within itself but separated from *b* by a disconformity and from *c* by a nonconformity. The lack of metamorphism of *e* by the granite indicates the granite was uncovered by erosion before *e* was deposited. If the intrusion is dated at 200 million years and bed *e* is Jurassic on fossil evidence, the nonconformity is late Triassic.

Cross section *C* shows a conformable sedimentary sequence deposited in the order *a* through *d,* from oldest to youngest; followed by folding and erosion producing the surface *e;* and then intruded by dike *f* to produce volcanic rock *g*. The rock *c* is dated as Eocene, so the age of the volcanism is less than about 60 million years.

WEATHERING AND EROSION

People make adjustments in their dress and manner as they move from one environment to another; similarly, rocks change both physically and chemically as they adjust to new environments. The group of changes that characterizes a rock's adjustment to conditions at the earth's surface is termed **weathering** and includes all of the physical and chemical modifications of the initial material that gradually convert rock into soil. Weathering of rock is commonplace because most rocks form at some depth within the earth at higher temperature and pressure and in a different chemical environment than is found at the surface. When these rocks are brought to the surface by outpourings of lava, deep **erosion,** or uplift, a whole family of physical and chemical reactions come into play.

The breakdown of rock material produces both dissolved salts and fragments of various sizes. These *products* of weathering may remain in place as residual materials but are more or less readily removed and carried by such agents as running water, wind, or glacial ice in the ever-present processes of erosion.

Figure 3.4 Examples of relative age

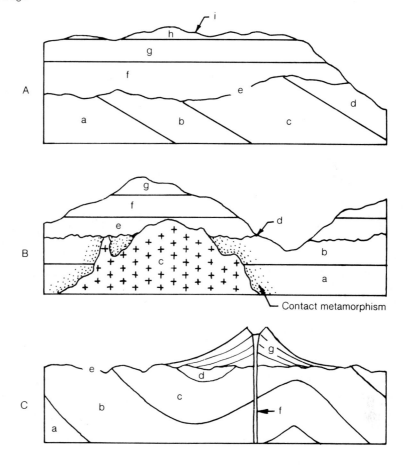

Contact metamorphism

KEY

A. Older sedimentary rock layers (*a–d*) separated from younger layers (*f–h*) by an erosional unconformity (*e*). The present surface (*i*) is a second unconformity in the making.

B. Older sedimentary rock layers (*a–b*) intruded by younger igneous rock (*c*). Erosion surface (*d*) then cut both igneous and older sedimentary rocks. Finally, sedimentary rock layers (*e–g*) were deposited above the erosion surface.

C. Sedimentary rock layers (*a–d*) were deformed by folding and eroded to surface (*e*). Later volcanism through conduit (*f*) deposited a cone of lava and pyroclastics (*g*).

Bedded sediments, western Kentucky

Unconformity, Tom Sauk Power Station, Missouri

Sheeting in granite, Cape Ann, Massachusetts

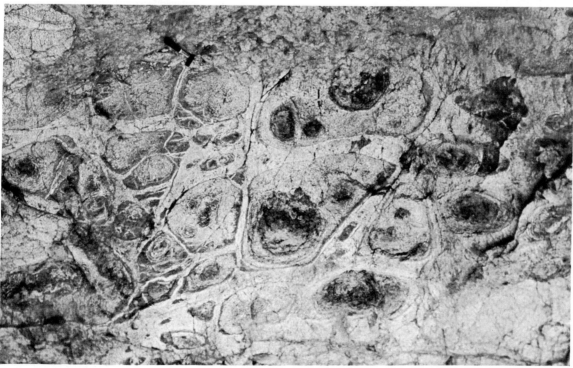

Concentric weathering in basalt, Virginia

Figure 3.5 Frost wedging

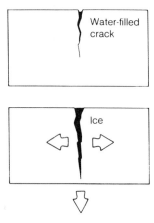

Figure 3.6 Development of sheetings

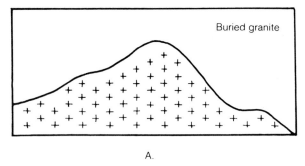

A.

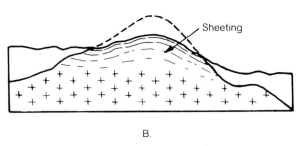

B.

KEY

A. Before erosional unloading
B. After unloading

Physical Weathering

Physical weathering breaks up or *comminutes* (pulverizes) rock without altering the composition of the rock particles. Frost action, a particularly effective means of physical weathering, produces localized stress in a rock that contains water in cracks or pores. Since water expands when it freezes, frost can cause the mechanical disruption of the rock (fig. 3.5). **Abrasion,** resulting from the physical contact between rock fragments in transport that grind against one another and against the rocks over which they pass, is a significant means of comminution. Physical weathering may also take place as the result of *pressure release*. If, for example, a granite intrusive body is emplaced at depth and later unroofed and unloaded by erosion, the accompanying release of pressure allows the granite to expand but, being brittle, it also fractures forming cracks parallel to the earth's surface. The result is a slabby surface called exfoliation or **sheeting,** which is characteristic of homogenous rock that has been unloaded (fig. 3.6).

Many other orientations of cracks may be developed when rocks are unloaded, and in fact, not only granite but nearly all rocks show tensional cracking so that exposed rock surfaces are typically broken into more or less regularly shaped blocks. These cracks are called **joints** from their similarity to the joints in masonry work. Typically, the spacing of joints decreases with depth, but incipient (unopened) joints will be present in deeper zones and may be usefully exploited in excavation and quarrying. The disruption of rock by jointing is of great importance in engineering because it changes the continuity of a rock mass and hence its bulk mechanical properties as well as providing openings for subsurface water flow.

Chemical reactions involving rocks and minerals also cause changes in their volume and so disrupt the material and comminute the rock. Chemical reactions may thus lead to physical weathering.

Chemical Weathering

The new surfaces produced by abrasion or fracturing (two per crack) generate new sites for chemically reactive agents to attack and dissolve or alter the rock. This chemical weathering, although less visible than physical weathering, is quantitatively the more important process in rock weathering. Chemical weathering processes may precede or follow the processes of physical weathering, but most usually

Table 3.3 Principal Products of Rock Weathering

Primary Minerals and (Major Elements)	Weathering Products			
	Unchanged solids	New solids	Colloids	Dissolved ions
Quartz (Si)	Quartz (Si)	—	—	—
Feldspar (K, Na, Ca, Al, Si)	—	Clay (Al, Si)	Si, Al	K, Na, Ca
Ferromagnesian minerals (Ca, Mg, Fe, Al, Si)	—	Clay (Al, Si)	Si, Al	Ca, Mg
		Iron oxides (Fe)	Fe	

The weathered product is a mixture in various proportions of iron oxides, clay minerals, and quartz. Soils will also contain more or less organic debris.

accompany them with one process amplifying the other in a synergistic fashion.

The earth's surface abounds in chemically active agents such as water, carbon dioxide, free oxygen, and organic acids. Alone or in combination these reagents can attack and transform all but the most resistant of minerals. The general trend of the chemical changes that take place is expressed by the *rule of LeChatelier,* which states that any shift in factors affecting the equilibrium of a system will cause the system to change in such a way as to neutralize the effect; that is, a decrease in pressure will lead to the production of phases of larger volume, or an increase in heat will cause *endothermic* (heat absorbing) reactions to take place. For example, the generation of carbon dioxide, CO_2, from a carbonate-containing mineral is promoted by decreased pressure because of the larger volume of the gas phase, $CaCO_3$ (solid) $\rightleftharpoons$ CaO (solid) + CO_2 (gas), and a reaction such as $2\,CO + O_2 \rightleftharpoons 2CO_2$ + heat goes to the left by raising the temperature.

Some broad generalizations as to the course of chemical weathering may be drawn from observation. Primary igneous minerals such as quartz and, to a lesser extent, muscovite mica are stable under surface conditions and tend to be concentrated as residual materials in weathering products. Feldspars alter to *secondary solids* such as sericite (a

variety of muscovite mica) and various clay minerals (e.g., **kaolinite** or **montmorillonite**). The ferromagnesian minerals olivine, hornblende, and biotite mica yield, respectively, numerous *alteration products* among which green talc, chlorite, serpentine, and epidote; white calcite, clay minerals (again), and secondary quartz; and brown to red oxides and hydrated oxides of iron are common. The final products of the chemical decay of an aggregate of these primary igneous minerals will be such resistant solid particles as quartz and muscovite; secondary solids such as clays, aluminum hydrates, and hydrated iron oxides; colloidal silica; and the **dissolved ions** Na^+, K^+, Ca^{2+}, and Mg^{2+}. A much simplified scheme of weathering products is shown in table 3.3. These solids and colloids are more or less thoroughly mixed with organic debris at the earth's surface resulting in a distinctive new product of great importance in civil engineering—soil.

The earliest stages of chemical weathering are dependent upon the local chemical environment, which may not even be the same for different parts of the same outcrop. The first effects in many instances are the **oxidation** of ferrous iron to the ferric state, with the consequent development of brown to red coloration, and the removal in solution of calcium, sodium, and to a lesser extent, magnesium. On the other hand, long continued chemical weathering

Figure 3.7 Water molecule, H_2O

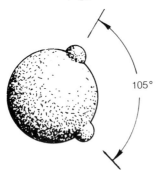

105°

tends to a control of the processes by the climatic factors of temperature and rainfall with the result that the products converge to a few widespread soil types.

The interaction of water with mineral solids is the cardinal factor in chemical weathering. The water molecule, H_2O, is asymmetric with two hydrogen ions embedded in the oxygen ion at an angular separation of 105° (fig. 3.7). Because the positive and negative charges are not aligned, the molecule has a positive (hydrogen) and negative (oxygen) side (i.e., it is dipolar).

Water molecules, in the process of **hydration,** are attracted to ions that are part of a crystal structure, or in the process of *dissolution,* they may completely isolate an ion by clustering about it. Should the charge on a mineral surface be particularly high, water molecules may be dissociated to form positive hydronium and negative hydroxyl ions by the process of **hydrolysis:**

$$2 \, H_2O \xrightarrow{\text{hydrolysis}} (H_3O)^+ + (OH)^-.$$

In the weathering of some minerals, the attachment of hydroxyl ions to mineral surfaces on the walls of a crack results in an increase in volume that propagates the crack and provides still more reactive surface.

Many minerals, particularly feldspars, tend to consume more hydrogen than hydroxyl ions during hydrolysis because $(H_3O)^+$ replaces the alkali metal in the crystal structure: $Na(AlSi_3O_8) + 2 \, H_2O \rightarrow (H_3O)(AlSi_3O_8) + Na^+ + (OH)^-$. Even more complex reactions transform the aluminum silicate

product into a clay mineral. The $(OH)^-$ buildup from such reactions may yield strongly reactive alkaline fluids that interfere with the weathering reaction. Neutralization by natural acids, principally carbonic acid, allows the reaction to proceed: $H_2O + CO_2 \rightleftharpoons H_2CO_3 \rightleftharpoons H^+ + (HCO_3)^-$ and $H^+ + (OH)^- \rightarrow H_2O$. Other sources of hydrogen ions are volcanic gases, which contain abundant hydrochloric acid, HCl; the oxidation of pyrite, $FeS_2 + 2O_2 \rightarrow FeSO_4$; and $FeSO_4 + H_2O \rightarrow H_2SO_4 + Fe^{2+}$; and decaying vegetation, which provides humic acids.

A related chemical action of great importance is **carbonation,** which is particularly effective in the destruction of carbonate rocks and the consequent development of cavernous conditions: $H_2O + CO_2 \rightleftharpoons H_2CO_3$; $H_2CO_3 \rightleftharpoons H^+ + (HCO_3)$, and $H^+ + CaCO_3 \rightleftharpoons Ca^{2+} + (HCO_3)^-$; therefore, $CaCO_3 \underset{\xleftarrow{}}{\xrightarrow{-H_2O-}} Ca^{2+} + (HCO_3)^-$. The products of the forward reaction are the calcium ion and the bicarbonate radical, which are in solution and if removed by flowing water leaving a limestone area result in cavernous ground (fig. 3.8). The reactions, however, are close to equilibrium, and deposition of $CaCO_3$, the mineral calcite, may be brought about by an increase in either the Ca^{2+} or bicarbonate concentration resulting in the formation of stalactites and stalagmites in caves and carbonate cement in sedimentary rocks.

TRANSPORT AND DEPOSITION

Weathered debris and exposed rock at the earth's surface are subject to erosive attack and transport by various agents of erosion and transportation whose tendency is always to move material downslope and laterally. The principal transporting agents, apart from humans, are direct gravitative action resulting in *mass movement* by falls, slumps, slides, and flows and gravity-induced *fluid flows* of air, water, or ice that erode and transport debris.

Gravitative mass movement includes all of those downslope phenomena in which a fluid phase is usually an important, but not necessarily a major,

Figure 3.8 Cave formation in limestone

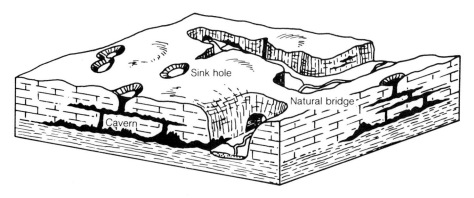

Valley glacier, Alaska

Fossil stream channel, Mexico

Alluvial fan, southern Spain

54   Geologic Considerations

Sand dunes, White Sands, New Mexico

component of the moving mass. Included are rock falls, which may accumulate as fragments, or scree, at the base of slopes and often form conical deposits of **talus**; slumps of poorly consolidated material to more stable attitudes; and surficial slides, flows, or creeps of soil or rock. Aside from natural causes, downslope mass movements may be triggered by poor design of engineering works and thus constitute a significant hazard (discussed at some length in chapter 13). Much greater total transport, though less striking, is provided by moving fluids. These act more or less continuously and, over time, transport huge quantities of dissolved and detrital materials. In the overall picture, mass movements act as short-distance feeders to the long-haul fluid agents. Sediment transport in streams is examined in chapter 5.

Different fluid agents are able to transport different sizes of particles; the largest wind-suspended particles being 0.2 mm in diameter, direct water-borne material seldom exceeding a few tens of centimeters in diameter, and ice being competent to carry blocks measurable in tens of meters. The total amount of solids, or *load,* that may be transported per unit volume of agent (as distinct from maximum particle size) also varies with the agent in the same order.

Moving water and wind are low-viscosity agents that more often move in turbulent than in laminar flow. Transport by them is through four recognizable mechanisms: solution, suspension, saltation, and sliding or rolling. If attention is focused on the base of the fluid agent at its interface with a solid substrate, some important aspects of the transport mechanism may be visualized. Solution or **corrosion** transfers atoms from solid to liquid whereas precipitation or adsorption has the opposite effect; abrasion, or **corrasion,** grinds away the substrate and reduces the size of moving particles. Fluid flow over an irregular bed generates **turbulence,** or random directional movements and velocities in the fluid, that increases with both bed roughness and fluid velocity. Fluid **shear** stress across bed particles causes them to *roll* or *slide,* and local low-pressure volumes give a lifting movement that may be sufficient to pop particles up from the bottom. Such particles may

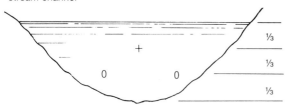

Figure 3.9 Loci of maximum velocity and turbulence in a stream channel

⅓

⅓

⅓

KEY

0 Thread of maximum turbulence
+ Thread of maximum velocity

then follow a trajectory through the fluid and come to rest at some distance downstream (i.e., they jump, or *saltate*) or they may be continuously buoyed and travel for considerable distances in *suspension.*

Turbulence, the great mover, increases toward the bottom of a moving fluid and, in channeled flow such as a stream, is further localized into two maxima in the active channel (fig. 3.9). It should be noted that this is positionally different from the maximum velocity locus within a stream.

The various mechanisms causing particle movement are sensitive to both the mass and the shape of fragments and hence are highly selective in the manner and distance of their transport. Sediment grains of similar hydrodynamic properties are thus selectively transported and deposited (i.e., they are *sorted*). Deposits of similar materials may result and the identity and nature of a transporting agent may often be determined by the distribution of sizes and the degree of rounding of the material in its deposits.

Wind can only move dry particles, so it is only an effective erosive agent in arid regions such as midlatitude deserts and polar regions. Only dust particles can be carried in suspension in air, and regions far downwind from source areas may receive thick accumulations of **loess** composed of fine airborne particles. Volcanic ash is another wind-transported material. Wind can move sand-size material by saltation, and sand grains tend to accumulate in *dunes* on the lee edge of deserts or behind beaches. Particles larger than sand remain in place in the source area as *lag gravel* or *desert pavement.*

Streams are more competent than wind and can carry larger loads but are less capable of fine sorting. Transported particles tend to be rounded by abrasion, but the intergranular water film cushions their impact and limits the size to which they may be reduced. Because of the wide range of flow conditions in the usual stream, due to variable discharge and the migration of its channel, the deposits that result from these fits-and-starts transports and depositions are complex. Within stream channels the general form of their depositions is interfingering lenses of bed load materials.

Moving ice, whether channelized as a mountain glacier or in an ice flood covering areas of subcontinental dimensions, has profound effects on topography in consequence of its strong abrasive action. Rock fragments plucked from the floor rasp away the bed rock, and blocks fall on the ice from adjacent cliffs and are borne away. The processes of load acquisition and transport are nonselective, and ice-related deposits are typically composed of unsorted, and often angular, debris ranging from clay to boulder size.

The downward and lateral movement of transported debris on land is toward local **base levels,** represented by interior basins, lakes, or main stream channels for tributaries, and eventually to the base level represented by the ocean shore. Conditions change abruptly when a base level is reached, and the deposition of sediment tends to be localized at this point: **deltas** develop at river mouths; lakes and basins fill; sediments are trapped in lagoons or behind barrier reefs or islands; and estuaries become silted. Movement of sediment, however, does not end at this point because downslope flows of dense, water-sediment mixtures, currents, or ocean waves continue its transport and sorting.

Much of the sediment load in lakes and oceans is carried along the shore-forming beaches, barrier islands, bars, and spits. Eventually, offshore movements remove the debris beyond the reach of effective wave and current action so that it accumulates as shallow-water sediments, which may be consolidated, or lithified, awaiting a drop in base level or a tectonic uplift that will again expose the sedimentary rock to the vicissitudes of weathering and the long journey to the depths of the sea. Many sedimentary rocks have evidently gone through such a cycle of erosion–transportation–deposition–lithification–erosion not once but many times because sorting and rounding processes are not efficient enough to account for the maturity observed in some sedimentary deposits if only one transit had taken place.

The complex system of surface–water movement downslope and along stream channels is not restricted to the earth's surface; it is related directly to both the atmosphere (through precipitation, evaporation, transpiration) and the subsurface. Open channels in the way of **caverns** exist locally and represent three-dimensional subsurface stream systems receiving water from and discharging it to the surface flows. As openings become smaller, however, resistance to flow increases and subsurface flow becomes sluggish and eventually stagnant.

Water falling upon and sinking into an exposed portion of the earth's surface percolates downward through a zone (variously called the **vadose zone,** the **phreatic zone,** or the **zone of aeration**) in which the soil and rock openings are not filled with water to a level, the **water table,** below which the rock is water saturated. Movements of water below the water table are in directions dictated by pressure gradients and are dominantly lateral but show much local variability.

Bodies of surface water are connected directly with the subsurface domain and, indeed, represent local conditions where the water table is above the surface of the land. Many aspects of this intimate connection of surface and subsurface water flow are important, and engineers will commonly have to deal with one or more aspects of groundwater complexities. Underground water is further discussed in chapter 6.

THE ROCK CYCLE

A common means of representing the cyclic nature of geologic phenomena is by means of a diagram. The *rock cycle* (fig. 3.10) is a particularly useful framework for study that organizes and summarizes the rocks and processes of physical geology. In the

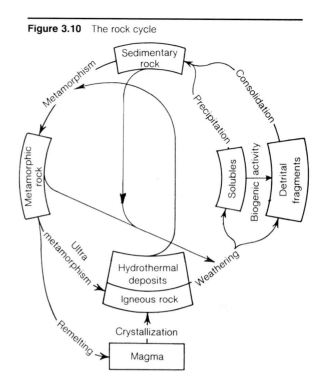

Figure 3.10 The rock cycle

diagram the *principal materials* are boxed and the *processes,* which move and transform them, are shown by arrows. Note that different paths or shortcuts are available and that many materials may spend long periods of time in subcycles; for example, the multiple cycles of lithification, weathering, transport, deposition, and relithification of sediments into sedimentary rocks.

Magma, which may be considered as a primary geologic material, serves as a starting point for the following brief step-by-step description of the rock cycle. Magma is a complex melt—a solution of silicate material formed in large pockets at a depth of many kilometers within the earth. Magma itself has never been observed directly, but *extrusive lava* is well known as are *igneous rocks,* which formed when magma that cooled and crystallized at depth was later exposed at the surface. Magma must have been formed in enormous quantities in the very early stages of the formation of the earth, and the generation and emplacement of magma, often in volumes measurable in hundreds or thousands of cubic kilometers, has continued throughout time. It is estimated that about 20 km³ of basaltic lava is erupted annually from the world's volcanoes and rift zones. On cooling, magma gives rise to the different kinds of igneous rocks whose composition is dominated by a relatively few silicate minerals.

When the magma in a large subsurface pocket has cooled, the minerals in the resultant igneous rocks are chemically stable only at the conditions of temperature, pressure, and chemical potential at which they formed. These rocks can be brought into a different environment near or at the earth's surface by mountain-building movements or by removal of the covering rocks by denudation. In the new environment, pressure and temperature are lower; oxygen, water, and carbon dioxide are more abundant; and the rock may be exposed to physical disruption and chemical change. The exhumed igneous rock and its constituent minerals will react to the new environment in a variety of ways, all tending toward equilibrium under the new conditions. Expansion because of lowered pressure results in open or incipient cracks, or joints; chemical reactions lead to oxidized, hydrated, and carbonated minerals that are typically of greater volume than the original, thus introducing more stress to produce more cracks providing more surface for more chemical activity.

The chemical and physical breakdown of a rock caused by these and other processes is called *weathering*. Working singly or together the physical and chemical processes cause rock to gradually decay and eventually become converted to a layer of *soil* composed of the more resistant minerals, mineralogic decay products, and various proportions of organic debris. All minerals of igneous rocks are not equally susceptible to physical and chemical weathering processes, and the intensity of the processes is also dependent upon climatic conditions. Weathering is thus a function of both a rock's mineralogic makeup and the locale.

Disaggregation of a rock by weathering generates dissolved ions and small particles, which are readily moved in the surface and near-surface environment. Underground and surface water, wind, glacial ice, waves and currents, and the direct action of gravity all provide means for *transport*. Because of the varied nature of the materials to be moved and the differing capability, or competency, of the agents, transport will tend to be selective, and like materials will travel and be deposited together. Transport over any distance will be "steady-by-jerks" with *deposition* and retransportation controlled by temporary changes in physical or chemical factors. Final deposition may be considered as taking place when the residence time is sufficiently long so that the material becomes *lithified* into a *sedimentary rock*.

The lithification of sediments takes place by direct crystallization, a consolidation of the grains by compaction, the deposition of a cementing agent in the grain interstices, or by some combination of these. In the process the interparticle void space is reduced and interstitial fluids are forced out. Sedimentary rocks typically show the consequences of their transport and depositional and lithification history in their more or less perfectly layered nature and tendency toward monomineralic composition.

Some sedimentary rocks are deposited in basinal areas or on continental margins that are tectonically active and may be carried downward by movements of the crust of the earth. If this happens, they are moved into environments that, with increasing depth, become hotter. Sequential changes in mineral composition and rock fabric occur and the rock changes form, or undergoes *metamorphism*. The changes may be local or regional, heat- or pressure-induced, or both. Ultimately, temperatures may rise to levels at which *melting* takes place and magma is generated. Alternatively, the metamorphic process may be such that an igneous-looking rock is formed by *ultrametamorphism* without going through a liquid magmatic stage.

The crust and upper mantle of the earth consist of a number of discrete plates, which are moved laterally by convective motions in the lower mantle at rates of a few centimeters per year. **Plate tectonics**—the generation, motion, and consumption of these large plates—provides a dynamic mechanism for the understanding of geological processes. **Subduction** carries an oceanic plate to the hot depths where metamorphism followed by selective melting occurs. The lighter liquids rise as intrusive and extrusive magmas and lavas with associated thermal and hydrothermal (hot water) effects. The conti-

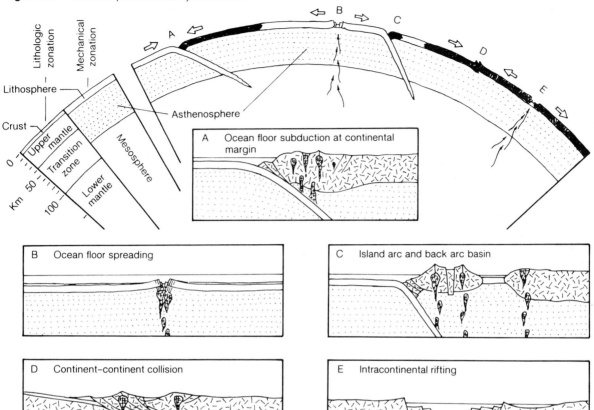

Figure 3.11 The outer portion of our dynamic Earth

Lithologic zonation

Mechanical zonation

Lithosphere

Crust

Upper mantle

Transition zone

Lower mantle

Mesosphere

Asthenosphere

Km 0 — 50 — 100

A — Ocean floor subduction at continental margin

B — Ocean floor spreading

C — Island arc and back arc basin

D — Continent–continent collision

E — Intracontinental rifting

nental plate is arched, broken by faulting, and crowned by a volcanic edifice exposing deep-seated rocks to the water- and oxygen-rich surface. In this region of considerable relief, weathering and erosion disaggregate and chemically modify the rocks and transport the sediment into an adjacent trench or spread it upon the continental block to be eventually lithified to sedimentary rock.

Analogous actions take place at *spreading centers* where rising magma differentiates to provide igneous products ranging from ultramafic (silica-poor) rocks near the base of the oceanic plate to basalt dikes, flows, and volcanic piles to effusive hydrothermal liquids. This differentiated sequence may then be transported by plate movement to convergent zones where it will be reworked thermally

in a downthrust subduction zone or uplifted and eroded in an upthrust zone of obduction.

The upper portion of figure 3.11 is a true-scale section through the outer portion of the earth. The smaller cross sections show some of the settings for rock formation. Section *A* might be the Pacific margin of South America with a marginal trench and plutons of igneous rock rising under the Andes Mountains. Section *B* represents an oceanic spreading center such as the Mid-Atlantic Ridge. Rising and differentiating magma at such a location generates the 5-kilometer-thick basaltic layer that underlies the ocean basins. Section *C* shows the complex relations to be found when an island arc, say Japan, develops outside a back arc basin separating it from the continental mainland. The arc area is characterized by volcanic rocks and basinal sediments in complex association. A continent–continent collision shown in section *D* might be represented by the Himalayan ranges and involves a thickened and complexly overthrust zone with sediments being shed from the uplifted region. Section *E* is representative of intracontinental *rifting,* as in the Red Sea.

Plate tectonics has provided geologists with a conceptual model for the timing and generation of earth surface features including the distribution of past and present oceans and mountain ranges, mineral deposits, climates such as glacial epochs, and the distribution of flora and fauna.

SUMMARY

Geologic processes follow a cyclic pattern, and in a broad view, no geologic material has become dominant in the whole 4.6×10^9 years of earth history. Also, since some materials typical of the interior portions of the earth are found at the surface there must be interaction between the interior and the surface. This broad view does not mean that local accumulation of a particular kind of rock material has not taken place but emphasizes the temporary nature of combinations, forms, location, and nature of aggregation on the geologic time scale. Permanency of the geologic situation may generally be assumed, however, for the anticipated life of an engineering project if care is taken not to upset the equilibrium of conditions that exist.

ADDITIONAL READING

Birkeland, P. W. 1974. *Pedology, weathering, and geomorphological research.* New York: Oxford University Press.

Boyer, R. E. 1971. *Field guide to rock weathering.* Boston, Mass.: Houghton Mifflin.

Carroll, D. 1970. *Rock weathering.* New York: Plenum Press.

Eyles, N., ed. 1984. *Glacial geology: An introduction for engineers and earth scientists.* Oxford: Pergamon Press.

Goldthwait, R. P. 1975. Glacial deposits. In *Benchmark papers in geology no. 21.* Stroudsburg, Pa.: Hutchinson Ross.

Keller, W. D. 1968. *The principles of chemical weathering.* Columbia, Mo.: Lucas Brothers.

Loughnan, F. C. 1969. *Chemical weathering of silicate minerals.* New York: American Elsevier.

Mohr, E. C. J., and Van Baren, F. A. 1954. *Tropical soils.* New York: Interscience.

Oxburgh, E. R. 1974. The plain man's guide to plate tectonics. *Proceedings of the Geologist's Association,* vol. 85. London: Geologist's Association.

Schulz, C. F., and Frye, J. C., eds. 1965. Loess and related aeolian deposits of the world. *Proceedings of the VIIth Congress, International Association for Quaternary Research,* v. 12. Lincoln, Neb.: University of Nebraska Press.

4

Soils

INTRODUCTION

Engineers, geologists, and agronomists are all deeply concerned with one or another aspect of soil, and each has developed particular but overlapping views of its formation and properties. Agronomists (pedologists) are primarily concerned with soil fertility, which includes its composition, zoning, texture, and water content. Geologists view soil as a component of the earth's surface zone of rock decay where rocks are transformed into surficial debris (regolith) by processes of physical and chemical weathering. A very large proportion of civil engineering practise is related in one way or another to the excavation, drainage, reconstitution, or building upon the soil; and geotechnical engineers deal principally with its physical and mechanical properties.

Soil is difficult to define but may be considered to be an unconsolidated to poorly consolidated and variable mixture of coarse-grained inorganic solids such as gravel or sand, silt, and clay admixed with more or less decayed vegetable debris and containing variable amounts of interstitial water. It is common for a soil to be vertically zoned and to change laterally in its constitution and thickness.

The complexity of soil, coupled with its importance as an engineering material, accounts for the emphasis that is given to it both here and in later chapters where such aspects as *soil index properties* (chapter 10), *soil mechanics* (chapter 11) and *slope stability* (chapter 13) are discussed.

In addition to solid constituents, soil contains water (as liquid, vapor, and sometimes ice) together with such gases as oxygen, nitrogen, carbon dioxide, and methane. The degree of water saturation and the rate of drainage markedly affect soil properties and are particularly important aspects of soils in projects for which soil stability or drainage is a factor. The upper limit of saturation—the *water table*—in the ground is not fixed but varies slowly in response to the balance of the **recharge** and **discharge** of groundwater, always lagging to some extent. Above the water table is a **capillary fringe** in which the soil pores are partially water-filled and, above the fringe, soil moisture is adsorbed on particle surfaces. Voids above the water table are gas-filled, principally with the atmospheric gases but in much different proportions than in the atmosphere. Oxygen is depleted and carbon dioxide is generated by the respiration of soil organisms while anaerobic decay of organic debris generates carbon dioxide and methane.

SOIL FORMATION

Soils are formed by the weathering of rocks (chapter 3), and their composition, texture, and zoning are the result of the interaction of six major factors: parent material, topography, climate, time, geologic conditions, and organisms. Additionally, the thickness of the soil zone is a function of the relative rates of the production of the soil and its removal by erosion.

A further complicating factor in the nature of soils is that they may have been formed in place on bed rock; on unconsolidated sediments transported to the soil-forming site by one or another geologic agent such as running water, wind, or glaciers; or the soil itself may have been moved away from its point of origin by downslope movement. In the first instance the soil will show gradational relations downward to its parent bed rock, but in the latter

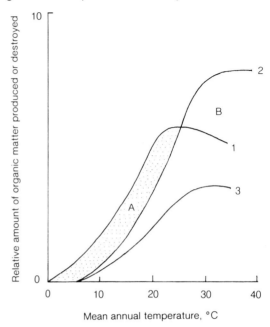

Figure 4.1 Plant production and decay

KEY

A. Humus accumulates
B. Humus absent
1. Rate of plant production
2. Rate of destruction (aerobic bacteria)
3. Rate of destruction (anaerobic bacteria)

two instances the underlying solid rock need have no genetic connection with the soil cover.

The nature and proportions of the inorganic and organic components in a soil depend initially upon the source rock, vegetation cover, and climate. Except under arid conditions, the incorporation within the soil of humic material derived from vegetable debris is principally influenced by temperature (fig. 4.1) while the stable mineral phases also depend to some extent upon the pH of the soil. Chemically alkaline environments (pH > 7) tend to contain montmorillonitic clay and carbonate minerals whereas acidic environments contain kaolinitic clay. Hydrated iron oxides are common in the midrange of pH (5 to 9), and such chemically resistant minerals as common quartz and rarer muscovite, rutile, and magnetite are present in soil in rough proportion to their abundance in the precursor rock. It should be noted that pH in the soil

Figure 4.2 Relation of principal soil groups to climate

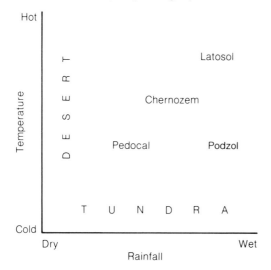

KEY

Desert soils Light-colored to red coarse-grained soils often containing calcareous materials and hardpan layers

Tundra soils Thin and often poorly drained acidic soils within the Arctic Circle and at high altitudes

Podzol soils Zoned soils (A, B, C horizons) rich in clay and colloids; common in eastern and central United States (humid regions)

Pedocal soils Lime-rich chestnut-colored soils of western Canada and northwestern United States; red and gray soils of western United States (arid regions)

Chernozem soils Organic-rich soils, rich in Ca, Mg, and K, sometimes with hardpan layer; found in central western North America

Latosol or laterite soils Variable tropical soils often enriched in aluminum and iron hydrated oxides forming hardpans

environment may vary vertically within a soil profile, thus providing the means for chemical or mineralogic stratification.

The initial stages of soil formation are dominated by the rock material vis-à-vis the soil-forming processes. With long existence, however, soils tend toward a few great *groups* (fig. 4.2) that are related to climatic zones and that show local variability related to rock type, topography, and weather conditions.

The extreme of soil formation is the development of **laterite,** a soil in which the upper portion is strongly depleted of all but clay minerals, iron oxides, and perhaps quartz and the lower zones are enriched in downward-transported iron and aluminum

Figure 4.3 Soil texture classes

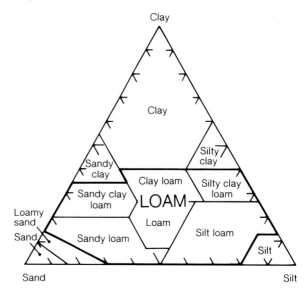

oxides and hydroxides. When well developed by an appropriate combination of rainfall, mean annual temperature, topography, and internal drainage such as characterize some present and past tropical regions, lateritic soils may be economic sources for both iron and aluminum.

SOIL DESCRIPTION AND CLASSIFICATION

Soil is a *mechanical mixture* of granular- and clay-sized inorganic particles, organic debris, and water, any of which may be absent or dominant. Additionally, these components may be uniformly or irregularly distributed and the granular fraction may be in open or close packing. Other than in rare circumstances soils are lacking in a significant organic fraction except in their upper portions.

Soils may range from dessicated to saturated and from impermeable to free draining with important consequences in their engineering properties. Chapter 6 deals with various aspects of groundwater in soils.

For agricultural purposes and in common usage, soils are described by utilizing grain size and texture as principal parameters (fig. 4.3) with soil

Figure 4.4 Soil zones

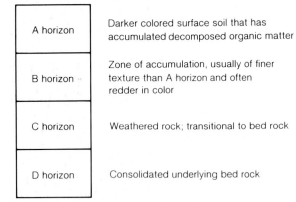

A₀ Vegetable litter
A₁ Humus-rich
A₂ Leached
B₁ Hardpan
B₂ Iron oxide-enriched

color, structure, and consistency (the tendency of a soil aggregate to hold together = **cohesion**) as secondary features.

A common feature of midlatitude soils is their *zonation* into horizons, or layers, of different composition, texture, color, and water content. These zones arise from increments to the soil from weathering of the rock below and dying vegetation added from above coupled with vertical internal movement of ions, colloids, and clay particles, often with the profile transected by a water table. These **soil zones, or horizons,** are commonly designated by the letters *A* to *D* as shown in figure 4.4; however, usage differs among disciplines and table 4.1 contrasts the accepted terms. The nature and perfection of zoning is different under different climatic regimes as shown diagrammatically in figure 4.5 wherein the chemical changes that are taking place in soil formation are also indicated. Table 4.2 provides some examples of soil profiles from different areas.

Figure 4.5 Soil zones in different climates

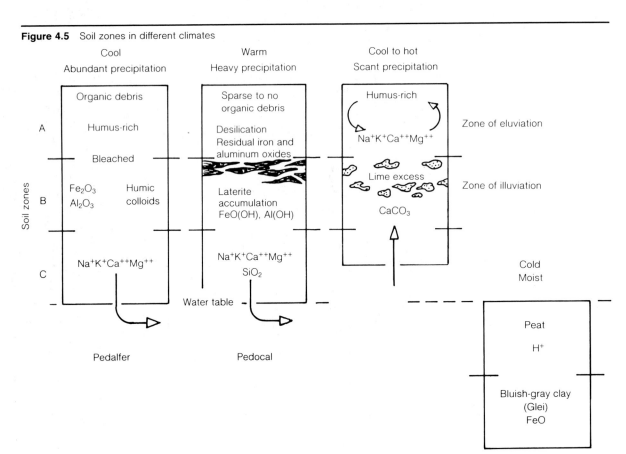

It is always of value when dealing with soils to know something of the environments in which they form since these impose different characteristics and hence engineering properties on these important materials. A tabulation of the principal *geologic settings* of soil deposits taken from the *U.S. Navy Soil Mechanics Design Manual NAVFAC DM–7.1* (1982) is provided in table 4.3, which gives a thorough digest from an engineering standpoint of the kinds and characters of soils.

Conceptually, a useful model soil for engineering purposes may be thought of as a granular aggregate whose pores are more or less filled by

Table 4.1 Soil Terminology

Engineering Usage	Pedologic Usage	Geologic Usage
Topsoil Surficial weathered layer of loose deposits capable of supporting plant growth	*A Horizon* Zone of vegetable litter and leaching	*Regolith* or *soil* Whole weathered profile from surface to unweathered rock
Subsoil Weathered layer of loose material between topsoil and unweathered rock	*B Horizon* Zone of accumulation *C Horizon* Transitional to bed rock	
Rock Hard and rigid deposits	*D Horizon* Bed rock	*Rock* Ancient deposits whether consolidated or not

Table 4.2 Typical Soil Profiles

Examples		Soils
Wyoming	A	Humus-rich
	B	Lime nodule accumulation
	C	Transitional to bed rock
Sierra Nevada	A	Sandy loam with roots
	B	Sandy clay
	C	Transitional through crumbly to fresh granodiorite
Kentucky	A	Humus-rich clay
	B	Clay with limestone fragments
	C	Transitional to fresh limestone
Maine	A_0	Forest duff
	A_1	Humus-rich light-brown sandy loam
	A_2	Ash-gray clayey loam
	B	Reddish-brown sandy compact clay
	C	Transitional through rotted granite retaining original texture to unaltered granite
The Netherlands	A_1	Humus
	A_2	Bleached-gray sandy alluvium
	B_1	Dark and compact hardpan precursor
	B_2	Banded and mottled zone of precipitated iron oxides
The Amazon basin	A_2	Yellowish-brown silty clay
	B_1	Hardpan of aluminum and iron hydrated oxides (laterite)
	B_2	Mottled clay
	C	Clayey siltstone
Cuba	A	Friable clay
	B_1	Iron and manganese oxide concretions
	B_2	Iron-rich clay
	C	Massive serpentine rock

Table 4.3 Principal Soil Deposits

Major Division	Principal Soil Deposits	Pertinent Engineering Characteristics
Sedimentary Soils *Residual* Material formed by disintegration of underlying parent rock or partially indurated material.	**Residual sands and fragments of gravel size** Formed by solution and leaching of cementing material, leaving the more resistant particles; commonly quartz.	Generally favorable foundation conditions.
	Residual clays Formed by decomposition of silicate rocks, disintegration of shales, and solution of carbonates in limestone. With few exceptions becomes more compact, rockier, and less weathered with increasing depth. At intermediate stage may reflect composition, structure, and stratification of parent rock.	Variable properties requiring detailed investigation. Deposits present favorable foundation conditions except in humid and tropical climates, where depth and rate of weathering are very great.
Organic Accumulation of highly organic material formed in place by the growth and subsequent decay of plant life.	**Peat** A somewhat fibrous aggregate of decayed and decaying vegetation matter having a dark color and odor of decay.	Very compressible. Entirely unsuitable for supporting building foundations.
	Muck Peat deposits that have advanced in stage of decomposition to such extent that the botanical character is no longer evident.	
Transported Soils *Alluvial* Material transported and deposited by running water.	**Floodplain deposits** Deposits laid down by a stream within that portion of its valley subject to inundation by floodwaters.	
	Point bar Alternating deposits of arcuate ridges and swales (lows) formed on the inside or convex bank of migrating river bends. Ridge deposits consist primarily of silt and sand; swales are clay-filled.	Generally favorable foundation conditions; however, detailed investigations are necessary to locate discontinuities. Flow slides may be a problem along riverbanks. Soils are quite permeable.
	Channel fill Deposits laid down in abandoned meander loops isolated when rivers shorten their courses. Composed primarily of clay; however, silty and sandy soils are found at the upstream and downstream ends.	Fine-grained soils are usually compressible. Portions may be very heterogeneous. Silty soils generally present favorable foundation conditions.
	Backswamp The prolonged accumulation of floodwater sediments in flood basins bordering a river. Materials are generally clays but tend to become more silty near riverbank.	Relatively uniform in a horizontal direction. Clays are usually subjected to seasonal volume changes.
	Alluvial terrace deposits Relatively narrow, flat-surfaced, river-flanking remnants of floodplain deposits formed by entrenchment of rivers and associated processes.	Usually well drained, oxidized. Generally favorable foundation conditions.
	Estuarine deposits Mixed deposits of marine and alluvial origin laid down in widened channels at mouths of rivers and influenced by tide.	Generally fine-grained and compressible. Many local variations in soil conditions.

From U.S. Navy Soil Mechanics Design Manual NAVFAC DM–7.1, 1982.

Table 4.3 Principal Soil Deposits (continued)

Major Division	Principal Soil Deposits	Pertinent Engineering Characteristics
	Alluvial-lacustrine deposits Material deposited within lakes (other than those associated with glaciation) by waves, currents, and organo-chemical processes. Deposits consist of unstratified organic clay or clay in central portions of the lake and typically grade to stratified silts and sands in peripheral zones.	Usually very uniform in horizontal direction. Fine-grained soils generally compressible.
	Deltaic deposits Deposits formed at the mouths of rivers.	Generally fine-grained and compressible. Many local variations in soil condition.
	Piedmont deposits Alluvial deposits at foot of hills or mountains. Extensive plains or alluvial fans.	Generally favorable foundation conditions.
Aeolian Material transported and deposited by wind.	**Loess** A calcareous, unstratified deposit of silts or sandy or clayey silt traversed by a network of tubes formed by root fibers now decayed.	Relatively uniform deposits characterized by ability to stand in vertical cuts. Collapsible structure. Deep weathering or saturation can modify characteristics.
	Dune sands Mounds, ridges, and hills of uniform fine sand characteristically exhibiting rounded and frosted grains.	Very uniform grain size; may exist in relatively loose condition.
Glacial Material transported and deposited by glaciers, or by meltwater from the glacier.	**Glacial till** An accumulation of debris, deposited beneath, at the side (lateral moraines), or at the lower limit of a glacier (terminal moraine). Material lowered to ground surface in an irregular sheet by a melting glacier is known as a ground moraine.	Consists of material of all sizes in various proportions from boulders and gravel to clay. Deposits are unstratified. Generally present favorable foundation conditions; but, rapid changes in conditions are common.
	Glaciofluvial deposits Coarse and fine-grained material deposited by streams of meltwater from glaciers. Material deposited on ground surface beyond terminal of glacier is known as an outwash plain. Gravel ridges known as kames and eskers.	Many local variations. Generally present favorable foundation conditions.
	Glaciolacustrine deposits Material deposited within lakes by meltwater from glaciers. Consisting of clay in central portions of lake and alternate layers of silty clay or silt and clay (varved clay) in peripheral zones.	Thin beds that are very uniform in a horizontal direction.
Marine Material transported and deposited by ocean waves and currents in shore and offshore areas.	**Shore deposits** Deposits of sands or gravels formed by the transporting and sorting action of waves on the shoreline.	Relatively uniform and of moderate to high density.
	Marine clays Organic and inorganic deposits of fine-grained material.	Generally very uniform in composition. Compressible and usually very sensitive to remolding.

Table 4.3 Principal Soil Deposits (continued)

Major Division	Principal Soil Deposits	Pertinent Engineering Characteristics
Colluvial Material transported and deposited by gravity.	**Talus** Deposits created by gradual accumulation of unsorted rock fragments and debris (scree) at base of cliffs. **Hillwash** Fine colluvium consisting of clayey sand, sand silt, or clay. **Landslide deposits** Considerable masses of soil or rock that have slipped down, more or less as units, from their former position on steep slopes.	Previous movement indicates possible future difficulties. Generally unstable foundation conditions.
Pyroclastic Material ejected from volcanoes and transported by gravity, wind, and air.	**Ejecta** Loose deposits of volcanic ash, lapilli, bombs, etc. **Pumice** Frequently associated with lava flows and mud flows, or may be mixed with nonvolcanic sediments.	Typically shardlike particles of silt size with larger volcanic debris. Weathering and redeposition produce highly plastic, compressible clay. Unusual and difficult foundation conditions.

Figure 4.6 Mechanical properties of soils

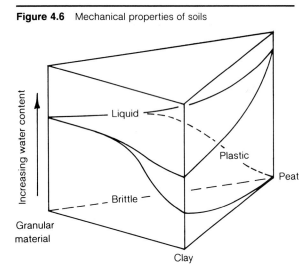

water, colloids, and clay minerals, or all three. Assuming the clay particles, colloids, and water to be uniformly distributed in the soil, its bulk mechanical properties will depend upon the degree of pore filling. If the pores are overfilled, the coarser grains must be separated and the bulk properties of the mass are those of a clay–water mixture: brittle, plastic, or liquid according to their proportions. On the other hand, filled or underfilled pores lead to the properties of a sand–water mixture without a significant plastic range but exhibiting some cohesion. These fundamental distinctions may be generalized as shown in figure 4.6 and are incorporated in the *Unified Soil Classification System* (table 4.4), which is widely used by engineers. Generally speaking, this system employs readily determined descriptive information together with such *index properties* as grain size and grading, compactness, and **Atterberg limits** data. Descriptions of the nature and measurement of soil index properties are incorporated in chapters 10 and 11.

Table 4.4 Unified Soil Classification System

Primary Divisions for Field and Laboratory Identification			Group Symbol	Typical Names	Laboratory Classification Criteria		Supplementary Criteria for Visual Identification
Coarse-grained soils More than half of the material finer than a 3-inch sieve is larger than the no. 200 sieve size.	*Gravel* More than half of the coarse fraction is larger than the no. 4 sieve size, about 1/4 inch.	*Clean gravels* Less than 5% of the material is smaller than the no. 200 sieve size.*	GW†	Well-graded gravels, gravel–sand mixtures, little or no fines.	$C_u = \dfrac{D_{60}}{D_{10}}$ greater than 4.‡ $C_z = \dfrac{(D_{30})^2}{D_{10} \times D_{60}}$ between 1 and 3.		Wide range in grain sizes and substantial amounts of all intermediate particle sizes.
			GP	Poorly graded gravels, gravel–sand mixtures, little or no fines.	Not meeting both criteria for GW		Predominantly one size (uniformly sorted) or a range of sizes with some intermediate sizes missing (gap-graded).
...do...	...do...	*Gravels with fines* More than 12% of the material is smaller than the no. 200 sieve size.	GM	Silty gravels, and gravel–sand–silt mixtures.	Atterberg limits below ''A'' line, or PI less than 4.	Atterberg limits above ''A'' line with PI between 4 and 7 is borderline case GM–GC.	Nonplastic fines or fines of low plasticity.
			GC	Clayey gravels, and gravel–sand–clay mixtures.	Atterberg limits above ''A'' line, and PI greater than 7.		Plastic fines.
...do...	*Sands* More than half of the coarse fraction is smaller than no. 4 sieve size.	*Clean sands* Less than 5% of material smaller than no. 200 sieve size.	SW	Well-graded sands, gravelly sands, little or no fines.	$C_u = \dfrac{D_{60}}{D_{10}}$ greater than 6. $C_z = \dfrac{(D_{30})^2}{D_{10} \times D_{60}}$ between 1 and 3.		Wide range in grain sizes and substantial amounts of all intermediate particle sizes.
			SP	Poorly graded sands and gravelly sands, little or no fines.	Not meeting both criteria for SW.		Predominately one size (uniformly sorted) or a range of sizes with some intermediate sizes missing (gap-graded).
...do...	...do...	*Sands with fines* More than 12% of material smaller than no. 200 sieve size.	SM	Silty sands, sand–silt mixtures.	Atterberg limits below ''A'' line, or PI less than 4.	Atterberg limits above ''A'' line with PI between 4 and 7 is borderline case SM–SC.	Nonplastic fines or fines of low plasticity.
			SC	Clayey sands, sand–clay mixtures.	Atterberg limits above ''A'' line with PI greater than 7.		Plastic fines.

*Materials with 5 to 12% smaller than 200 mesh are borderline cases designated GW-GM, SW-SC, etc.

†*G* gravel, *S* sand, *C* clay, *O* organic, *PT* peat; *W* well graded, *P* poorly graded, *M* mixture; *L* low plasticity, *H* high plasticity

‡See chapter 10 for laboratory classification criteria

Source: From *U.S. Navy Soil Mechanics Design Manual*, NAVFAC DM–7.1, 1982.

Table 4.4 Unified Soil Classification System (continued)

Primary Divisions for Field and Laboratory Identification		Group Symbol	Typical Names	Laboratory Classification Criteria		Supplementary Criteria for Visual Identification		
						Dry Strength	*Reaction to Shaking*	*Toughness Near Plastic Limit*
Fine-grained soils More than half of the material is smaller than the no. 200 sieve size. (Visual: more than half of particles are so fine that they cannot be seen by naked eye.)	*Silts* and *clays* (liquid limit less than 50).	ML	Inorganic silts, very fine sands, rock flour, silty or clayey fine sands.	Atterberg limits below ''A'' line, or PI less than 4.	Atterberg limits above ''A'' line with PI between 4 and 7 is borderline case ML–CL.	None to slight.	Quick to slow.	None.
	...do...	CL	Inorganic clays of low to medium plasticity; gravelly clays, silty clays, sandy clays, lean clays.	Atterberg limits above ''A'' line, with PI greater than 7.		Medium to high.	None to very slow.	Medium.
	...do...	OL	Organic silts and organic silt–clays of low plasticity.	Atterberg limits below ''A'' line.		Slight to medium.	Slow.	Slight.
...do...	*Silts* and *clays* (liquid limit greater than 50).	MH	Inorganic silts, micaceous or diatomaceous fine sands or silts, elastic silts.	Atterberg limits below ''A'' line.		Slight to medium.	Slow to none.	Slight to medium.
	...do...	CH	Inorganic clays of high plasticity, fat clays.	Atterberg limits above ''A'' line.		High to very high.	None.	High.
	...do...	OH	Organic clays of medium to high plasticity.	Atterberg limit below ''A'' line.		Medium to high.	None to very slow.	Slight to medium.
...do...	*Highly organic soils* . . .	Pt	Peat, muck, and other highly organic soils.	High ignition loss; LL and PI decrease after drying.		Organic color and odor, spongy feel, frequently fibrous texture.		

Figure 4.7 The deserts of the world

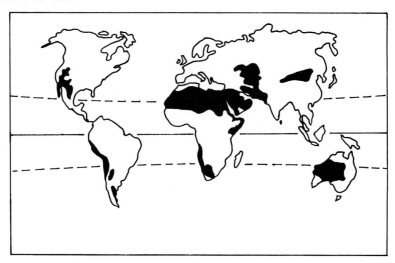

SOME SPECIAL SURFICIAL MATERIALS

Loess

Loess is an unusual surficial material formed by the deposition of windblown dust and often encountered downwind from arid regions. Desert regions are found today in the polar regions and in belts along the Tropics of Cancer and Capricorn. The Sahara, the deserts of Arabia and part of Iran, the Gobi, and the Sonoran desert of northern Mexico and the southwestern United States are found in the subtropical region of the Northern Hemisphere. The equivalent arid belt in the Southern Hemisphere includes the Kalahari desert in South Africa, the Atacama desert in Chile, the Patagonian steppe of Argentina and most of the Australian continent (fig. 4.7). Glaciation in the geologically recent past greatly enlarged the arid areas in northern North America, Europe, and Asia and promoted the development of loess deposits. Today, thick deposits of loess are found in the United States midcontinent (see map 5, appendix II).

Loess is formed when dust from successive falls is fixed in place by the growth of grasses whose root systems help to hold the loose mass together. The result is a lightly cemented, highly porous and permeable, pale yellowish to buff colored, unbedded sedimentary rock. Typical loess is characterized by the presence throughout of small calcite-lined capillary tubules, which appear to have been occupied by rootlets. The lower portions of a deposit often contain carbonate concretions.

A characteristic of loess is its capacity to stand stably in vertical or even overhanging walls, which is unusual for such a lightly cemented material. It may be readily excavated and drains well when undisturbed. The cohesion of loess is easily destroyed, however, and it is quickly worn by traffic or gullied by concentrated water flow such as might occur in ditches, along pipe lines, or beneath eave lines. Loess presents additional construction hazards because it cannot support heavy weights and is highly unstable when subjected to severe earthquakes.

Glacial and Glaciofluvial Deposits

Between about two million years and ten thousand years ago, the northern portions of North America and Eurasia were subjected to periodic advances of thick continental glaciers. These masses of ice transported and deposited huge quantities of rock debris, which in the glaciated regions often constitutes the principal surficial materials found today. Map 6, appendix II, gives the distribution of these deposits in the conterminous United States.

Figure 4.8   Periglacial deposition

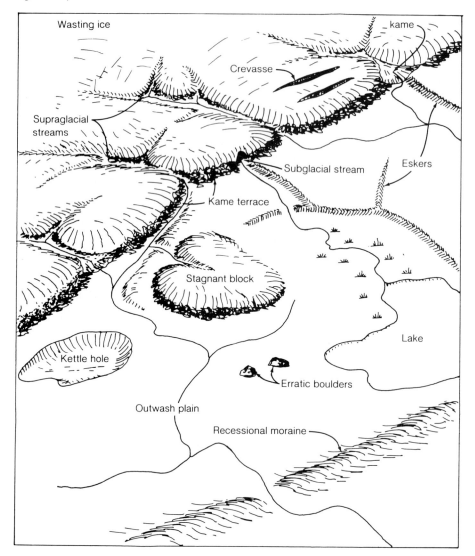

Two distinct kinds of sedimentary materials result from glacial action: the first is unsorted debris deposited directly by the ice and the second is that which has been reworked and sorted by glacial melt-water. The first kind, direct *ice deposits,* may contain particles ranging in size from clay (rock flour) to house-sized boulders that have been transported and deposited together, and the second kind comprises *glaciofluvial deposits* of sorted clays, sands, and gravels characteristic of ordinary stream action.

Both kinds of deposits will be found together in peri-glacial regions, and since each poses different engineering problems, their recognition is important.

The details of glacial transport and deposition are complex, so planners are well advised to seek knowledgeable assistance regarding the nature and distribution of glacial deposits. The general picture of sediment distribution may be gained by study of figure 4.8. In this figure, the rate of melting at the ice front is greater than the rate of ice flow so that

Figure 4.9 Glacial moraines in the Great Lakes region

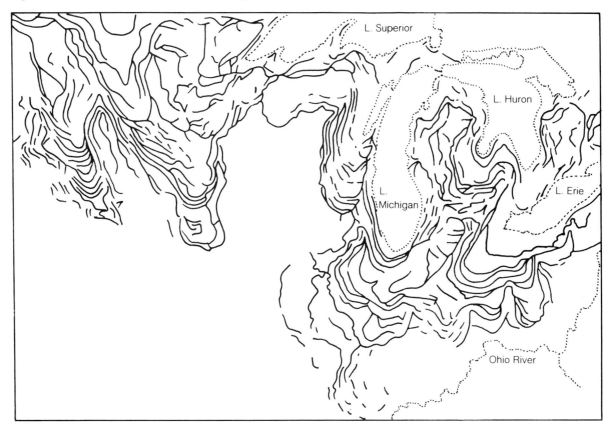

the glacial front is in retreat. The melt-back leaves a carpet of unsorted debris (*ground moraine*) over its abandoned ground that underlies the other deposits. Temporary halts allow a buildup of debris at the ice front forming a low ridge called a *recessional moraine* (the furthest advance is marked by a *terminal moraine*). Figure 4.9 shows the distribution of **moraines** in the Great Lakes region.

Streams will be present on, in, and under the glacier, and they will transport and deposit debris supplied from the ice and its moraines. These streams differ from ordinary ones only in that they flow in or on ice for part of their course and the supply of sediment typically exceeds their carrying capacity. In figure 4.8 supraglacial streams flow over the glacier in ice-walled valleys filled with rock debris. At the toe of the glacial front the abrupt change in gradient causes the stream to drop much of its load, which accumulates in an alluvial cone termed a **kame,** later to collapse when the supporting ice melts. Similarly, a stream and its sediment-filled channel may be gently let down on the ground moraine and, without support of the ice, the latter becomes a sinuous ridge called an **esker** marking the former course. The stream at this time is displaced laterally.

Sometimes ice hills may be cut off from the main glacier by stream action, stagnate, and waste away. During this time they are often surrounded by outwash sediments so that their former presence is marked by a depression called a **kettle hole.** Figure 4.10 illustrates kettle hole topography.

Glacially deposited erratic boulders, Silver Lake, California

Glacial outwash gravels, Maine

Figure 4.10 Kettle hole topography

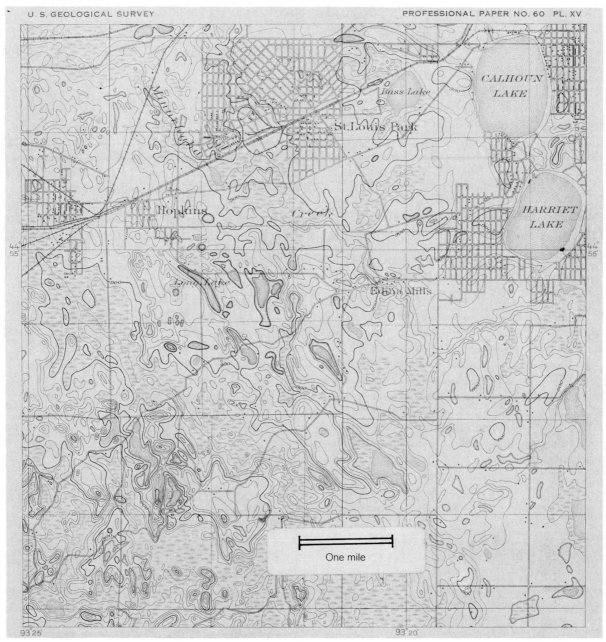

U. S. GEOLOGICAL SURVEY

PROFESSIONAL PAPER NO. 60 PL. XV

One mile

Part of the Minneapolis (Minn.) 1:62500 Quadrangle, U.S.
Geological Survey.

The abundance of sediment and the modification of topography in the outwash zone makes this a region of interrupted, often sluggish, drainage leading to the development of extensive shallow lakes and swamps. These become filled with fine sediments washed out of the glacial and glaciofluvial deposits, sometimes thick layers of **diatomaceous earth,** and eventually abundant plant debris.

Readvance of an ice sheet over its previous deposits is often marked by the presence of elongate, ellipsoidal hills called **drumlins.** These are typically comprised of unsorted debris. Figure 4.11 shows an area in which many drumlins are found. The ice movement was from the north northwest.

Peat

Decayed vegetable matter accumulates when the rate of decay is less than the rate of plant growth (see fig. 4.1), and the preservation of decayed material is promoted if it accumulates under water so that oxygen is excluded. These conditions are commonly met in the swamps, bogs, and **muskegs** of the relatively flat, poorly drained, glaciated areas of the United States, Canada, Europe, and Asia. Extensive bodies of partially decayed vegetable matter also accumulate in brackish or freshwater swamps along low-lying coasts such as are found along the eastern and southern shores of the United States. In this environment, a slow rise in sea level will allow thick layers of vegetable debris to accumulate in the marginal estuaries, lagoons, and deltaic swamps. After burial for some millions of years this material is transformed to lignite and eventually to coal. All of the vast coal fields of the world have been formed in swamps of this kind.

An accumulation of decayed vegetable matter is called **peat.** It may be derived from a variety of vegetation but is usually a product of sphagnum moss in northern locations, grasses and reeds in central regions, and mangrove or cypress in subtropical areas. Peat varies from a light, spongy material in its upper layers to a dense, brown or black substance at the bottom of deep swamps. Roots and tree trunks are often present.

In its natural state, peat contains 90%–95% water, which is only reduced to 88%–91% by

drainage. Air drying reduces the water content to about 25% at which point its specific gravity is about 0.5 (hand cut) or 0.8–1.0 (macerated). Fine-grained, detrital material such as clay is always present admixed with the peat, usually increasing in amount downward.

Expansive Soils

One class of clay minerals formed by rock weathering, especially by the breakdown of volcanic ash, has the property of ready uptake and loss of water molecules between the sheets of atoms that comprise its structure. The mineral class is the *smectites* with *montmorillonite* as the most common species. This mechanism of clay expansion is described in some detail in chapter 1 and some of the mechanical consequences are examined in chapter 11. If these clay minerals accumulate in a soil or are concentrated into beds by transport and deposition, the resulting mass will swell or shrink, according to the availability of water, with disastrous results to structures on or within it. It is termed a **vertisol** by pedologists and **expansive soil** or *swelling* soil or clay by engineers. Measured free-swelling ranges are from 50% to 2 000% and the expansive pressure generated is in the range of 15 000 kg/m² to 50 000 kg/m². This pressure is well in excess of the load imposed by small buildings, which will, in consequence, be uplifted and rotated by the expanding soil. A less obvious but potentially more dangerous failure is that of buried utility lines.

Expansive soils are greater hazards and responsible for more damage than earthquakes, volcanic eruptions, landslides, and floods combined. The presence of dangerously expansive soil at an engineering site will probably be known to local builders or engineers. A characteristic microtopography of irregular ridges termed **gilgai** develops on undisturbed ground, and soil and mechanical testing should, of course, be used when assessing soil conditions at any project site of significant size. The more reliable tests are the use of a *consolidometer,* whereby the expansive pressure caused by introduction of water to a confined sample is directly measured, and *Atterberg limit* determinations (chapter 10) in which either a plasticity index greater than

Figure 4.11 Drumlins

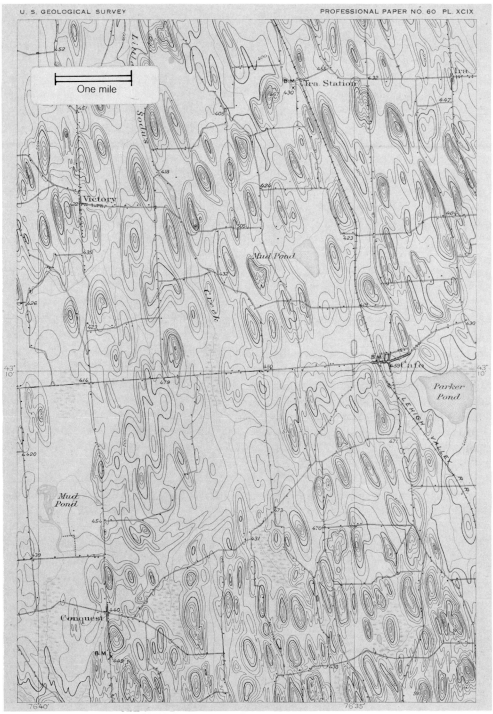

One mile

Part of the Weedsport (N.Y.) 1:62500 Quadrangle, U.S.
Geological Survey.

Figure 4.12 Percent swell as a function of dry weight and liquid limit

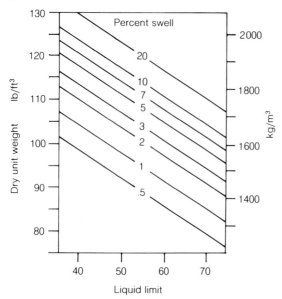

15 or a percent swell above 3 (fig. 4.12) indicates significant potential swelling.

Expansive soils have a worldwide distribution and are particularly dangerous in semiarid regions because of the wide variations in soil moisture that occur there. In the United States they are particularly common in Alaska, California, the Rocky Mountain and central plains areas, and around the Gulf Coast.

SUMMARY

The engineering characteristics of soils are covered later in chapters 10 and 11. Here the geologic features of soils including soil formation, description, and classification have been emphasized.

Soil is the skin of the earth. It is formed by the action of air and water on subjacent rock and reflects this ancestry of rock and climate in its composition and morphology. Many soils are zoned into recognizable horizons, some have special origins or compositions (loess, peat, and glacially derived soils, for example), and all have pertinent engineering characteristics as set forth in table 4.3.

ADDITIONAL READING

DeSitter, L. U. 1964. *Structural geology.* 2d ed. New York: McGraw-Hill.

Dregre, H. E. 1976. *Soils of arid regions.* New York: Elsevier Science.

Glazovskaya, M. A. 1983, 1984. *Soils of the world,* vols. 1 and 2. Trans. C. M. Rao. Rotterdam: A. A. Balkema.

Hough, B. K. 1969. *Basic soils engineering.* 2d ed. New York: Ronald Press.

Jarrett, P. M., ed. 1983. *Testing of peats and organic soils.* American Society for Testing Materials, STP 820. Philadelphia: American Society for Testing Materials.

Krynine, P. D., and Judd, W. R. 1957. *Principles of engineering geology and geotechnics.* New York: McGraw-Hill.

Lambe, T. W., and Whitman, R. V. 1979. *Soil mechanics.* New York: John Wiley & Sons.

Leggett, R. F. 1961. *Soils in Canada: Geological, pedological, and engineering studies.* Toronto: Toronto University Press.

Lotkowski, W. M. 1966. *The soil.* Chicago: Educational Methods.

Marshall, C. E. 1964. *The physical chemistry and mineralogy of soils.* New York: John Wiley & Sons.

Mitchell, J. K. 1976. *Fundamentals of soil behaviour.* New York: John Wiley & Sons.

5

The Hydrologic Cycle
and Sediment Transport

INTRODUCTION

Water flowing as surface streams represents the principal agent for erosion and transport of rock debris. Hills are slowly worn down, valleys carved, and the detritus of weathering is carried to lakes or the ocean. Running water is even, or rather especially, effective in carving desert landscapes because of the intensity of the rare rainstorms and associated floodwaters.

Soil and rock are porous so that water soaks into the ground, permeating it and moving slowly in response to pressure differences. Groundwater provides a water supply through wells penetrating the saturated zone, is responsible for seepage into excavations, modifies the physical properties of soil, and because of upward capillary movement, is the cause of hardpans and salt buildup within soils. Deep groundwater is hot because rock temperature increases downward; this water may be tapped as a geothermal energy source.

Both surface water and groundwater freeze when temperature falls. Ground ice markedly changes the physical properties of soil and rock;

Figure 5.1 Variation in rainfall, runoff, and temperature for the conterminous United States

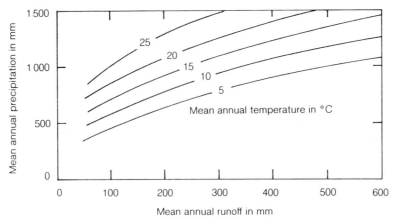

Modified from W. B. Langbein, U.S. Geological Survey Circular 52, 1949.

volume changes and the ability of the mass to conduct water is nullified. Seasonal freezing and thawing thus cycles surficial material through annual changes that can result in breakage of buried pipes, heaving of foundations, or flooding because of the sealing of surface materials against water infiltration.

Very deep zones of permanently frozen ground exist today in far northern latitudes as a legacy of glacial conditions in the Pleistocene epoch. These regions pose unique and particularly difficult engineering problems.

CLIMATIC FACTORS AND THE WATER BUDGET

The *supply of water* to a region is closely related to climatic factors, which fix the amount and distribution over the year of rain, sleet, or snow *runoff* and the amount of *infiltration* into the subsurface; and return to the atmosphere by *evaporation* or by *transpiration* of plants.

Surface runoff by sheetwash and channeled flow in streams delivers water to temporary storage in lakes and the ocean. The ratio of runoff to precipitation, however, varies widely and depends on many factors. The slope of the surface is of obvious importance. The amount and duration of rainfall and

its distribution throughout the year are important since light showers tend to be totally absorbed by the ground whereas sudden torrential rainstorms and extended periods of heavy rain saturate the upper portion of the ground, and a large proportion of the rainfall makes its way to streams by surface routes. Also, melting of snow above frozen ground will produce a very high proportion of runoff to infiltration. Additionally, rocks and soil vary greatly in **permeability** depending on their degree of lithification, particle size, packing, and sorting. In particular increased amounts of clay in the soil and the type and amount of vegetation cover can markedly modify the amount of adsorption and hence affect runoff. Figure 5.1 and table 5.1, which provide summary data relating temperature, precipitation, and runoff, illustrate but a few of the many pertinent relationships. Surface runoff is very important and the total annual discharge for the conterminous United States is huge; gauging station data indicates that the rivers deliver approximately 1 375 cubic kilometers of runoff water to the sea each year.

Infiltration of water into the subsurface, directly from precipitation or through the bottoms of streams and lakes, *recharges* the store of underground water, which in most instances is being continuously *discharged* through springs, wells, and into

Table 5.1 World Distribution of Runoff

Continent	Area, km²		Annual Runoff, mm
Europe (including Iceland)	10 372		260
Asia	45 336		170
Africa	31 972		203
Australia and New Zealand	8 541		76
South America	19 280		450
North America	21 925		295
Greenland, Canadian archipelago	4 164		180
Malayan archipelago	2 811		1 600
TOTAL	144 401	WEIGHTED AVERAGE	270

Figure 5.2 The hydrologic cycle

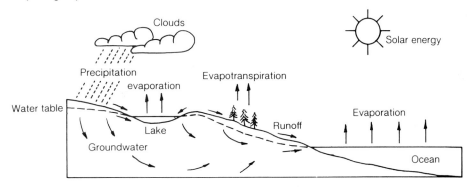

streams and lakes. Groundwater movement is typically steady but very slow whereas recharge from precipitation tends to be locally rapid and aperiodic. In consequence, the very large subsurface volume of joints and pores available for groundwater *storage* is maintained at a roughly full state, and the amount of groundwater in storage is as much as 4 000 times that available as fresh water on the surface. Temperature and wind, together with the nature of surficial materials and slope, are particularly important in fixing the amount of infiltration by such means as modifying evaporation rates, sealing soil pores by freezing, or controlling the amount of vegetation present.

The **hydrologic cycle** (fig. 5.2) shows the general movements of water above, on, and under the earth's surface but gives no indication of rates. These may, however, be determined at any point since nec-

essarily $R = D \pm \triangle S$, where $R =$ inflow or recharge, $D =$ outflow or discharge, and $\triangle S =$ the change in storage. The inflow items include surface and subsurface inflow, precipitation, and imported water. The outflow items include surface and subsurface outflow; consumptive loss, including evaporation and transpiration; and also exported water. Changes in storage items include changes in surface pondage, changes in the amount of groundwater, and changes in soil moisture. Many of the factors associated with the hydrologic cycle can be measured easily, but some are difficult to assess. Rainfall and stream runoff are measured at many U.S. Geological Survey monitoring stations throughout the country, but groundwater storage and flow and evapotranspiration are much more difficult to determine. An indication of the complexity of the problem of water balance may be seen by a study of

Table 5.2 Average Annual Water Budget, Tucson Basin, 1936–1963

	Annual Mean	
	Acre-feet	$m^3 \times 10^6$
Surface water		
Inflow, Santa Cruz River	68 000	83.9
Outflow		
Surface flow	17 000	21.0
Infiltration	51 000 ⌐	62.9
Underground water		
Inflow		
Infiltration	51 000 ◄	62.9
Underflow in tributary valleys	17 800	21.9
Inflow along edge of basin	31 000	38.2
Sewage effluent return	8 300	10.2
Industrial return	9 000	11.1
Outflow		
Underflow, Santa Cruz River	10 000	12.3
Evapotranspiration	15 500	19.1
Irrigation, public supply, industrial pumpage	176 700	218.0
Total inflow	134 100	165.3
Total outflow	219 200	270.4
Storage depletion	85 100	105.0

Data from E. S. Davidson, *Geohydrology and water resources of the Tucson Basin, Arizona.* U.S. Geological Survey Water-Supply Paper 1939-E, 1973.

table 5.2, which is a simplified accounting of inflow, outflow, and storage in the hydrologic basin around Tucson, Arizona. The net storage depletion of this basin has important socioeconomic implications for the continued growth of this area.

EROSION AND DENUDATION

The portion of the water budget constituting runoff is the most important of the agents producing *erosion* or **denudation** of the earth's surface. Rocks weathered by physical and chemical processes are broken down into small particles that are *transported* towards the sea in surface streams. This **suspended** and **traction** load, in turn, wears away the stream bed, and as the bed is lowered and the banks are steepened, more debris can creep, or be washed, into the stream. The rate of denudation or erosion varies with the slope of the land, its composition, its protecting vegetation cover, and atmospheric and climatic conditions. The relationship of precipitation to *sediment yield* for different vegetative coverage on land is summarized in figure 5.3. In addition

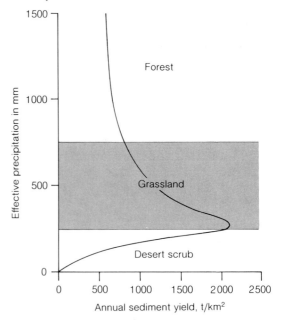

Figure 5.3 Relation of precipitation, vegetative cover, and sediment yield

CASE HISTORY 5.1 Soil Erosion

Figure 1 Erosion along highway

The photograph shows an interstate highway under construction in the state of Georgia. The structural and concrete work have been completed for the overpass, the roadbed has been graded and paved, and the embankment formed. The material in the embankment and the drainage ditch is obviously susceptible to erosion by runoff, and its gullied appearance is typical.

The exposed bank has been left too long after fill and grading without stabilization by grass, crown vetch, straw, or other protection. The bank will have to be regraded before it can be effectively stabilized. Erosion of this type can take place in a few days or months depending upon rainfall and soil material. Stabilization should thus closely follow construction.

Table 5.3 Rates of Denudation in the United States

Drainage System	Area km²	Runoff, m³/s	Denudation, cm/10³ year
Colorado River	637	.65	16.5
Pacific slope	303	2.27	9.1
Western Gulf of Mexico	829	1.56	5.3
Mississippi River	3 238	17.56	5.1
South Atlantic (East Gulf)	736	9.20	4.1
North Atlantic	383	5.95	4.8
Columbia River	679	9.77	3.8

to clastic or sediment debris, dissolved solids are also transported in the stream runoff.

The average denudation rate for the United States as a whole is 6.1 cm per thousand years (table 5.3). Human activities have had a marked effect on the rate of erosion and denudation, and the precipitation to runoff ratio and many flash floods and other changes in stream behavior are undoubtedly due in part to the sealing of large areas of ground by such things as housing, roads, parking lots, or airports. Such areas have almost a 100% runoff, and furthermore, the precipitation is converted immediately to runoff and channeled very rapidly by drainage systems to streams whose levels, following a storm, rise at a far greater rate than if the runoff came from an area of natural vegetation. The rates of erosion, denudation, and runoff are thus greatly affected by the type of vegetation cover, land usage, and rainfall rates and periods. A study by Putnam (1972) showed that under average climatic conditions for the eastern United States a 20 km² drainage area would yield about 0.35 m³ of sediment per year if wooded, 3 m³ if farmland, 13 m³ if an urban area and 190 m³ if the soil was not vegetated.

Drastic changes in the erosion rate of an area due to artificial factors can cause silting and other disequilibrium situations in streams. The rate of erosion may be greatly increased in some agricultural areas by some cultivating and tilling practises. At construction sites and in areas of mining and quarrying the erosion rate will increase unless the disturbed ground is reclaimed and replanted with grass or other vegetation.

FLOW CHARACTERISTICS IN OPEN CHANNELS

The importance of channeled flow of streams as a means of both erosion and transport makes it important to examine some phenomena related to the flow of water.

In general, fluid flow is *laminar* only at low velocities and the free flow of wind, ocean currents, and natural streams is typically *turbulent* or transitional from laminar to turbulent conditions. Natural examples of laminar flow are found in the movement of groundwater and glaciers. In laminar flow, the fluid particles move with constant spacing in straight or gently curving paths, which can be observed by injecting dye into the flow, a technique that also shows the particle trajectories in turbulent flow (fig. 5.4). Examination of the figure shows that water discharge is higher for laminar flow and local pressure differences (path separation) are characteristic of turbulence.

The common velocity measured for a stream is the mean velocity, U, usually measured at $1/3$ depth below the surface (fig. 5.5). This is an average of the velocities at vertically spaced points within the flow, which are designated by V in hydraulics calculations and measured by probes consisting of Pitot tubes or propeller flow meters.

Experiments have shown that the conditions that distinguish laminar and turbulent flow can be determined by the *Reynolds Number, R:*

$$R = \frac{\text{velocity} \times \text{depth} \times \text{density}}{\text{viscosity}}.$$

Figure 5.4 Paths of individual water particles in laminar and turbulent flow

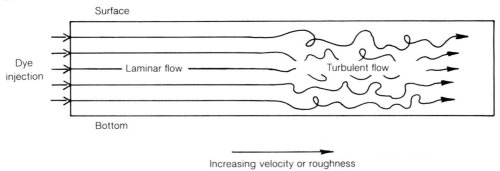

Figure 5.5 Location of maximum potential or actual turbulence within a channel

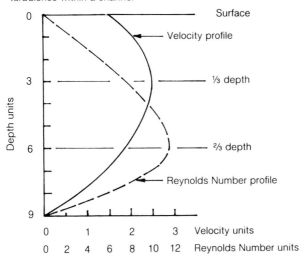

The Reynolds Number corresponding to a particular velocity profile is shown in figure 5.5 and may be seen to reach maximum values at about 2/3 depth, thus fixing the location of maximum potential or actual turbulence. In SI (International System) units, the critical value of the Reynolds Number separating laminar and turbulent flow of water lies between 500 and 5 000. Since the viscosity of the fluid is dependent on the temperature and suspended load, these become factors in the determination of the Reynolds Number. Velocity will approach zero and flow will be laminar in the region

close to the channel boundaries known as the **boundary layer.** Both the velocity and turbulence increase rapidly away from the boundaries with maximum turbulence concentrated in two symmetrically disposed threads at 2/3 the channel depth (see fig. 3.9).

In uniform velocity flow in a stream, the downstream gravitational component balances the frictional effects generated by turbulence and friction against bed, banks, and suspended particles. The discharge of a channel, Q, is thus the quantity of water passing through a given cross section, A, in a unit time and equals the area of cross section times the mean velocity of flow, V_m:

$$Q = AV_m$$

Apart from *gradient*, the mean velocity is influenced by the *roughness* of the channel as determined by the type and size of material of the walls and bed of the channel and the frictional drag of the *wetted perimeter* (the length of the bed in cross section wetted by the flow). Figure 5.6 depicts three cross sections of channels having the same cross-sectional areas but different wetted perimeters. Channel A has the smallest wetted perimeter and offers the least frictional resistance to the flow. It may be noted that the cross-sectional shape having a minimum wetted perimeter, whether boxlike, triangular, or semicircular is always a 2:1 ratio of width to depth. The relationships between the parameters of gradient, velocity, channel characteristics, and sediment load

Stream gauging, Potomac River

Figure 5.6 Channel cross sections of equal area

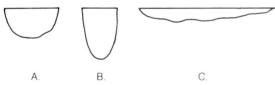

A.   B.   C.

KEY

A.   Smallest wetted perimeter
B.   Intermediate wetted perimeter
C.   Largest wetted perimeter

Modified from W. W. Rubey, U.S. Geological Survey
Professional Paper 218, 1952.

are obviously complex, and close observation and measurement of a stream is essential to establish the engineering geology of erosion, deposition, sewage disposal, bridge and dam design, or flood-control works.

SEDIMENT TRANSPORT IN STREAMS

Natural waters are never completely pure. Precipitation as rain or snow contains dissolved gases from the atmosphere, and at the earth's surface water dissolves solids from soil and rocks. These *dissolved substances* eventually find their way into stream waters. The amount of dissolved material depends

Table 5.4 Compositions of Natural Waters

	Dissolved Substances (values in parts per million)							
	HCO$_3$	SO$_4$	Cl	Na	Mg	Ca	Fe	SiO$_2$
Amazon River	17	3	1.7	1.8	1.1	4.3	.05	7.0
Mississippi River	101	41	15	11	7.6	34	.02	5.9
Colorado River	183	289	113	124	30	94	.01	14
Jordan River (Israel)	238	174	473	253	71	80	—	—
Seawater	140	2 649	18 980	10 556	1 272	400	.02	4
Great Salt Lake	—	6 680	55 480	33 170	2 760	160	—	—

Source: *The Encyclopedia of Geochemistry and Environmental Sciences,* IVA, R. W. Fairbridge, ed. © Van Nostrand Reinhold Company, New York. Adapted by permission.

on the climate, season, geological setting, and rock types. The principal dissolved materials are bicarbonate, sulfate, chloride, sodium, magnesium, calcium, iron, and silica (table 5.4).

Estimates of the total dissolved sediment load delivered to the ocean by the rivers of the United States are of the order of 0.27×10^9 tonnes per year, and all the rivers of the world are estimated to deliver 3×10^9 tonnes to the oceans annually. Most of these salts are stored in the oceans or salt lakes from which they are recovered by geologic or human-devised evaporative processes. Ancient seas and lakes have yielded the great commercial deposits of salt, gypsum, and borax; other materials such as magnesium and bromine are today being recovered directly from brines.

The transport of solids by a stream is determined by the flow characteristics of the stream, which vary in time. The solid portion of the stream load consists of solid or colloidal particles carried in suspension or dragged along the bottom. Also, particles may saltate (jump) from the bed and follow a more or less extended downstream trajectory before again coming to rest. The very fine grained solids and colloids will remain in permanent suspension while the bed load moves by fits and starts. Consequently, the suspended and bed load portions of the sediment load move at different rates. The relation of suspended load transport to stream discharge for two rivers is shown in figure 5.7.

Figure 5.7 Stream discharge and suspended load

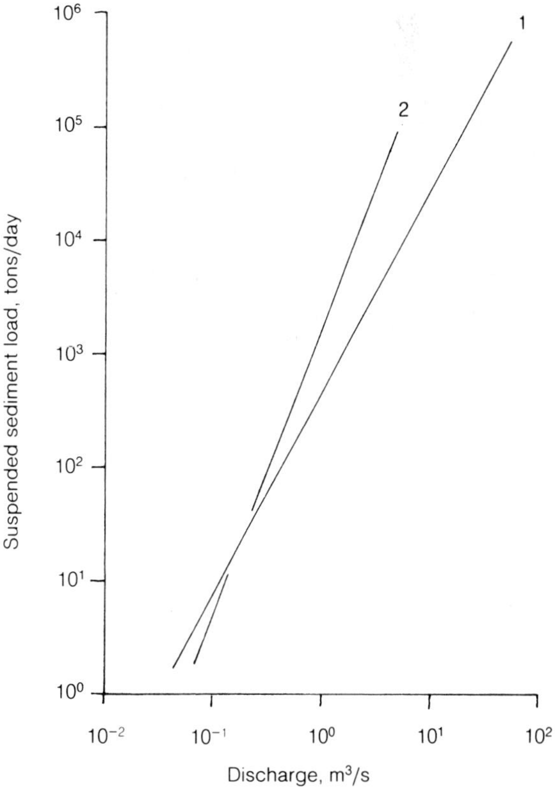

KEY

1. Powder River, Arvada, Wyoming
2. Rio Puerco in flood, New Mexico

Modified from L. B. Leopold and L. B. Miller, U.S. Geological Survey Professional Paper 282A, 1956.

The Hydrologic Cycle and Sediment Transport 87

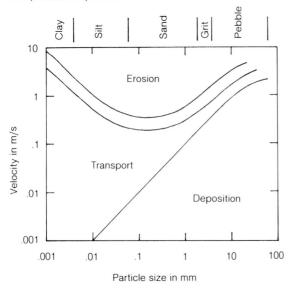

Figure 5.8 Relations between velocity, particle size, erosion, transport, and deposition

Water flowing in an open channel is subject to the two principal forces of gravity and friction. The gravitational force is exerted vertically downward and if the surface on which the water is resting is inclined, as in natural channels, there is a tangential component of the gravitational force that propels the water downslope. Friction between the water, the channel boundaries, and the particles resting on the stream bed tends to resist the downslope movement of the water. This resisting force is called the *bed shear stress*. At some point after initiation of water motion its bed shear stress will overcome the resistance of bed load particles to motion, and this incipient movement of bed particles is called the *threshold* of sediment transport. General relationships of transport, erosion, deposition, and particles size are given in figure 5.8.

The threshold of sediment transport is not a well-defined parameter since there is no one point at which all the particles of the same size and density move at the same time as the velocity of flow over a stationary bed is increased. Experimental data can be obtained by studying water flow in flume tanks with or without sediment or simulated sediment. A typical laboratory flume is equipped with discharge and point-velocity measuring equipment together with sediment introduction, measurement, and retrieval systems. The general relationship between particles and forces of the bed is shown in figure 5.9.

An indication of the complexity of the entrainment process may be gathered from the results obtained by Vanoni (1960) in tests of various rating curves (fig. 5.10). He measured the critical stress for entraining sediment particles in a boundary layer by using complete velocity profiles and a burst frequency to determine critical movement. He noted that when the threshold of movement was approached the motion of the grains occurred in bursts or gusts produced by the turbulence of the flow. By observing a small area with special apparatus he could count the number of bursts over a convenient period of time, and for certain sand sizes, it was also possible to estimate the average number of grains in motion during a burst. By plotting the calculated sediment discharge data against water discharge values he was able to establish the reliability of other sediment discharge formulae that had been proposed. A wide scatter is apparent in the calculated results as well as in the actual measurements. It is apparent that the slopes of some of the discharge curves are consistently steeper, and it is these three curves (Laursen, Meyer Peter-Muller, and the Einstein bed load function) that yield results closer to the slope of the data points. Further evaluation of transport formulas is widely presented in engineering literature on hydraulics.

In summary, estimates of the rate of sediment transport in natural channels are generally based on formulae that have been derived from laboratory data. However, actual correlations between laboratory results and field measurements are too few to justify recommendation of any particular formula as being more accurate than others.

STREAMS

The runoff water derived from precipitation flows downhill following the course of lowest altitude and forming a body of moving water called a stream. Small streams ultimately combine their discharges to form *stream systems,* which may attain great size

Figure 5.9 Forces on submerged particles

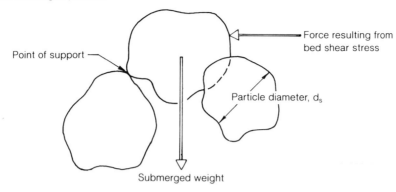

Figure 5.10 Sediment-rating curves for the Colorado River at Taylor's Ferry Curves labeled with investigators' names. See Vanoni (1960) for further references.

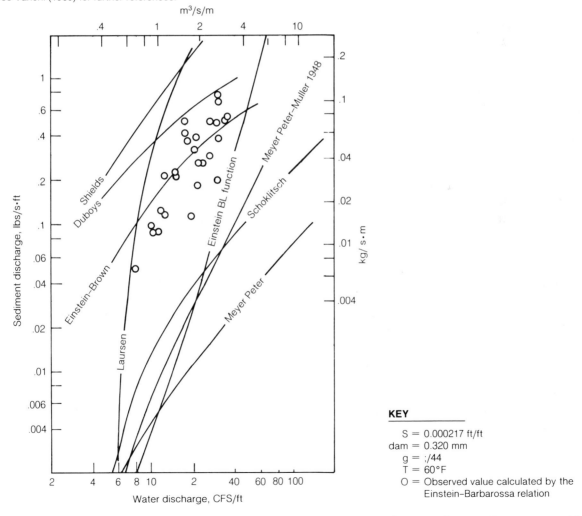

KEY

$S = 0.000217$ ft/ft
dam $= 0.320$ mm
$g = ;/44$
$T = 60°F$
O $=$ Observed value calculated by the Einstein–Barbarossa relation

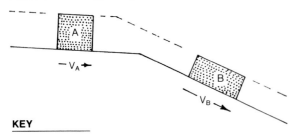

Figure 5.11 Energy and gradient

and complexity. Fairly large streams are called rivers. For example, the Amazon River system of South America or the Mississippi River system of North America drain major portions of continents. Rivers are of great importance for water supply, navigation, irrigation, flood control, erosion, and sedimentation and have consequently been the subject of extensive studies by engineers, hydrologists, and geologists.

The *energy* of flow of a stream is derived from gravity and is related to the slope of the stream bed. The downstream component of gravity is proportional to the sine of the slope angle, which is virtually the same as the tangent of the slope angle since most streams flow at low gradients. The tangent of the slope angle is used in hydraulic engineering calculations. A simple means of visualizing the relation of energy to gradient is to consider a greatly simplified channel in longitudinal section (fig. 5.11) in which discharge is constant at all points; that is, the velocity × cross-sectional area (= discharge volume) is constant for regions *A* and *B*. Since the kinetic energy may be approximately equated with the momentum, (= ½ mV^2) and the *kinetic energy* in the two regions is related as the squares of the velocities in the two regions:

$$\frac{KE_B}{KE_A} = \frac{\frac{1}{2}\,mV_B^2}{\frac{1}{2}\,mV_A^2} = \frac{V_B^2}{V_A^2}.$$

If the water in a flowing stream did not experience friction of the water particles against one another and the bed and banks of the streams, it would theoretically attain great velocity. In nature, however, *friction* is very effective and most streams flow at fractions of a meter to a few meters per second. The greatest velocity ever measured for a river reach in the United States was 7 m/s in the Potomac River during the flood of March 1936.

The *gradient* of a stream is the slope of its bed and decreases along its course from the headwaters to the mouth producing a characteristic, longitudinal, stream *profile* steepening exponentially upstream (fig. 5.12). The cross-valley profiles change with the topographic location or *tract* of the stream and are designated as the mountain tract, valley tract, plain tract, and delta tract. The general relationship of the tracts of a stream to the stream profile are given in figure 5.12.

The **mountain tract** is usually characterized by steep gradients and steep-sided V-shaped valleys, undissected upland, and limited **strath** development. The **valley tract** has less steep walls and usually a **flood plain** on the inside of the bends; uplands are well drained and relief is maximum. The **plain tract** has a very broad cross-valley profile with a broad flood plain and low divides. The stream has a gentle gradient and typical **meanders.** Many rivers enter the delta tract as they approach the sea or large lakes. The **delta tract** is a region of sediment deposition with constantly changing stream courses and a series of branches called *distributaries*, which are not always well developed.

A stream erodes its valley by a combination of downcutting and lateral wasting of the valley walls. The *base level of erosion* is the theoretical limit of this downcutting. Stream erosion may be controlled by a temporary base level if the stream enters a lake or encounters hard or resistant bed rock that delays the downcutting; the latter situation produces a waterfall or rapids.

Streams in the mountain tract exert most of their energy in a downward direction (fig. 5.12) and tend to have relatively straight courses in plan. In longitudinal section, however, they are seen to be comprised of a fairly regular series of *pools* and

Figure 5.12 Generalized tracts of a stream

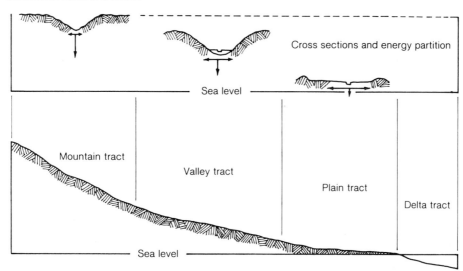

Cross sections and energy partition

Sea level

Mountain tract

Valley tract

Plain tract

Delta tract

Sea level

Mountain tract, Chihuahua, Mexico

Valley tract, Virginia

Lower valley tract, Rio Grande, Boquillas, Mexico

Figure 5.13 Pool and riffle relations

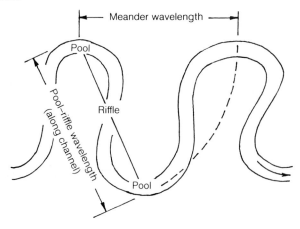

Figure 5.14 Source and depositional areas of sand in a meandering stream

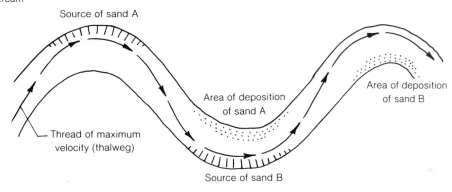

riffles whose spacing or wavelength increases regularly with stream width. As the stream approaches base level, its erosive energy is gradually directed from vertical to horizontal, and the pool and riffle system becomes horizontally looping, quasi-sinusoidal in plan and is termed *meandering*. The original pool and riffle wavelength is preserved and incorporated in the meander wavelength as shown in figure 5.13.

A balance of erosion and deposition is involved in all rivers. In a meandering river, erosion takes place principally at the outside of the bends and deposition at the inside with the source and deposition areas for sand on meander beds as illustrated in figure 5.14. It is the erosion of the outer sides of

the bends and deposition on the inner that cause the migration of the meanders both outward and in a downstream direction with effects such as shown in figure 5.15. These changes in course are most commonly accomplished during times of flood when a river tends to overflow at the narrowest part of a meander in order to straighten its course and the bypassed meander is turned into a backwater, bayou, or cutoff meander. The new course is called a *chute*.

The wavelength of a meander is proportional to the width of the river and is also proportional to the radius of curvature of the meander. These relationships (fig. 5.16) also apply to the Gulf Stream and to meanders in glacial ice.

Figure 5.15 Erosion and deposition in a meandering stream

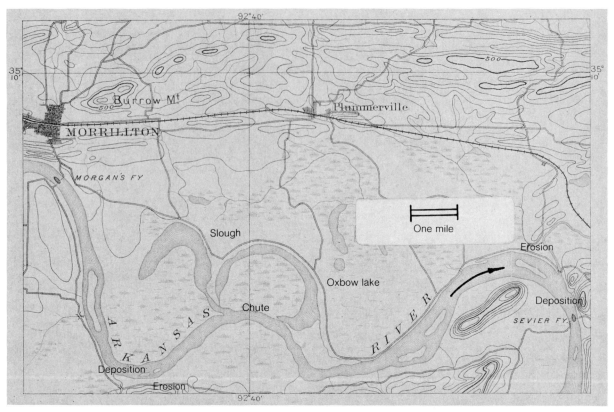

Part of the Morrillton (Ark.) 1:62500 Quadrangle, U.S. Geological Survey.

Natural **levees** form during periods of flood when a river overflows its banks and sediment is deposited close to the bank line (fig. 5.17). These levees may hold the level of the river above that of the surrounding valley floor and cause the flood plain to slope away from the river forming marshy areas called *back swamps.*

In some streams, sediment deposits form in the main channel and cause the stream to divide around islands, which develop from bars in the stream bed. The bars are usually longitudinal and develop from material too coarse for the stream to move except during periods of flood. This coarse material, in turn, traps finer sediment and may be reinforced by vegetation to form a semipermanent island. When many depositional islands form, the stream is divided into a ramifying set of channels and such *braided channels* are characteristic of streams carrying heavy sediment loads or represent a portion of the stream course where the banks consist of poorly consolidated material that is easily erodable.

Stream Deposits

The size of particles that can be transported in running water (its competency) varies exponentially with its velocity, or more properly its turbulence. This, coupled with the widely different discharge of a stream as it experiences times of flood and low water, results in time-controlled erosion, transport, and deposition of selectively sized particles. *Stream channel deposits* may thus be anticipated to be fairly well sorted, but to occur in interfingering lenses of finer and coarser grained materials.

The turbulent flow common in the upstream tracts of rivers is competent to move gravel and perhaps cobbles although it is more likely that very

Figure 5.16 Relationships between meander length, channel width, and radius of meander curvature

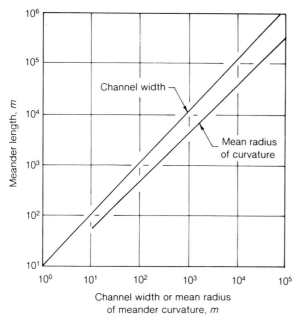

Channel width or mean radius of meander curvature, *m*

Redrawn from L. B. Leopold, M. G. Wolman, and J. P. Miller, *Fluvial Processes in Geomorphology.* © 1964 W. H. Freeman and Company. Used with permission of W. H. Freeman and Company.

Figure 5.17 Development of natural levees

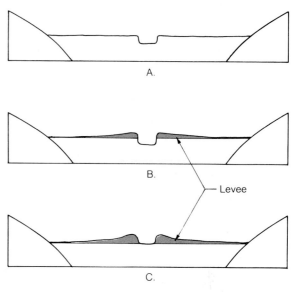

KEY

A. Cross section of stream channel and flood plain.
B. The stream overflows its banks and deposits sediment close to the bank line as levees.
C. Successive flooding increases the height and breadth of the levee.

coarse material is moved by undermining during periods of high flow. The competency generally decreases downstream to gravel, sand, and silt-size material, which characterize most river channel and flood plain deposits.

Deposition in active stream channels is of a temporary nature and sediments deposited at a low-flow stream stage may well be moved (and replaced) at the next high-flow stage. However, stream deposits will not only be found in the active channel of a stream but also comprise the abandoned flood plains and terraces beside the stream that represent its earlier location and activity.

Final deposition of river-borne sediments occurs as a **delta** when a river discharges into a body of standing water where its velocity is checked and its sediment load deposited. This situation can occur where a tributary stream enters a main channel but is most common where a stream enters a lake, impoundment, or the sea. The largest deltas in the world form at a sea edge. The delta deposits of the Mississippi River are over 1 000 kilometers in length and have been building from the vicinity of Cairo, Illinois, since Mesozoic time. The drainage basin includes the mid-continent of the United States from the Appalachian to the Rocky mountains and provides a constant supply of sediment to build the delta into the Gulf of Mexico. The main channel in the delta area splits into a number of distributary channels, often separated by swamps, and the course of the distributaries is constantly changing as they meander across the soft deltaic sediments.

The general cross section of the sediments in a delta shows three distinct sets of beds. The *bottomset* beds are the first deposited, conform to the general slope of the sea floor, and extend farthest out into the sea. These are overlaid by the *foreset* beds built out with a greater angle of initial dip and filling most of the space between the sea floor and surface. These are overlaid, in turn, by nearly horizontal

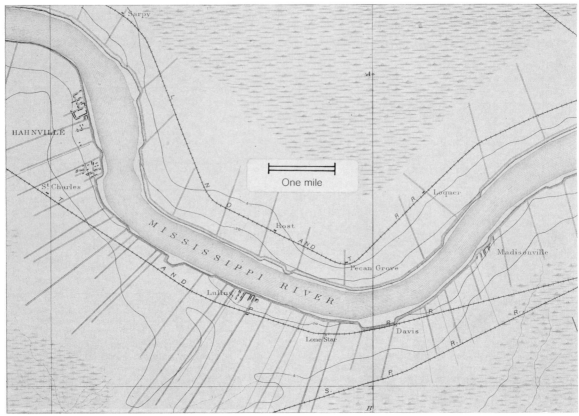

Natural levees and back swamps, Mississippi River

Part of the Hahnville (La.) 1:62500 Quadrangle, U.S. Geological Survey.

topset beds deposited at or close to sea level. The bottomset and foreset beds are generally deposited in a marine environment but the topset beds may be deposited in a partly marine and partly nonmarine environment, the kind of environment in which coal-forming plants grew to be later buried and transformed to lignite and coal.

SUMMARY

Water is the essence of life on our planet. The hydrologic cycle describes the way in which water is moved between the three great reservoirs of atmosphere, surface water, and groundwater. The first is the province of the meteorologist, but surface water and groundwater are part of the geologic milieu of earth process and material. (Groundwater is discussed in chapter 6.)

Stream systems of rivers and tributaries drain the water falling on the earth's surface, and running water is the principal agent of sediment transport. It is responsible for most of the erosion and transport of both dissolved and solid debris to places of temporary and final deposition.

The nature of streams, the mechanics of their flow, and their transport characteristics are of crucial importance in the design and construction of many engineering projects.

ADDITIONAL READING

Davis, W. M. 1902. Base level, grade, and peneplanation. *Geographical Essays,* vol. 18. Boston, Mass.: Ginn & Co.

Einstein, H. A. 1950. *The bedload function for sedimentation in open channel flows.* U.S. Department of Agriculture Soil Conservation Service Technical Publication 55. Washington, D.C.: U.S. Department of Agriculture.

Gilbert, G. K. 1914. *The transportation of debris by running water.* U.S. Geological Survey Professional Paper 86. Washington, D.C.: U.S. Government Printing Office.

Hjulstrom, P. 1959. *Transportation of detritus by running water in recent marine sediments.* Tulsa, Okla.: American Association of Petroleum Geologists.

Judson, S., and Ritter, D. F. 1964. Rates of regional denudation in the United States. *Journal of Geophysical Research,* v. 69, no. 16. Baltimore, Md.: Johns Hopkins Press.

Leopold, L. B., Wolman, M. G., and Miller, J. P. 1964. *Fluvial processes in geomorphology.* San Francisco: W. H. Freeman.

Macklin, J. H. 1948. Concept of the graded river. *Bulletin Geological Society of America,* v. 59. Boulder, Colo.: Geological Society of America.

Middleton, G. V., and Southard, J. B. 1978. *Mechanics of sediment movement.* SEPM Short Course no. 3. Society of Economic Paleontologists and Mineralogists. Tulsa, Okla.: Binghampton Press.

Mutlu Sumer, B., and Müller, A. 1983. *Mechanics of sediment transport.* Rotterdam: A. A. Balkema.

Rubey, W. W. 1938. *The force required to move particles on the bed of a stream.* U.S. Geological Survey Professional Paper 189E. Washington, D.C.: U.S. Government Printing Office.

——— 1952. *Geology and mineral resources of the Hardin and Brussels quadrangles* (Illinois). U.S. Geological Survey Professional Paper 218. Washington, D.C.: U.S. Government Printing Office.

6

Groundwater and Frost

INTRODUCTION

Water in the atmosphere and on the earth's surface is familiar. Water also exists in quantity within the pores and crevices of rock and sediment beneath the surface. Indeed, more underground water is to be found than is present in the combined volume of the world's lakes and rivers.

This groundwater is not only valuable as a water source but as a component of geologic materials is an important contributor to their mechanical properties. Further, it must be dealt with in the many engineering situations in which excavations penetrate the zone of water-saturated materials.

The free movement of water underground is impeded by friction as it percolates through the constricted passageways within rocks; however, the rate of groundwater movement, although usually small, is continuous in response to differential pressures. These pressures arise from both differences in fluid head (elevation) and in adjustments within a rock or sediment that result in changes in pore volume.

POROSITY AND PERMEABILITY

Soil, sediment, and rock contain **voids**—spaces, cracks, and pores—that are always filled with fluids (usually air or water), and these fluid-filled voids may constitute a significant portion of the total rock volume, thus serving as *fluid reservoirs*. If the pores are connected, the fluids move more or less easily through the solid framework under the influence of differential pressure. This storage and movement of water within the ground is a particularly important aspect of geology and engineering and is dealt with in the following pages. To limit the discussion, **connate fluids,** which are encountered in deep drilling as salty formation waters, natural gas, or petroleum, are excluded. Such fluids are trapped in closed systems in the earth and do not participate in circulation that will eventually bring them to the surface.

The **porosity** of a rock or a sediment is the ratio of the pore volume to the total volume expressed as a percentage (see chapter 10). Since the mineral and rock grains constituting rocks vary in size and shape, the pore space also varies (fig. 6.1). Most clastic sedimentary rocks have porosities between 10% and 45% although some poorly consolidated sediments have porosities as high as 80% and recently deposited muds, or slurries, may hold up to 90% water by volume. The porosities of sediments decrease as they become consolidated into solid rock. Igneous rocks are at the other extreme of the porosity scale with unfractured granite, gabbro, and obsidian having essentially zero porosity. In general, a porosity of less than 5% is regarded as low, from 5% to 15% is medium, and over 15% is considered high.

Permeability, or **hydraulic conductivity,** is the ability of a rock to conduct a liquid and therefore determines the yield of water from rock or water-bearing sediment. A rock that carries significant amounts of underground water is called an **aquifer;** the best aquifers will normally have the highest permeabilities and differential pressure due mainly to gravity. Gravity supplies the energy to cause underground water to flow. A rock of very low permeability is termed an **aquiclude.**

Permeability is related to the size and connectivity of pore passages. Pore passage diameters vary

Figure 6.1 Porosity in geologic materials

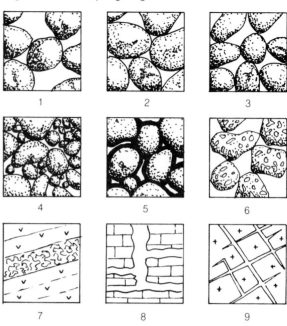

KEY

1. Open packing, porosity higher
2. Close packing, porosity lower
3. Good sorting, porosity higher
4. Poor sorting, porosity lower
5. Cementation, porosity reduced
6. Porous grains, porosity increased
7. Porous zone between lava flows
8. Solution joints in limestone
9. Fractures in crystalline rock

roughly with particle diameters and it is common practice to relate the hydraulic conductivity to particle size. The approximation, following Hazen (1893),[1] is that $K = 100 D_{10}^2$ where K is the hydraulic conductivity in cm/s and D_{10} is the diameter of particles at 10% finer weight (as discussed later in chapter 10). The concept is that finer material occupies the interstices between larger grains and permeability will thus be a function of the size, size distribution, shape, and manner of packing of particles with a dominant role being played by the finer fraction. Permeability, however, is strongly affected

1. Hazen, A., 1893. *Some physical properties of sands and gravels.* Mass. State Board of Health, 24th Ann. Report for 1892.

Figure 6.2 Permeability of some common sediments

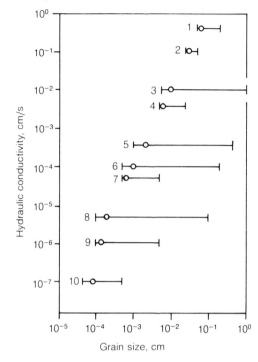

KEY

Grain size
1. Uniform coarse sand
2. Uniform medium sand
3. Well-graded sand and gravel
4. Uniform fine sand
5. Well-graded silty sand and gravel
6. Silty sand
7. Uniform silt
8. Sandy clay
9. Silty clay
10. Clay (30–90% clay sizes)

D_{10} value indicated by open circle

by slight variations in particle size and size distribution (gradation), so highly accurate measurements on a few samples generally have less value than more tests of lesser accuracy. Figure 6.2 illustrates the range of hydraulic conductivity that may be expected in some common sediments and will provide approximate coefficients if testing is not practical. The regularity of permeability coefficient changes with D_{10} value should be noted.

GROUNDWATER MOVEMENT

When a hole is drilled or a well or a mine shaft sunk into the ground it will usually penetrate a layer of soil, sediment, or rock containing water, and in such a well or a vertical mine shaft there is a level to which the hole will fill with water. This level identifies the level of the **water table.** Below the water table the rocks are saturated, and so the water table is defined as the upper surface of the zone of saturation (fig. 6.3). The water table is also the locus of points in

the groundwater at which pressure is equal to atmospheric pressure. Above the water table is the **vadose** (Latin meaning shallow) **zone** extending upward to the earth's surface. The vadose zone is subdivided into the region of soil moisture (closest to the surface), an intermediate belt, and a **capillary fringe** lying immediately above the water table. The thickness of the capillary fringe depends on the size of pores with smaller pores resulting in greater capillary uptake and a thicker fringe. The region of soil moisture contributes water to vegetation or to the atmosphere by evaporation.

The maximum amount of water that may be held in the vadose zone is termed **field capacity** and is the weight of water per unit volume retained against gravity divided by the total weight of material. In the capillary fringe above the water table, water motion is upward at rates controlled by pore size and departure from field capacity of the vadose zone. Water percolation in permeable materials

Figure 6.3 Groundwater zones

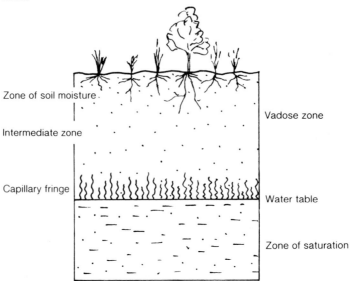

Zone of soil moisture

Intermediate zone

Capillary fringe

Vadose zone

Water table

Zone of saturation

Figure 6.4 Distribution and movement of water underground

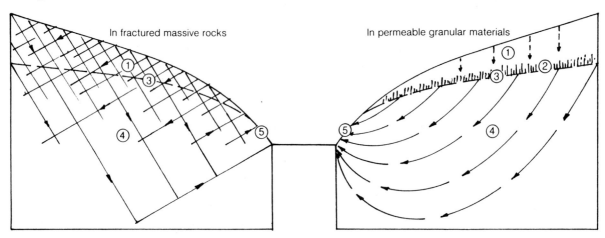

In fractured massive rocks

In permeable granular materials

KEY

1. Vadose zone
2. Capillary fringe
3. Water table
4. Percolation in saturated zone
5. Discharge zone

Figure 6.5 Some features of water underground

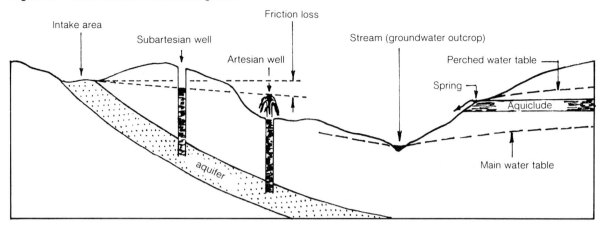

below the water table is controlled by hydraulic gradients and is typically along convex-downward arcuate paths. Figure 6.4 shows the relations of the zones and water movement for groundwater in uniformly permeable material and in otherwise impermeable but fractured rock.

The shape of the unconfined groundwater surface, or water table, is typically a subdued reflection of the topography, rising under hills and falling toward valleys. In more detail the surface is also a function of permeability with lesser slopes related to greater permeability. The water table is sometimes at or above the land surface where it is represented by the water surface of lakes, streams, or swamps and its "outcropping" may be marked by springs. In addition to the various physical constraints that control the position of the water table, its configuration will vary in time as the balance of recharge, discharge, and storage is changed by climatic factors or by withdrawal of water through wells or through springs.

Since water underground is present in several distinct settings, its quantity and motion are subject to different hydraulic parameters. In porous surficial materials water filters downward in the zone immediately below the surface to some lower level where its vertical movement is blocked by impermeable or water-saturated material. Its presence in partially air-filled pores results in a chemically and biochemically active system. When variations in permeability are present underground, as is geologically realistic, groundwater in permeable horizons (aquifers) may be constrained by less permeable or impermeable barriers (aquicludes) so that its free surface cannot conform with the pressure or **potentiometric surface** representing its unconfined static level. Such confined groundwater will, of course, rise to its potentiometric level if relieved by a fracture or well in the aquiclude (fig. 6.5). A system of this kind is called **artesian** (after the type locality in Artois, France) if water rises above ground level and subartesian if the rise is less. Some water and other liquids such as petroleum have been trapped by aquicludes deep within the earth for long periods of time and are typically in an artesian condition as shown by their rise when penetrated by a well.

Darcy Relation

Since water below the water table is not stationary, it is appropriate to consider the factors affecting its movement. Subsurface water movement is nearly always laminar because the Reynolds Number (chapter 5) rises above the threshold value for turbulent flow only in the relatively large openings offered by pores in gravel coarser than 2 mm, open cracks, or solution channels. The factors determining the rate of movement of water along or under the water table are the permeability and the *hydraulic gradient,* which is the ratio between the difference in elevation, or head, and the distance measured along the slope of the water table. In nature groundwater gradients are fairly low, varying from 1 to 10 m per 1 000 m. Since groundwater movement is slow and dominantly laminar, typically

Table 6.1 Effective Velocity of Groundwater Percolation Under a Gradient of 1%

Material	V_{eff} in m/day
Clay and gumbo	0 — .001
Silt, fine sand, loess	.02
Medium sand	.35
Medium to coarse sand and sandy gravel	~ 1.9
Gravel	~ 9.0

From C. F. Tolman, *Ground Water.* © 1937 McGraw-Hill Book Co.
Reprinted by permission of the McGraw-Hill Book Co.

a few meters per day to a few meters per year, its movement follows the law known as the **Darcy relation,** which was formulated by Darcy, a French hydrologist who in 1856 showed that laminar flow velocity and discharge are proportional to the hydraulic gradient. The hydraulic gradient is $\sin \theta$ or h/l where h is the head, l is the length of the flow path from intake to discharge and θ the slope angle. The law states that $V = K\, h/l = K \sin \theta$ where V is the percolation velocity and K is the hydraulic conductivity related to the properties of the rock. Darcy did not attempt to segregate the factors that control K but used experimental measurements as currently favored. Values of K include properties of the medium such as pore size, pore connectivity, and fluid adsorption; properties of the fluid such as its density and porosity; and properties of the system such as temperature.

Permeameters based on the Darcy relation may be used to determine P, which is defined as the rate of flow through a unit cross section under a vertical (100%) gradient. The general relations are $K = Qlt/TAh$ where Q = quantity of discharge, l = column length, t = temperature correction, T = time, A = cross-sectional area, and h = head.

The rate of percolation is $Q = K\, h/l = K \sin \theta$ where Q is the discharge. Q is directly related to percolation velocity, V, since a discharge of 1 m³/day through a 1 m² cross section equals a velocity of 1 m/day. However, because water occupies only the pore space and all of the water in the pores is not moving, the volume of water passing through a unit cross section must be divided by the *effective*

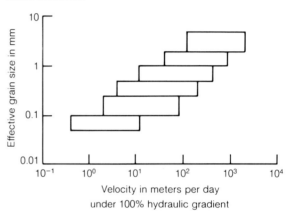

Figure 6.6 Approximate percolation velocities for well-sorted sediments

porosity to obtain the effective velocity of percolation, thus

$$\text{Effective velocity} = \frac{K \sin \theta}{\text{effective porosity}}.$$

Table 6.1 gives values of effective velocity of groundwater percolation through various materials under a 1% gradient.

The effective porosity is a function of the size, shape, packing, and sorting of the grains. For uniformly sized material the relation of grain size and the percolation velocity are shown in figure 6.6, but as is always the case, laboratory or field measures should be made for the site of interest. Field measurements of effective velocity are usually accomplished by *pumping* and recovery *tests* and occasionally by various *water tracing* techniques employing dyes or fabric brighteners. Other means

Figure 6.7 Groundwater flow

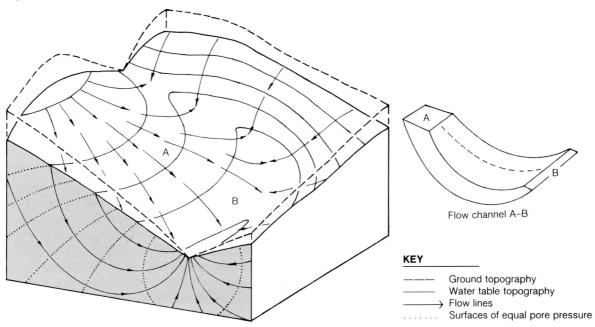

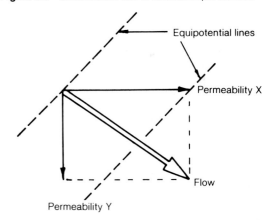

Flow channel A–B

KEY

– – –	Ground topography
———	Water table topography
——→	Flow lines
. . . .	Surfaces of equal pore pressure

are to introduce an electrolyte such as common salt in a well and monitor the change in electrical conductance in surrounding test wells or to use radioactive tracers.

Flow Nets

The flow of groundwater below the water table is along downward-looping paths from recharge area to discharge area as shown on the face in figure 6.7. In the figure the ground topography (dashed) has been removed and the water table is shown as a contoured surface. *Flow lines* on this surface diverge from mounds and converge to trenches. Assuming the saturated ground to be equally permeable in all directions, the flow lines will be perpendicular to equipotential lines represented by the contours. Flow will also be perpendicular to equipotential surfaces (surfaces of equal pore pressure) below the water table. These surfaces are indicated by dotted lines on the face of the block. Flow from any recharge area on the water table surface, say from *A* to *B*, is within a *flow channel* bounded by *flow surfaces* (as

Figure 6.8 Groundwater flow in an anisotropic medium

Equipotential lines

Permeability X

Flow

Permeability Y

shown in the inset whose shape is dictated by the position of equipotential surfaces).

If permeability is not isotropic, as is often the case, groundwater flow in an anisotropic medium will not be at right angles to an equipotential line or surface (fig. 6.8).

Figure 6.9 Flow net around a cofferdam

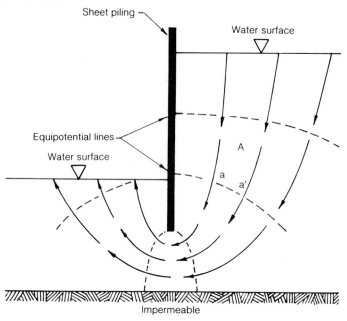

In quantitative studies of groundwater flow, a *flow net* is constructed with flow lines so spaced that each channel they bound has an equal discharge flow rate and each flow channel is subdivided into elements by lines of equal potential drop. As an example of a flow net, consider figure 6.9, which represents a common engineering hydrological situation. The flow elements are of approximately equal area (volume) so the discharge velocity of each element equals Q divided by the end length of the element (i.e., $a - a'$ for element A in the figure). The velocity thus changes as the flow lines converge or diverge. Flowing water transmits a frictional drag on the solids through which it moves that is proportional to channel size and flow rate. This is termed *seepage force* and may have considerable engineering significance; for example, increased discharge velocity might cause *piping* beneath the piling in the figure or the upward movement on the downstream side might be large enough to equal the submerged weight of the sediment generating a liquid soil-water suspensoid and a quick condition (see chapter 11).

WATER TABLE POSITION

The water table is usually penetrated by humans for one of two reasons. The first is to make an excavation for an engineering purpose such as a foundation, road cut, or pipe burial, or to mine to extract economic materials. The second is to recover water for domestic, agricultural, or industrial uses. In all such cases the water table is artificially lowered and the greater the lowering of the water table the more water flows into the well or excavation until some equilibrium controlled by Darcy's law is attained. The need to stop or control groundwater flow greatly affects the engineering economics of many situations.

A fundamental aspect of the position of the unconfined water table is that it represents the level at which water will stand at one atmosphere pressure. The effects of noncoincidence of the real water surface and an imaginary potentiometric surface are everywhere met in dealing with water underground. Some of these phenomena are described in the following paragraphs.

Figure 6.10 Cone of depression

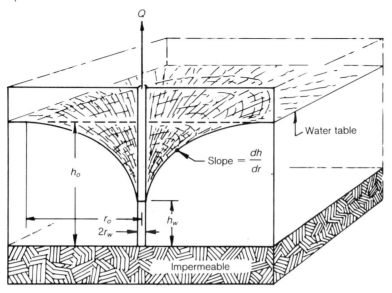

Derivation of equations for a cone of depression

Water withdrawal from the well causes steady, radially inward flow to the well from the unconfined aquifer. The cross section through which the water is flowing at some distance from the well is a cylinder centered on the well of radius r ($r < r_o$) and height h ($h < h_o$) above the well bottom. The water table slope at this radius is dh/dr. Therefore, the discharge from the well is $Q = 2\pi\,Prh\,dh/dr$ where P is the coefficient of permeability. Integrating with limits of r_o and h_o

$$Q = \pi P \frac{h_o{}^2 - h_w{}^2}{\log_e (r_o/r_w)} = \pi P \frac{(h_o{}^2 - h_w{}^2)}{2.3 \log_{10} (r_o/r_w)}$$

This equation fails to accurately describe the drawdown curve near the well, but values of Q for given heads are good. Usual values for r_o are 150–300 m.

In the case of a bore or well for the recovery of water it is desirable to recover the maximum amount of water with the minimum lowering of the water table and with the minimum depth of the well from a supply that is naturally controlled by the permeability of the aquifer. Activities such as drawing down a reservoir or pumping a well cause first a rapid lowering of the potentiometric surface followed by a prolonged *dewatering* of the material stranded above this surface. This is a potential cause of slope failure as described in chapter 13. Around a pumped well the lowered pressure surface has a general conical shape. With continued steady pumping a balance of discharge and recharge is achieved when the water surface conforms to the pressure surface, and increased groundwater flow down the steepened gradient balances that removed by pumping. The geometric relations and derivation of pertinent equations for this **cone of depression** are given in figure 6.10.

In artesian systems the potentiometric surface is above the upper level of saturated permeable material confined below an aquiclude. Completion of a cased well under these conditions (fig. 6.11) causes an immediate rise in the water level to the potentiometric surface. Should this point be at or above ground level a flowing well or fountain will result. Confinement of water underground by manufactured barriers serving as aquicludes results in artesian conditions, which may be very complex and can involve multiple pressure surfaces as suggested by the relations of figure 6.12. When a free water surface is maintained above the local water table, as might be the case for a river in flood or in an irrigation ditch, water seeps through the stream bed and forms a groundwater mound on the water table (fig. 6.13).

Figure 6.11 Potentiometric surface in artesian and subartesian flow

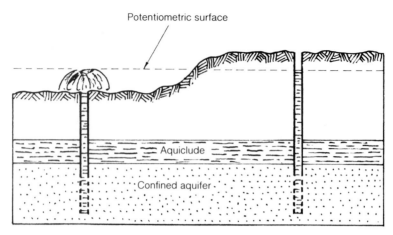

Figure 6.12 Multiple potentiometric surfaces

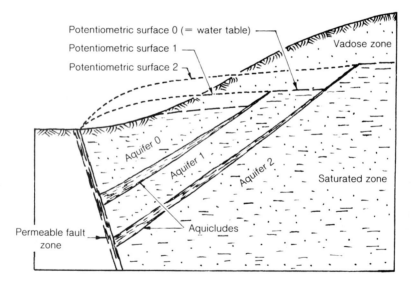

Figure 6.13 Leakage from surface channel forming a water table mound

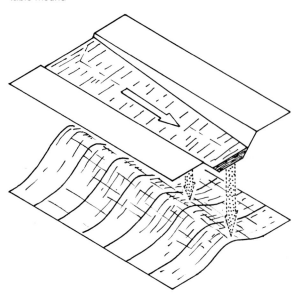

SOME SPECIAL GROUNDWATER CONDITIONS

Karst Regions

Limestone rocks are composed of carbonates which are soluble to varying degrees in acid. Groundwater usually contains carbonic acid from carbon dioxide dissolved from the atmosphere and humic acid derived from the decay of plants in the soil. Underlying carbonate rocks can dissolve and form mazes of underground caverns, tunnels, and drainage systems. The ground surface over such regions characteristically shows **sinkholes** and disappearing streams (fig. 6.14) and is called **karst topography,** after the Kras area of Yugoslavia where such topography is well developed. Unless the engineer is aware of these features, land subsidence, severe foundation problems, dam leakage, and other engineering problems can be encountered. Solution channel systems can be hundreds of kilometers in

Blocked sinkhole, Kentucky

Stalactites and stalagmites, Carlsbad Caverns, New Mexico

Figure 6.14 Karst terrain

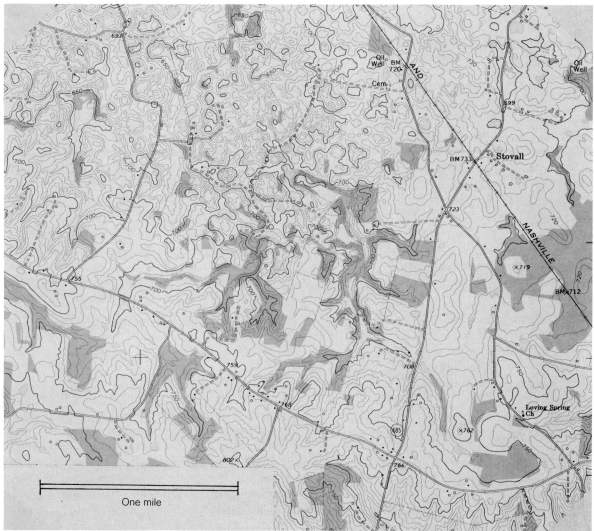

One mile

Part of the Park City (Ky.) 1:24,000 Quadrangle, U.S. Geological Survey.

length and the surface openings are almost impossible to close by concrete grouting or other means. Karst topography can usually be recognized on aerial photographs, it is usually well known to local inhabitants, and site details may be examined by drilling. Analysis by geological and geophysical means is required in less obvious situations.

Groundwater in Coastal Areas

An interesting and important groundwater phenomenon occurs in permeable rocks and sediments in coastal areas. Fresh water derived from landward recharge is brought into contact with salt water derived by infiltration from the ocean, and since they have slightly different densities and do not mix

Figure 6.15 Salt–fresh groundwater relations at a coastline

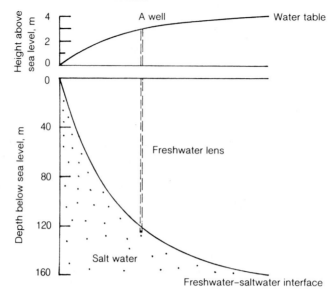

readily, a rather sharp interface is present. The general geometry is shown in figure 6.15. The freshwater table is in its usual position and the location of the freshwater–saltwater interface is fixed by the freshwater head above sea level in the ratio of about 40:1; that is,

$$\text{density} \left(\frac{\text{fresh water}}{\text{salt water} - \text{fresh water}} \right)$$
$$= \frac{1.000}{1.025 - 1.000} = 40.$$

The apparent abundance of fresh water in the coastal zone indicated by figure 6.15 is misleading in that drawdown from a well will remove the weight of a cone of fresh water around the discharge point causing a 40:1 rise in the saltwater surface with consequent saltwater intrusion into the well. For example, a well at *A* would encounter 3 + 120 m of fresh water. When pumped, however, the freshwater – saltwater interface would cone upward by 40 m for each one meter of drawdown from the surface. Saltwater intrusion would thus occur when the downward coning reached a depth of only three meters.

GEOTHERMAL ENGINEERING

The earth has a natural **geothermal gradient** increasing in temperature toward the center at a rate of 20–30°C per km, and hence a great quantity of heat energy is stored within the crustal and deeper layers of the earth. This heat energy can become even more concentrated in zones of igneous rock intrusion where magma temperatures range up to 1 200°C. Igneous areas decline in thermal activity with time, but even in the dying stages there may be sufficient subterranean heat remaining to raise the temperature of groundwater and produce the steam and hot water that has long been used to generate electricity (Geysers area, California) and to heat whole cities such as Reykjavik in Iceland.

In World War II Japanese troops cut off from supplies on Rabaul Island in the Pacific Ocean used volcanic heat to evaporate seawater for salt and pure water. Today there are operating *geothermal steam power* plants in New Zealand, Japan, Italy, the Philippines, Mexico, Russia, the United States, and other areas. Many other geothermal areas are under study. Potential areas in the United States are to be found in the western states including Alaska and Hawaii.

CASE HISTORY 6.1 *Geothermal Energy*

Figure 1 Geyser in Yellowstone National Park

Figure 2 Geothermal power plant

The two scenes depict the uncontrolled natural escape of steam and water from a geyser in Yellowstone National Park, a typical geothermal "hot spot" area, and the controlled use of the steam to generate electricity in a geothermal power plant.

The geyser periodically erupts and showers the surrounding area with steam and boiling water; the materials dissolved in the groundwater precipitate as a deposit of geyserite around the vent that shows clearly in the photo. This dissolved material is a particular problem in the design and operation of geothermal plants. The cooled geyser water infiltrates underground to be reheated and the periodic eruptive process is repeated.

In a geothermal power plant the emission of steam is controlled and harnessed to be fed into generating turbines to produce electric power. The used water is usually returned to the ground and reused making the process efficient from an engineering standpoint, environmentally acceptable, and relatively cheap.

Potential geothermal areas can be located by infrared satellite or aerial surveys.

Figure 6.16 Setting for geothermal power

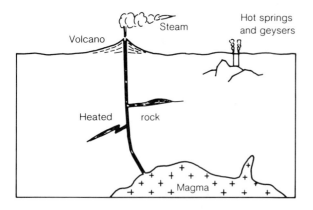

Figure 6.17 Supergene enrichment of an ore body at the water table

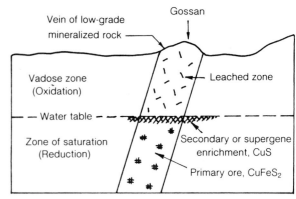

A generalized setting for geothermal power sources is given in figure 6.16. Even if no water is naturally present, water may be circulated through heated dry rocks by injection and recovery wells. Thermal energy will be harnessed on a large scale in the future by the use of heat exchangers or by steam used directly to drive generators for electric power.

SUPERGENE ENRICHMENT

Metalliferous mineral deposits containing ores soluble in the vadose zone may be secondarily enriched at the groundwater table by **supergene enrichment** through downward transport and reprecipitation. For example, in figure 6.17 primary chalcopyrite, $CuFeS_2$, an ore of copper, above the water table is oxidized in the presence of water to yield copper ions and the sulfate radical in solution and solid iron oxide:

$$4\ CuFeS_2 + 19\ O_2 \xrightarrow{\text{water}}$$
$$4\ Cu^{2+}\ 8\ (SO_4)^{2-} + 2\ Fe_2O_3.$$

The copper and sulfate ions migrate downward leaving porous and iron-stained rock (**gossan**) behind as an indicator of their former presence. At and below the water table the cupriferous solutions undergo reduction and precipitate as a copper sulfide such as chalcocite, CuS, forming a layer enriched in copper:

$$Cu^{2+} + (SO_4)^{2-} \rightarrow CuS + 2\ O_2.$$

Other minerals may be enriched by selective solution, vertical transport, and reprecipitation in the zones of aeration and saturation. Particular minerals may thus be concentrated at the water table or be found as residual deposits in the zone of aeration from which other minerals have been leached. This is the general mechanism that has formed the great iron, nickel, manganese, and aluminum (bauxite) deposits of the world whose forms are controlled by present or past water tables.

EVAPORATION

The migration of dissolved materials may be upward rather than downward under conditions of high aridity when the zone of saturation is losing water by *evaporation.* Under such conditions various salts will be deposited at the top of the capillary fringe where the transporting water is evaporated. This mechanism will lead to **hardpans** or **duricrusts** within the soil or poison it for agricultural uses. It is a very serious consideration for irrigation schemes, which are typically developed in arid regions. Examples are the poisoning of irrigated areas in California and the Nile Delta by salt carried upward with evaporating groundwater. In extremely arid areas commercial deposits of saline minerals may be formed by this process.

FROST ACTION

The freezing and thawing of groundwater in surficial geologic materials is a subtle hazard that must be considered anywhere that winter temperatures fall below the freezing point of water. *Frozen ground* is a concretelike solid rather than a cohesive or cohesionless mass and hence reacts quite differently to stress than when in its unfrozen state. There is a significant *expansion* in volume when water-saturated earth freezes, and thawing ground may well be a liquid suspensoid. These changes, following an annual cycle, can adversely affect the foundations of roads and structures, break buried utility lines, or cause slopes to fail.

The freezing point of water, nominally 0°C, is lowered by both pressure in or on the water and by the presence of dissolved material. Because freezing and thawing of soils is seasonal, temperature is variable in time and both its increase and decrease work downward from the surface. Water content is variable within the soil, especially above and below the water table, and water in small pores is subjected to greater pressure than in large pores because of its stronger attraction to pore walls (capillarity). Additionally, water expands about 9% on freezing and its crystallization is exothermic. Combining these variables into a general description of the history of a *freeze-thaw cycle* in soil is obviously difficult and not completely understood.

Consider the freezing of water in a saturated granular aggregate at the 0°C isotherm: freezing begins within the pores but not on the higher-pressure pore surfaces, the latent heat of crystallization of the exothermic freezing opposes the reaction, and overcooling is needed to maintain the ice. The transformation of water to ice results in a volume change, further pressurizing the freezing site and forcing still liquid water to move away.

Growing *ice crystals* within the soil mass increase its water volume by about 9%, but measured expansions of frozen ground range upward of 20% which fact, coupled with the observed presence of ice lenses within frozen ground, requires that additional water be drawn into the freezing site. A suggested means to do this is to increase the capillarity of soils above the water table by reducing their pore size through a partial filling with ice.

Another possible means for the movement of water into a zone of freezing begins with the formation of a near-surface ice zone following a drop in air temperature. Water in contact with the lower surface of this ice zone will freeze while the ions adsorbed on particle surfaces in a water film remain in solution, thus increasing their concentration locally. This has the effect of drawing in water from below to reduce the concentration, and the cycle repeats. So long as the upward flow and freezing rate are in balance water will be drawn upward, will freeze to cause **heaving,** and on melting the soil mass may become a suspensoid because of the presence of excess water.

In clay-rich soils the amount of adsorbed ions is usually large but the permeability is low, whereas the reverse is true for coarse-grained soils. As a result, neither type satisfies the conditions for heaving and the consequences of groundwater freezing are generally not objectionable. Silty deposits, on the other hand, are very susceptible to frost heaving. The phenomenon of **frost heaving,** which is the increase in volume of frozen ground, is enhanced when there is a high water table from which water may be pulled by capillarity into the freezing zone above. The soil capillarity is naturally high because of a critical percentage of silt-sized particles and the permeability of the soil is good, thus allowing easy migration of water. Silty soils possess these characteristics and are the materials most likely to heave. The particle size critical to the development of heaving is 0.02 mm; if the amount of particles of this size is less than 1%, no heaving is likely, but if the amount is more than 3% in nonuniform soils or more than 10% in uniform soils considerable heaving may occur (fig. 6.18). Heaving is also dependent upon the rate of freezing since a rapid hard freeze will seal the pores and stop the upward drawing of capillary water to the freezing zone. Conversely, slowly dropping average temperatures promote heaving.

Frost heaving, Alaska

Figure 6.18 Frost heaving as a function of particle size

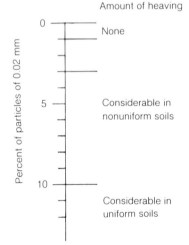

Amount of heaving

Percent of particles of 0.02 mm

0 — None

5 — Considerable in nonuniform soils

10 — Considerable in uniform soils

The effects of frost heaving are everywhere to be seen in those latitudes subject to annual freezing. Water in cracks of rocks in natural outcroppings or exposures freezes, expands, and wedges the rock apart. These fragments when released by thawing litter the rock surface and slide or fall, if free to do so, and thus make work around rock faces especially dangerous and highway cleanup in road cuts necessary in the springtime. More subtle effects of frost heaving are the slow migration of boulders down slopes or the stones that "grow" in fields. Freezing soil expands carrying enclosed boulders toward the surface; thawing, however, allows soil particles to fill in below the boulder leaving it slightly nearer the surface and, after many cycles, it rests partially exposed and "floating" in the soil. If this process occurs on a slope, the motion of the boulder is outward in freezing, but vertically downward in the thawing process, thus giving a downslope component.

With decreasing temperature the various effects that oppose the complete freezing of interstitial water are overcome and the soil mass attains a rigid and expanded condition. Freezing continues gradually downward so long as ambient air temperatures remain below the freezing point. The low thermal conductivity of the soil mass, however, makes the freezing and thawing lag significantly behind temperature changes in the air. The frost is not "out of the ground" and soil temperatures not high enough for plant growth until well after the average air temperature is above the freezing point.

The depth that freezing may reach varies from point to point on the earth from zero at sea level in tropical and subtropical regions, to 1 to 2 m for much of the northern conterminous United States, to over 400 m in the permanently frozen ground (**permafrost**) that underlies much of the polar regions. The depth is rather variable on a local scale because of the differences in soil conditions, the quality of pore water, the amount of insulating cover such as snow or surface water, and the presence of population centers, which raise the local temperature. Local building codes establish an official frost depth, and utility lines and the footings of structures should be below this depth in order to avoid the effects of frost heaving.

When frozen soil melts, the process proceeds from the surface downward and the water released cannot initially percolate through the still-frozen ground below. The thawed soil mass is saturated and can be unstable. Mudslides are a common result of this condition, especially if rain occurs and the mass becomes a **suspensoid.** This partially thawed condition is of especial engineering concern in the construction and maintenance of unpaved roads. It may be alleviated by using coarse-grained, free-draining material for the roadbed and providing adequate drainage ditching.

Permafrost

During Pleistocene time ice sheets of continental dimensions covered large portions of the northern hemisphere and in North America extended southward into the area of the United States. When they finally melted back 10–15 thousand years ago they left a legacy of sedimentary debris in the form of glacial and glaciofluvial deposits that had been carried by the glacial ice and meltwater. Deposits of loess are also related to this geologic process. (See

Figure 6.19   Distribution of permafrost in the Northern Hemisphere

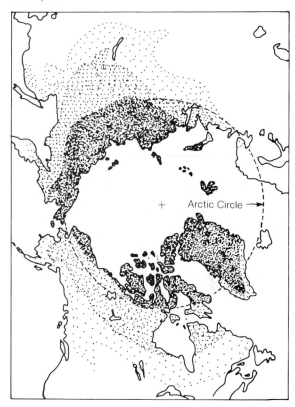

KEY

Continuous permafrost
Discontinuous permafrost
Sporadic permafrost

chapter 4 and appendix II.) Farther north above about latitudes 55°–60° much of the ground has remained permanently frozen since the glacial period and this *permafrost* poses a special set of problems to engineering activity.

Figure 6.19 shows the distribution of continuous, discontinuous, and sporadic permafrost in the northern hemisphere. Frozen ground extends to depths of 300–450 m as a solid mass to the north and then shallows and becomes discontinuous southward. Above the permafrost is a surface zone, the **active zone,** which annually thaws and freezes.

Figure 1 Permafrost effect on railroad

The photo depicts a light-duty, branch railroad line in Alaska that was built to connect a copper mine with a main railroad. The line was obviously designed for light traffic and employed little or no ballast beneath the railroad ties and the track is of low-poundage rail.

Movement of the active zone above permafrost has resulted in the collapse of foundation material and completely destroyed the grade of the track. The vegetation growing between the ties and the condition of the rails indicate that the company did not consider the line a great engineering success and chose another means of ore transport.

Figure 6.20 Soil frost depth

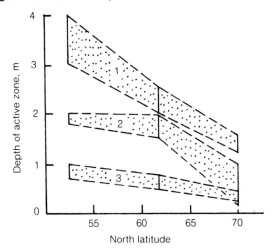

KEY

1. Sandy soil
2. Clayey soil
3. Peaty soil

Figure 6.21 Soil profile in permafrost

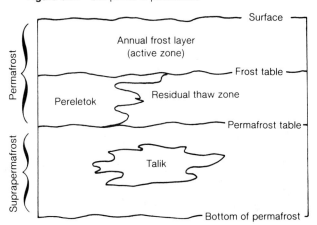

This active zone, when thawed, is often waterlogged because downward movement of the water is inhibited by subjacent ice and there is little topographic relief in these arctic areas. The thickness of the active zone varies with latitude, soil type, vegetation, and snow cover but is seldom more than a few meters (fig. 6.20).

The soil profile in regions of seasonally or perennially frozen ground is both complex and variable in time. Figure 6.21 illustrates some of the terminology used for permafrost regions. Definitions of these and other related terms follow:

Annual frost layer (active zone) The top layer of ground subject to annual freezing and thawing

Excess ice Ice in excess of the fraction that would be retained as water in the soil voids on thawing

Frost table Depth of annual thawing

Ground ice A body of more or less clear ice within frozen ground

Ice wedge A vertical wedge-shaped mass of ice

Muskeg Poorly-drained organic terrain, consisting of a mat of vegetation overlying peat

Pereletok Frozen layer at the base of the active layer that remains unthawed for one or two seasons

Permafrost Perennially frozen ground

Residual thaw zone Layer of unfrozen ground between the permafrost and the annual frost zone. Absent when annual freezing extends to permafrost

Suprapermafrost The entire layer of ground above the permafrost table

Talik Unfrozen zone within permafrost

Soil conditions within the region of permanently frozen ground lead to the development of a number of characteristic features including *ice wedges* and *patterned ground, pingos,* and *solifluction* on slopes. Thermal contraction during the arctic winter causes vertical cracks to form in a polygonal pattern reminiscent of the cracking in a layer of drying mud but in a much larger mesh. Re-expansion during the following summer results in an upturning of the permafrost strata and the cycle, repeating over many seasons, generates vertical, veinlike masses of foliated ice in the upper 10 m or so. These **ice wedges** are arranged in a polygonal

Figure 6.22 Patterned ground

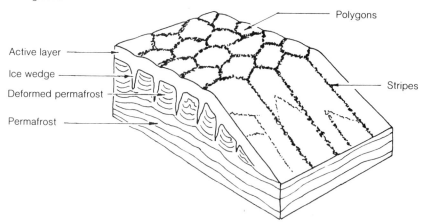

Active layer
Ice wedge
Deformed permafrost
Permafrost
Polygons
Stripes

pattern whose mesh is 10–100 m. Differential thawing and freezing of a surficial layer containing ice wedges leads to the sorting and small movement of surface debris and results in the formation of *patterned ground,* a conspicuous feature of vast areas in arctic terrains. If the ground is flat the typical patterns are circles, nets, or polygons, but downslope movement on slopes of only a few degrees will cause the stone polygons to be drawn out into lobes or stripes (fig. 6.22).

Solifluction is the downslope movement of thawed, oversaturated soils with typical rates of a few centimeters per year. Movement is most rapid at the surface and usually dies out at a depth of less than a meter. The term has occasionally, but improperly, been used to describe slow downslope movement of soils outside the arctic regions.

A remarkable feature of arctic terrains is the formation of isolated ice-cored hills up to 45 m high and 600 m in diameter. They are called **pingos** and about 1 500 have been identified in Canada alone where they are restricted to thick alluvial, deltaic, or glaciofluvial sands. Pingos originate when excess pore water from a confined body of saturated soil is injected beneath an impervious sheet of frozen ground. The cover is arched upward and the water frozen to a core of ground ice.

Engineering activities have and will be conducted under arctic conditions and some useful guides for engineering projects on permafrost are epitomized by the words *investigate, vacate, separate, insulate,* and *cooperate.* Some detailed maps of permafrost locations have been made and air photographs can be used to good advantage since near-surface permafrost retards tree growth and mounds and potholes often develop upon it. Test pitting and drilling should be conducted on local sites and samples carefully tested, especially for grain size and water content. Local, well-drained soils affording excellent construction sites may often be found, commonly on the narrow ridges of glacial deposits or dunes.

When permafrost is encountered, the best advice is to vacate the site if possible. If it is not possible to avoid the frost, separation from it by building above grade on piles coupled with insulation may serve. The effects of shading and reflected sunlight on the ground must be considered.

The best, easiest, and often the cheapest way to deal with permafrost is to cooperate with it. Aside from designs that recognize the unique properties and instabilities of frozen ground, the possibility of using ice for sealing or support should not be overlooked. Deliberate freezing by the circulation of refrigerants through unfrozen ground has been used to stabilize ground for foundations or shaft-sinking and to form impermeable dam cores.

Construction in permafrost regions is complicated by the need to control the seasonally active layer and maintain the integrity of the permafrost both during and after the construction phase. It is sometimes possible to remove the active zone and replace it with a coarse and permeable fill, but preferred practise is to insulate the road or structure above the nominal ground surface. The problem of carrying the Alaska pipeline across seasonally active areas was solved by elevating it above the ground on well-footed supports. Care must be taken during construction to leave undisturbed, insofar as possible, the natural *insulating cover* of snow or vegetation. Disruption of this cover leads to the development of thaw pits, which eventually enlarge to extensive lakes.

Successful methods for road construction include the avoidance of cuts, placement of coarse subgrade material (not derived from beside the road) directly upon the vegetation, and using deep, narrow drainage ditches well away from the road. Particularly unstable situations may require the installation of underdrains, replacement of silt by coarser material, or insulation with gravel, sand, or vegetable mulch.

Small, unheated buildings may generally be placed directly on the ground or be supported on posts or piles. Heated buildings, however, must be elevated leaving an insulating air space below. Support may be by posts footed on gravel mats or by piles footed in permafrost. When piling is used a well or collar around each pile should be built to allow for movement in the active layer. Excessive melting of the active layer may be controlled by keeping the area around the building free of snow and shading the base by an apron or fencing on the sunny side.

In summary, most engineering problems in permafrost areas are caused by the thawing of ground that contains large quantities of ice, and the most serious difficulties are encountered with soils containing a significant silt fraction. Figure 6.23 (on p. 122) provides a means of correlating the unit dry weight of a soil, its ice volume, and water content.

In the figure a silt ($G = 2.70$) having a dry unit weight of 1 522 kg/m³, and containing 60% excess ice has a total porosity (proportion of ice volume in the sample) of 78%, a dry weight of the frozen sample of 609 kg/m³, and a water content of 114%.

SUMMARY

The upper portion of the earth is more or less porous and permeable. Water enters the interstices between solid rock-forming particles and completely saturates the rock below the water table. The water table is marked by the surfaces of lakes and rivers above the ground and by the level of standing water in wells beneath the surface.

The movement of groundwater is, with rare exceptions, both slow and laminar; flow nets relating flow direction to pressure differences are particularly useful in this context. The balance of pressure (head) differences and permeability is expressed by the Darcy relation; the seepage of groundwater into wells, artesian flow, and spring discharge are phenomena that show that groundwater moves to regions of lower pressure.

Water-containing soil is very different mechanically from frozen soil, and in regions subjected to alternate freezing and thawing, special consideration must be given to this fact. The expansion of water on freezing is of particular importance. High latitude regions have permanently frozen ground extending to considerable depths with only the upper portion melting seasonally.

The vadose zone above the water table, through which water seeps downward and is also drawn upward by capillarity, is especially active chemically. Here the soil-forming chemical reactions take place, hardpans and duricrusts form, and supergene enrichment may take place.

Figure 6.23 Relations of soil weight, ice volume, and water content

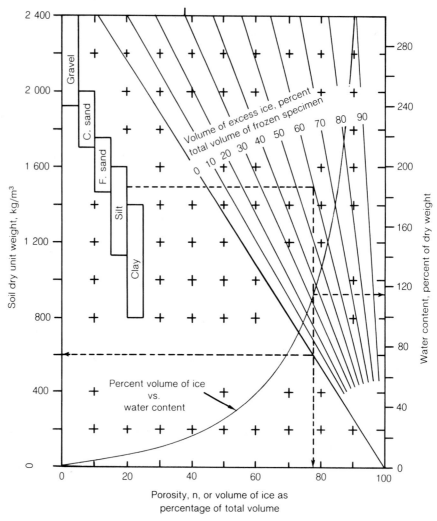

KEY

------ Silt, dry unit weight 1 522 kg/m³, 60% excess ice; total porosity 78%, dry weight of frozen sample 609 kg/m³, water content 114%

ADDITIONAL READING

Beck, B. F., ed. 1984. *Sinkholes: Their geology, engineering, and environmental impact.* Rotterdam: A. A. Balkema.

Black, R. F. 1954. Permafrost: A review. *Bulletin Geological Society of America,* v. 65. Boulder, Colo.: The Geological Society of America.

Brown, R. J. E. 1970. *Permafrost in Canada.* Toronto: University of Toronto Press.

Davis, S. N., and DeWeist, R. J. M. 1966. *Hydrogeology.* New York: John Wiley & Sons.

Flint, R. F. 1971. *Glacial and quaternary geology.* New York: John Wiley & Sons.

Franciss, F. O. 1985. *Soil and rock hydraulics.* Rotterdam: A. A. Balkema.

Freeze, R. A., and Back, W. 1983. *Physical hydrology.* Florence, Ky.: Van Nostrand Reinhold.

Freeze, R. A., and Cherry, J. A. 1979. *Groundwater.* Englewood Cliffs, N.J.: Prentice-Hall.

Hopkins, D. T. M. et al. 1955. *Permafrost and groundwater in Alaska.* U.S. Geological Survey Water Supply Paper 489. Washington, D.C.: U.S. Government Printing Office.

Jennings, J. N. 1971. *Karst.* Cambridge, Mass.: M.I.T. Press.

Matthess, G. 1982. *The properties of groundwater.* Trans. J. C. Harvey. New York: John Wiley & Sons.

Meinzer, O. E. 1923. *The occurrence of groundwater in the United States.* U.S. Geological Survey Water Supply Paper 489. Washington, D.C.: U.S. Government Printing Office.

Muskat, M. 1951. *The flow of homogenous fluids through porous media.* New York: McGraw-Hill.

Powers, J. P. 1981. *Construction dewatering: A guide to theory and practise.* New York: John Wiley & Sons.

Price, M. 1985. *Introducing groundwater.* Winchester, Mass.: Allen & Unwin.

Réthátic, L. 1983. *Groundwater in civil engineering.* New York: Elsevier Science.

Rinehart, J. S. 1980. *Geysers and geothermal energy.* New York: Springer-Verlag.

Todd, D. K. 1980. *Groundwater hydrology.* 2d ed. New York: John Wiley & Sons.

Tolman, C. F. 1937. *Groundwater.* New York: McGraw-Hill.

U.S. Army. 1952. *Arctic construction manual TM5–560.* Washington, D.C.: Department of the Army.

Washburn, A. L. 1980. *Geocryology: A survey of periglacial processes and environments.* New York: John Wiley & Sons.

Woods, K. B. 1973. *Proceedings Permafrost International Conference.* Washington, D.C.: Building Research Advisory Board, National Research Council.

7

Oceans and Shorelines

INTRODUCTION

Saline ocean waters cover more than 70% of the earth's surface today, and study of sedimentary rocks on land shows that most of the earth's surface has been at one time or another under the sea. Ocean (and lake) shorelines have long been the loci of engineering activities related to coastal navigation, harbor facilities, shore protection, and groundwater supply. The ocean floors are currently under active study, but only in the past several decades has real progress been made in understanding their topography, structure, and composition. At present most of the engineering problems associated with the ocean are confined to the coastlines and relatively nearshore regions. In the near future greater attention will have to be given to the deeper regions of the oceans where recovery of oil, mining of minerals, disposal of waste, and other activities will require refined engineering techniques.

A few broad features of the oceans will be discussed here, but the main treatment will be concentrated on nearshore and coastal features.

SALINITY

The remarkably constant *salt content* of open ocean water results from a balanced system of chemical contributions from the land and losses by evaporation, precipitation, and organic activity. The *primary input* is dissolved salts produced by rock weathering, which have been transported to the ocean by surface and groundwater flow. Some substances are removed from ocean water by direct chemical precipitation and others, for example, Ca^{2+}, HCO_3^-, and silica are continuously extracted by organisms to make their tests, shells, and skeletons. Major *losses* occur when large bodies of sea water are cut off and evaporate to become extensive and valuable deposits of salt. Additionally, some *cyclic salts* are evaporated from the sea surface and returned to the land with rainfall or snow.

If the content of a few of the dissolved species in river water and ocean water is taken from table 5.4, some of the important aspects of seawater chemistry may be highlighted. Sodium and chlorine maintain an approximate 1:1 ratio but are enriched in the ocean by factors of 10^2–10^4, sulfate is enriched by factors of 10^1–10^3, and bicarbonate by factors of 10^0–10^1. The bicarbonate–sulfate ratio is changed from generally bicarbonate-dominant fresh water to sulfate-dominant seawater.

Salinity in the ocean, although comparatively constant, varies with depth, latitude, and other factors such as evaporation rates, the influx of fresh water in the equatorial zones due to high rainfall, and the melting of ice in the polar zones. Evaporation in partially enclosed areas of the ocean such as the Mediterranean Sea causes a marked increase in salinity while dilution by a major river may yield brackish water for long distances seaward.

CURRENTS AND TIDES

Mass movement of ocean water occurs at all scales in response to tidal forces, the earth's rotation, wind stress, waves, and density differences arising from differential temperature or salinity. The *surface currents* of the oceans (fig. 7.1) are giant, counter-rotating cells north and south of the equator, which are more or less deformed by the shapes of the bordering continents. Currents leaving the polar regions such as the Humbolt and Benguela currents are cold water flows while those leaving tropical regions such as the Gulf Stream and Kuroshio Current are warm. Ocean currents are classified by the U.S. Navy Hydrographic Office as follows:

> Currents related to density distribution
> > Thermal differences
> > Salinity differences
>
> Currents caused by wind stress
> Tidal currents
> > Rotating
> > Reversing
> > Hydraulic
>
> Currents caused by surface waves
> > Deep water
> > > Mass transport
> >
> > Shallow water
> > > Onshore
> > > Longshore
> > > Rip

Tides, a major factor in coastal engineering situations, have a periodic movement controlled by the positions of the sun and moon relative to the earth and also by the rotation of the earth on its axis. Aside from direct observation of celestial bodies, the ebb and flow of ocean tides is probably our most accessible means of recognizing the complexities of celestial mechanics.

The *gravitative pull* of the moon and sun on the earth slightly distorts its rocky mass and more strongly its water envelope. The greater mass of the sun exerts the stronger pull, about 167 times that of the moon, but because of its distance from earth the difference in gravitative attraction on the near and far portions of the earth is negligible so there is no perceptible diurnal change in tidal height. The radius of the earth (6 378 km) is, however, an appreciable fraction of the 384 400 km radius of the lunar orbit so the moon's gravitative pull is significantly different at different points on and within the earth. Referring to figure 7.2, the average attractive force at the earth's center is

$$F_o = \frac{mass_{earth} \times mass_{moon}}{orbital\ radius^2} = \frac{k}{R^2}$$

Figure 7.1 Oceanic surface currents

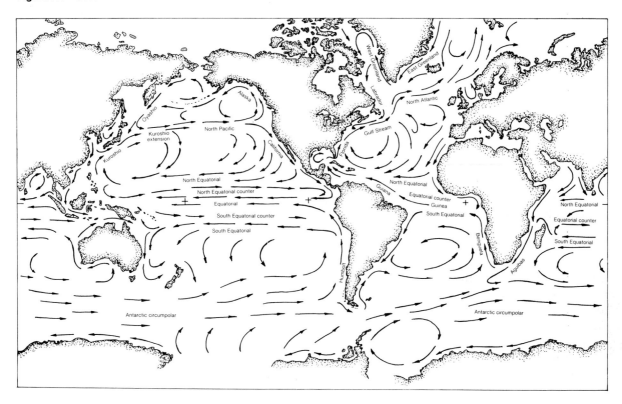

Figure 7.2 Differential lunar gravitative forces on the Earth

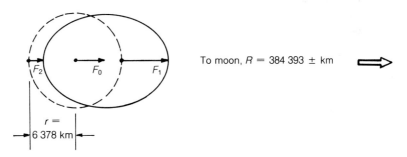

To moon, $R = 384\ 393 \pm$ km

and the forces at the near and far sides with respect to F_o are different by about 3%:

$$\frac{F_1}{F_o} = \frac{R^2}{(R - r)^2} = 1.034, \text{ and}$$

$$\frac{F_2}{F_o} = \frac{R^2}{(R + r)^2} = .968.$$

If it is imagined that F_o shortens the earth–moon distance by one unit, then F_1 and F_2 cause shortening of 1.034 and .968 units respectively, and a spherical hydrosphere is drawn out as a prolate ellipsoid having its major axis along the earth–moon line. The orientation of this ellipsoid is fixed by the vector sum of the moon and sun positions, and as the earth spins beneath the two bulges, a given point on the surface experiences a twice-daily tide. The rocky body of the earth is little affected by these deforming forces because its elastic response time is too slow; the hydrosphere, however, exhibits a pronounced tidal bulge.

Solar attraction adds or subtracts to the lunar forces as a function of the relative positions of the earth, moon, and sun (fig. 7.3) causing the tidal bulge to increase in height with sun–moon alignment whereas nonalignment of sun and moon results in *retardation* (lagging) or *acceleration* (priming) of the tidal peak. The orbit of the moon is inclined at 5.12° to the plane of the ecliptic (fig. 7.4), which adds still another harmonic to tidal motion experienced at a given point.

This concept of tidal bulges under which the earth rotates is, of course, grossly oversimplified since no account is taken of the effect of seafloor topography or continental barriers and their configuration. It may be anticipated that tidal cycles and their attendant currents will show a large amount of *local variability*. Tide records are available for most shore locations and this data, coupled with safety factors to accommodate higher water levels generated by onshore winds, should provide the basis for design criteria.

At most places the average interval between successive high or low tides is $12^h\ 25^m$, but there is considerable variation in this interval. Usually the highest (spring) tides reach their maximum a day or so after the new and full moon and lowest (neap) tides a day or so following the quarters. Spring tides are generally higher at the time of the equinoxes and lowest at the solstices. It is unusual for the two tides on each day to be equally high and low. Tides can vary from 12 to 20 m between high and low in areas such as the Bay of Fundy in Canada to just a few cm in other parts of the world, principally because of the configuration of the shoreline.

Any body of water has a natural period of oscillation, which depends on its dimensions, and a standing tidal oscillation will be set up when this period corresponds to the period of the tide-producing forces. The application of **Coriolis forces** to such a system results in clockwise current flow (Northern Hemisphere) around resonant nodes, and to a greater or lesser degree, such nodes and currents charac-

Figure 7.3 Sun, moon, and tide

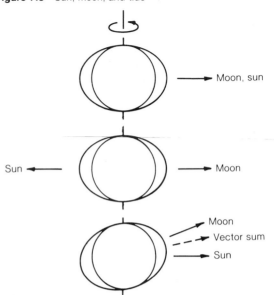

terize the tidal motions in all of the world's oceans and seas.

The *currents* caused by the rise and fall of the tides are rarely of sufficient strength along open coasts to pick up bottom material although they may transport wave-suspended particles. In semienclosed water bodies such as bays and estuaries, however, tide-generated currents may attain sufficient velocity to erode and transport a considerable amount of material. It may be remarked that an **estuary** is a particularly complex coastal environment because of such tidal effects coupled with fresh water and sediment influx. It is also a common locale for dredging, shoreline protection, and the construction of piers, bridges, and tunnels.

ESTUARIES

Estuaries have been much used, both directly and indirectly through their transfer of river water to the open ocean, as a means of waste disposal. The hydromechanics of estuaries, however, are so complex that detailed knowledge obtained by observation and study must be applied if this means of waste dispersion is to be employed. Waste disposal in estuaries, although a common practice, is not to be recommended.

Figure 7.4 Earth–Moon relations of importance in tidal motion

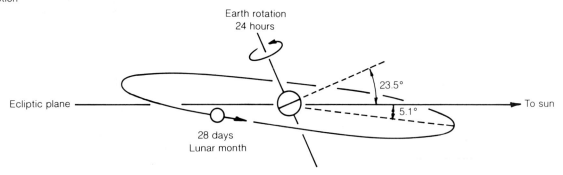

Among the more important aspects of estuaries are the tidal motions, which induce turbulent flow, and the presence of a more or less saltwater-contaminated layer of fresh water delivered by the river. Fresh water being lighter spreads over the surface of a saltwater body with variable but sometimes with little or no entrainment of salt water from below. Entrainment that does occur increases the volume of the freshwater layer and its discharge towards the mouth. Increased river flow causes the wedge of saltwater intrusion in an estuary to move downstream and generally results in a more rapid circulation and mixing of water. The time required to replace the existing fresh water in an estuary at a rate equal to the river discharge, the *flushing rate* of the estuary (following Dyer, 1973. Used with permission of John Wiley & Sons, Inc.), should be

$$T = \frac{Q}{R} = \frac{V + P}{P}$$

where T = flushing time, Q = total volume of river water accumulated in the estuary, R = river flow, V = low tide volume, and P = intertidal volume (the tidal prism).

Calculations for real estuaries show this simple relation to give flushing times that are too low, principally because complete mixing is assumed. Corrections may be applied by considering the fraction of fresh water in an estuary or estuary segment,

$$f = \frac{S_s - S_e}{S_e},$$

where f = freshwater fraction, S_s = salinity of undiluted seawater, and S_e = mean salinity in the es-

Figure 7.5 Outfall relations in an estuary

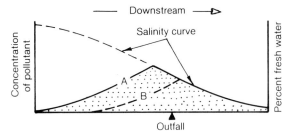

tuary. Then flushing time, total volume of river water accumulated in the estuary, Q, and the river flow, R, may be related as

$$T = \frac{(f)\, V_e}{R}$$

where V_e = estuary volume. Measured flushing times of modest-sized estuaries (e.g., the New York Bight) are in the order of 6–12 days while that for a large estuary such as the Bay of Fundy is about 76 days.

Tidal mixing will distribute a nondecaying pollutant both up- and downstream with a maximum concentration near the discharge point. Prediction of its steady-state distribution may be made using salinity distribution information if the pollutant acts in the same way as fresh or salt water. Figure 7.5 shows the kind of distribution (curve *A*) that would be present if the pollutant is strongly partitioned to fresh water. It may be seen that upstream pollution would be markedly reduced and downstream pollution little changed by moving the outfall seaward (curve *B*).

WAVES

Concentric dynamic envelopes of air and water enclose the earth and interact to produce important consequences including the transfer of energy from winds to the water surface. The energy in the waves that are generated is equally divided between *kinetic* energy associated with the orbital motions of the water particles and *potential* energy associated with the variations in elevation of the water surface. Of greatest interest and importance in civil engineering problems are the *shallow water waves* that develop at shorelines and whose energy is responsible for hydraulic wedging, erosion, and sediment transport.

Water waves have the form of a *trochoid,* the curve generated by a point on the spoke of a rolling wheel. For the ideal, wind-generated, open-ocean wave whose size depends upon wind speed, wind direction, and the distance over which the wind acts (*fetch*) the *wavelength, amplitude, velocity, and period are closely interrelated:*

$$L = VT \text{ where } L \text{ (in m)} = 0.64\ T^2, \text{ and}$$

$$V = \sqrt{\frac{gL}{2} \tan H \frac{2\pi d}{L}}$$

where L = wavelength (horizontal crest–crest distance), V = velocity, T = period, H = amplitude or wave height (vertical crest–trough distance), d = water depth, and g = gravitational constant.

In *deep* water d/L is large and $\tan H \frac{2\pi d}{L}$ approaches unity, therefore, for deep water

$$V = \sqrt{\frac{gL}{H}}.$$

For most purposes the water may be considered deep for wave calculation purposes if $d \gg L/2$ since if $d/L = 0.5$, $\tan H \frac{2\pi d}{L} = 0.9963$. In *shallow* water when $d \ll 0.5\ d/L$, $\tan H \frac{2\pi d}{L}$ is approximately equal to $\frac{2\ d}{L}$ and $V = \sqrt{gd}$. The relations of wavelength, period, and water depth are plotted in figure 7.6 and those for wave velocity, period, and water depth in figure 7.7.

Waves do not represent a mass water motion but rather the *orbital* motion of individual water particles. In deep water the circular orbit of a surface particle has the diameter of H, and below the surface the orbital size decreases rapidly to a point of no motion at a depth equal to the wavelength. If the depth of the water is less than this value, the lower orbits are first deformed into horizontal ellipses and then destroyed by bottom drag at a depth of $L/2$. In these elliptical orbits the landward acceleration under the wave crest is greater than that

Figure 7.6 Relations of wavelength, period, and water depth

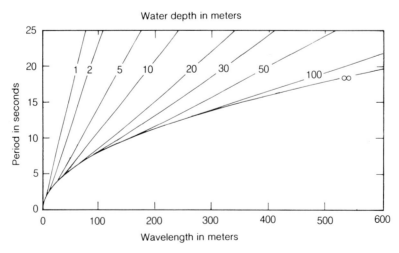

Water depth in meters

Wavelength in meters

seaward under the trough (fig. 7.8). Thus, when a wave approaches a shelving shore a point is reached at a depth equal to one-half the wavelength when the free motion of the deepest orbiting particle is blocked by the bottom. From this point shoreward, energy is transferred from the wave to the bottom, the wave is forced to steepen, and volume requirements cause a shortening of wavelength (fig. 7.9).

Figure 7.7 Relations of wave velocity, period, and water depth

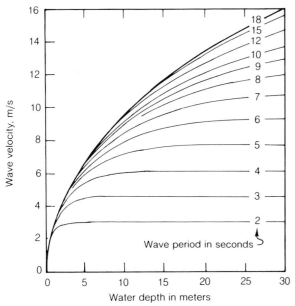

Reprinted from pp. 1–14 of *Technical Report No. 2*, Beach Erosion Board, U.S. Army Corps of Engineers, 1942, 43 pp.

Eventually, at a depth of about 4/3 H, the steepening causes the internal angle of the wave cusp to fall below 120° and the wave becomes unstable and topples forward as a *breaker*. In the absence of wind the form of a breaking wave is directly related to the profile of the beach and the steepness of the wave (the limiting ratio of wave height to depth of water below its trough being about 0.78). With steepest wave and flattest bed slope a breaker forms that gently spills forward at the crest without a well-defined breaking point. On a medium slope the breaker plunges forward with a roar and has an ill-defined forefoot of foam. Where the slope is greater than about 1:10, the breaker surges on the beach with a collapse of the crest, and *standing waves* (**clapotis**) may be formed by the reflection of a portion of the wave energy.

The energy of deepwater waves, which must be dissipated against the shore, may be estimated from measurement of their amplitude, *H,* and either wavelength, *L,* or period, *T.* To a first approximation, the total energy per unit distance along the crest per wavelength for deep water is $E = WH^2L/8$ where W = weight per unit volume of water. H may be estimated by observing waves at their instant of breaking when they are 1.4 times their deepwater amplitude, and L found by timing a number of breakers to find the period in seconds and using the relation

$$L = \frac{gT^2}{2}.$$

Figure 7.8 Features of waves

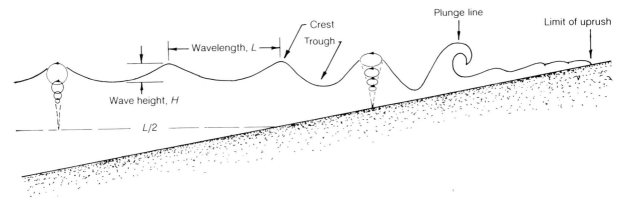

Figure 7.9 Change in wavelength and wave height as a function of water depth

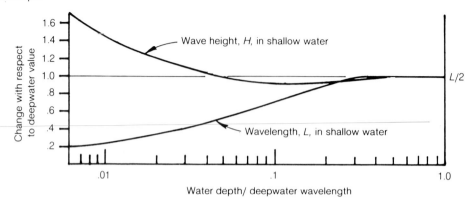

Figure 7.10 Deepwater wave-forecasting curves

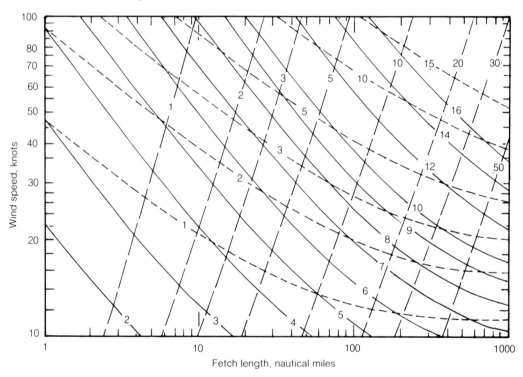

KEY

------- Significant wave height, meters
_____ Significant period, seconds
— — — Minimum duration, hours

Figure 7.11 Deepwater clapotis

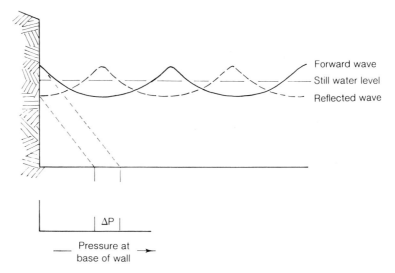

— Forward wave
— Still water level
- - - Reflected wave

Pressure at → base of wall

Forecasting of deepwater waves and, by extrapolation, their shallow-water characteristics may be accomplished by using empirical forecasting curves (fig. 7.10) developed by Bretschneider (1952). The average amplitude of the highest one-third of the waves and their average period have been used to establish "significant" parameters.

It should be noted that although wind-generated waves (such as have been described) dominate the sea surface, other kinds exist. These include earthquake-generated waves, seiches, and internal waves. **Earthquake**-generated waves are of two kinds, shock waves similar to those produced by a depth charge and those produced by submarine landslides, volcanic action, or fault movements. These latter **seismic sea waves,** or **tsunamis,** tend to have a very long wavelength, sometimes in excess of 500 km, and long periods of 10 to 60 minutes. They do not break on approaching a shore, but give rise to an extremely turbulent and dangerous mass movement on shore and will form large **bores** in rivers. Disturbances of the kind that cause seismic sea waves may also set a body of water sloshing back and forth in oscillations whose period depends largely on the dimensions of the moving mass of water. Such oscillations are called *seiches*.

The **wave form** of an ordinary wave may be thought of as the interface between air and water, two fluids of different density. Waveform interfaces

Figure 7.12 Wave diffraction at vertical structures

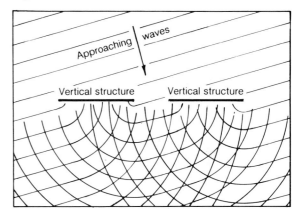

also form within water bodies that are internally stratified because of differences in temperature or salinity. These *internal waves,* like surface waves, have a high energy content, and the design of structures that penetrate stratified waters must take them into consideration.

Wave Reflection and Refraction

The impingement of waves on the seabed or other obstacles results in changes in the shape of the wave and its direction of travel. *Reflection* from cliffs, breakwaters, or seawalls sets up clapotis as shown in figure 7.11. *Refraction* occurs at corners and ends of vertical structures (fig. 7.12).

Figure 7.13 Changes in form and orientation of a wave approaching a shelving shoreline

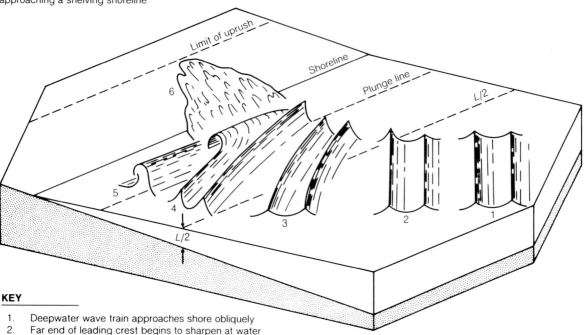

KEY

1. Deepwater wave train approaches shore obliquely
2. Far end of leading crest begins to sharpen at water depth of one-half its wavelength, $L/2$
3. Drag causes crestlines to refract toward shoreline
4. Oversteepening causes wave to topple forward
5. Breaker moves parallel to shore
6. Forward rush of water up the shore

The changes in wave form and orientation that occur as a deepwater wave approaches a shelving shoreline are complex since different portions of the wave may be in water of different depth. Some of these changes are illustrated in figure 7.13.

If the shore slope is uniform and the shoreline straight, the **energy distribution** along the shore is uniform, as shown by the spacing of the dashed lines in figure 7.14. However, wave trains approaching curving or irregular shores expend more of their energy on those portions of the shore most nearly perpendicular to the direction of wave train movement.

This is illustrated for a simple geometric case in figure 7.15 for wave trains approaching perpendicular to the shore in segment *ab*, at 60° for segment *bc*, and 30° for segment *cd*. The energy content of the deepwater waves is constant, but the different attack angles of the waves on the shore cause the energy expenditure per unit length of shore to be in the ratio of the sines of the attack angles, in this instance 1.00: 0.86: 0.50. Generally speaking, wave energy is concentrated on headlands and submarine ridges and dissipated by divergent wave trains in bays and over submarine valleys (fig. 7.16). Wave refraction diagrams can be constructed for given directions of approach and wavelength for any stretch

Figure 7.14 Energy distribution along a straight beach

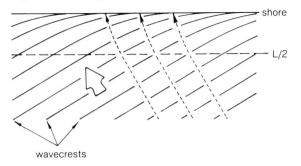

Figure 7.15 Energy distribution as a function of attack angle

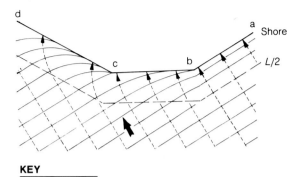

KEY

ab Perpendicular to shore
bc At 60°
cd At 30°

Figure 7.16 Concentration and dispersion of wave energy

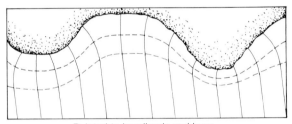

Related to headlands and bays

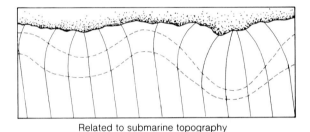

Related to submarine topography

KEY

– – – – – – – Submarine contour, outermost = L/2
————————— Wave normals, equally spaced in deep water

of coast where bottom contours are known. Munk and Taylor (1947) and Johnson *et al* (1948) describe particular methods.

Currents are generated by shoaling waves and include flow from zones of high energy concentration to areas of divergence: *longshore currents* close to and paralleling the shore and *rip currents* that return excess water mass seaward in narrow zones.

Real shores, of course, are typically irregular in plan with headlands and bays, offshore islands and estuaries, shoals and channels. These irregularities coupled with changes in *L*/2 (wave base) related to tidal changes and differing wavelengths of impinging waves make the theoretical study of wave energy dispersion unaccompanied by careful on-site studies essentially meaningless.

Sandy beach, Yucatan, Mexico

Gravel beach, Maine

Estuary, Carnavon, western Australia

Eighty-mile beach, western Australia

Tidal flats, Monrovia, Liberia

Cliffed shore, California

Figure 7.17 Beach nomenclature

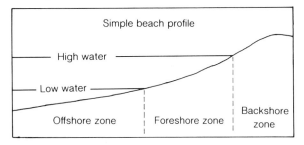

Simple beach profile

High water

Low water

Offshore zone | Foreshore zone | Backshore zone

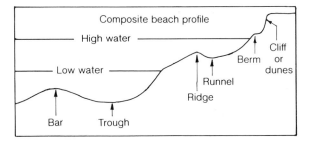

Composite beach profile

High water

Low water

Cliff or dunes

Berm

Runnel

Ridge

Bar | Trough

Figure 7.18 Suspension of sediment with respect to the plunge line of a breaking wave

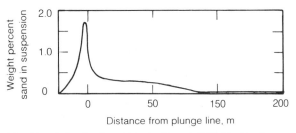

Distance from plunge line, m

After *Beach Erosion Board Interim Report 1933.* Washington, D.C.

Figure 7.19 Longshore drift

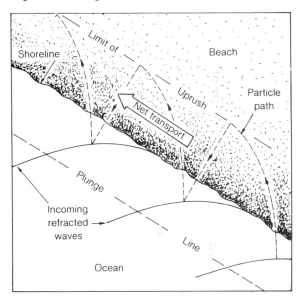

Sediment Transport

Except for certain cliffed shores, the edge of large water bodies is commonly marked by a more or less wide, stable, and complex zone termed a **beach.** The shape of a beach and the nature of its constituent sediments represent a dynamic balance between oceanic and terrestrial forces and both its shape and sediments may change as that balance changes. By their nature, beaches are a protective mechanism for shoreward installations, and the presence of a wide, stable beach is the best protection of an engineering work on the seacoast. Nomenclature appropriate to the description of beaches is given in figure 7.17.

The greatest dissipation of wave energy occurs where the wave breaks and the resultant turbulence in the breaking zone suspends and winnows the bottom sediments (fig. 7.18). There is a tendency for coarser materials to accumulate along the *plunge line* while finer sediments migrate shoreward to build the beach or seaward to form *offshore bars*. Along sandy beaches one or more bars are typically formed just outside the plunge line (the low tide plunge line if the tidal range is large). The formation of an outer

bar rising above the bed may cause waves to break farther offshore, then reform to break again near the beach.

The particles thrown into suspension by breaking waves are carried parallel to the beach in two zones separated by the plunge line. Inshore of this line the particles are carried obliquely up the beach by the uprush but return directly downslope to the shoreline and beyond (fig. 7.19). In consequence, there is a net *migration* of particles along the shore. Seaward of the plunge line, and to depths of not over about 12 m, particles are moved by the

Figure 7.20 Coastal erosion and deposition

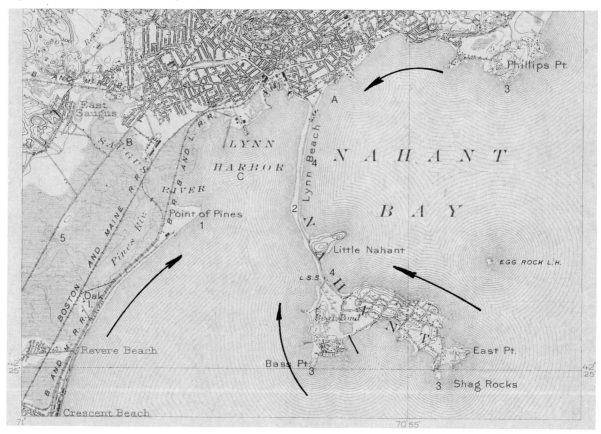

Part of the Boston Bay (Mass.) 1:62500 Quadrangle, U.S. Geological Survey.

KEY

1. Spit
2. Tombolo
3. Headland
4. Beach
5. Saltmarsh

A. Seaweed accumulation
B. Solid waste dump site
C. Silted harbor

current generated by the longshore component of the refracting waves. These twin rivers of sand, called **longshore drift,** flow steadily until blocked by an obstruction or by entering deeper water. The amount of transport measured at a few points on the east and west coasts of the United States ranges from 2 to 10×10^5 m³/yr. A particularly useful relationship appears to be that $Q = 210\ E^{0.8}$ (Caldwell, 1956) where Q = sand movement in cubic yards per day and E = the intensity of longshore energy in millions of footpounds per foot of beach per day.

The amount of run-up of water and suspended particles and thus the amount of longshore drift

Storm Washover. The pond behind the storm beach ridge on the map silted up and was turned into a nine-hole golf course. Cobbles and gravel were robbed from the beach ridge to provide solid fill on which the tees and greens were built. The height of the ridge in its natural state was adequate to protect the low-lying ground behind, but when lowered by the removal of material for fill it was overtopped by storm waves. An expensive masonry apron wall had to be constructed to protect the area.

shoreward of the plunge line is controlled by the height and steepness of the deepwater wave, the effects of refraction and shoaling on the wave, the permeability of the beach sediments, the nature of the sedimentary material, and changes in sea level caused by wind and tide. Generally, greater rates of drift will be associated with unstable beach profiles. Longshore drift of water and sediment has important effects on coastal engineering; any fixed structure built on the beach or seaward from it can interrupt this flow of sediment with disastrous results and sediment can build up at unwanted points and erode from other central locations (fig. 7.20).

It may easily be seen that, although the qualitative features of wave movement are constant, details will vary with both changes in the waves such as direction of approach and wavelength and changes in the shoreline such as tidal level and shoreline gradient. Further, the configuration of a shoreline, particularly one with a sandy or shingle foreshore, is not independent of but largely determined by the average character of the impinging waves. The marked changes in *beach profiles* between summer and winter and the general cuspate form of sandy beaches may be laid to this cause.

CASE HISTORY 7.1 Coastal Engineering

Figure 1 Coastal engineering

The coastal scene at Ocean City, Maryland, is typical of the highly developed eastern coast of the United States. A combination of coastal engineering, architectural design, land reclamation, and highway construction has produced valuable beachfront, recreational and residential facilities.

The Ocean City development is on a coastal barrier island separating the Atlantic Ocean on the left from the bay in the right background of the photograph. The view clearly shows the delicate balance between the two marine environments, which are separated by a narrow barrier of land only 20 feet or less in elevation above sea level. This barrier can easily be breached by storms or hurricanes.

The beach and the oceanfront property are protected by beach groins to stabilize the sand. The high-rise buildings are constructed on concrete pilings to minimize flood damage from storms and high tides. A housing development on land reclaimed from the backbay is depicted on the middle right in the photograph.

Coastal engineering projects must consider erosion, environmental and design characteristics appropriate to the location.

Figure 7.21 Seasonal beach profiles

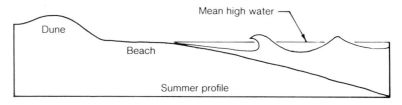

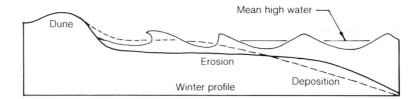

COASTAL ENGINEERING AND PROTECTION

The actual coastline, or interface between water and land, creates important engineering problems. Demands for recreational areas, high-priced residential zones, port facilities, and many other activities are fast consuming the suitable coastline areas in the world. Most coastlines are in some general state of equilibrium (erosional, depositional or static). However, abnormalities such as dredging, artificial structures, storms, cyclones, or hurricanes can rapidly unbalance the sediment–water equilibrium, usually with disastrous and costly results. Protection against these situations should always be in the coastal engineer's mind in the design and construction of buildings, bridges, harbors, and marinas.

Protection of a coastal area usually involves either the creation of protective structures or the use of natural barriers in the form of dunes, reefs, or offshore islands. In all cases the aim is to dissipate and control the enormous energy of waves, currents, and tides and to minimize their modification of the coastline.

Most shorelines have configurations and profiles that vary with seasonal changes, which bring about a change in the prevailing wind direction or

the intensity of wave attack. A simple example is shown in figure 7.21 and is typical of the Atlantic coast of the United States. The beach erodes in winter to produce offshore deposits in shallow water, often as one or more submerged bars, and then redeposition of the beach occurs during the summer months. If the beach is a long one the eroded sand may be deposited down current and replaced by the sand from an up-current source. For a pocket beach between two headlands, however, the eroded sand will be replaced from the offshore deposits.

The principal structures used for shoreline protection are *breakwaters, inlet jetties, groins* and *seawalls*. In figure 7.22 the prevailing wind and the longshore current are from south to north, so the sand transport along the shore is northwesterly. Two types of **breakwater** are shown. The first breakwater is offshore, parallel to, and provides an area of semi-sheltered water toward the shore without interrupting the longshore drift of sand. The second breakwater is a shore-anchored type often used for marina construction. Although it provides more sheltered water for small craft, it causes sand to pile up on its south side—sand that eventually bypasses the breakwater and causes shoaling of the basin—and it contributes to the erosion of the shoreline just north of it. The inlet entrance is protected by fixed

Figure 7.22 Shoreline protection

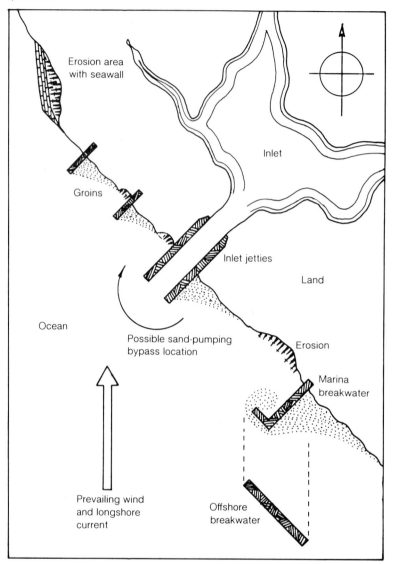

jetties (with similar depositional–erosional effects to be discussed later in more detail). **Groins**—fixed structures set perpendicular to the beach that are constructed of sheet piling, wood, stone, or concrete—trap sand moving along an erodable beach and help stabilize it. Their disadvantage is that they create an area of erosion down current to the north by interrupting the equilibrium of sand flow along

the shoreline. This effect will remove the beach sand and undermine property foundations. A preventative measure for this situation would be the construction of a seawall to dissipate wave and current energy. **Seawalls** are constructed of sheet piling, rock riprap, concrete walling, or angular concrete blocks (fig. 7.23). Apronlike seawalls are intended to dissipate the energy of oncoming waves, and vertical

Figure 7.23 Seawalls

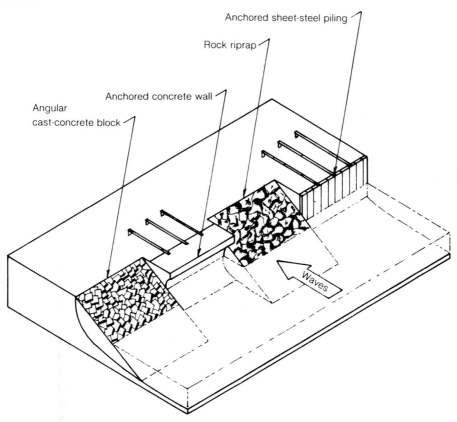

Anchored sheet-steel piling

Rock riprap

Anchored concrete wall

Angular
cast-concrete block

Waves

or concave seawalls are designed to deflect them. The apron type is generally preferred as being cheaper to install and maintain and is not as easily undermined by waves as are walls.

Nonstructural protection methods involve the stabilization of the beach and dune area by introducing vegetation, restricting building and other construction work to areas behind the beach, or dredging. Dredging is used not only to remove material from offshore or from a channel, inlet, or harbor to deepen water for navigation but to provide material to restore an area of erosion. The latter process is known as *beach nourishment* and, at best, is a costly and temporary method of control.

Inlets cause special problems for the coastal engineer since they are sites of delicate equilibrium and are often used to provide access to the ocean from sheltered water such as bays, harbors, and marinas. Rivers usually make their own inlets, which may or may not be navigable, and may have to be dredged or stabilized with fixed structures such as jetties. Artificial inlets almost always need to be stabilized. River flow, tides, and tidal currents are important factors since all of the sediment transported inward on one flood tide has to be removed on the ebb or the inlet will shoal. Hence, there must be an equilibrium between water movement and consequent sediment transport through the inlet as a periodic function of tidal movements with the steady discharge of water and sediment from the stream superimposed.

Figure 7.24   Flow conditions at Jupiter Inlet, Florida

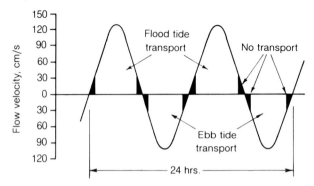

A typical relationship of flow velocity, sediment transport, and time is found in Jupiter Inlet, Florida (fig. 7.24). Sediment is transported inward on the flood and outward on the ebb tides separated by short interim periods of no transport. If jetties are used to stabilize an inlet they serve as a groin and usually create a sand accumulation on the upstream side of the jetty relative to the direction of longshore drift and an area of scour on the downstream side (fig. 7.25). In addition, deposition and consequent shoaling take place offshore from the inlet while inside the inlet more deposition occurs due to the reduced flow velocities of the tidal movements. Either depositional locus could render the inlet impassable to navigation. The judicious and controlled use of dredged holes at these points can create vortex scour action and minimize this sedimentation (Moore, 1970). The problem of deposition and erosion of the adjacent beaches can be partially solved by sediment by-passing, which involves pumping the sand across the inlet. Such a system is in operation at Lake Worth inlet, Palm Beach, Florida, and in other areas; by-passing has never provided a complete solution to the problem, however.

It should be obvious from the foregoing that coastal protection measures, though often necessary, do not always function in the planned manner. The interaction of the ocean with its shoreline is always complex, and any modifications of the shore are apt to bring about unanticipated and deleterious consequences. Careful planning, including model experimentation, should always be conducted and avoidance of modification practised whenever possible.

Figure 7.25   Conditions at Jupiter Inlet, Florida

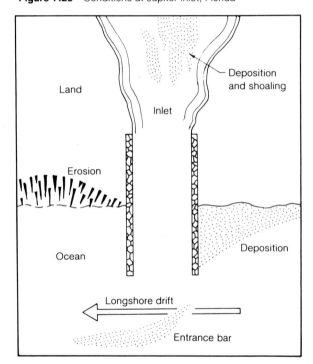

CLASSIFICATION OF BEACHES AND COASTS

There is no universally accepted set of terms or classification scheme that applies to beaches or coasts. The following can serve, however, as models. The nomenclature for simple beaches is given in figure 7.17.

Following King (1959), beaches may be classified as follows:[1]

Shingle beaches
Sand beaches
   Tidal
      Smooth profile
      Ridged profile
   Tideless
      Barred profile
         Straight bars
         Crescentic bars
      Smooth profile

Submarine bars, never exposed above water level, and exposed barrier beaches or islands often exist seaward from the beach while lagoons and salt marshes are found on the landward side.

Coasts may be classified by a number of schemes.

Valentin (1952) classified coasts as follows:[2]

Advancing coasts
   Coasts of emergence
   Outbuilding coasts
      Organic deposition (mangrove, coral)
      Marine deposition (barrier islands, cuspate forelands, spits)
      Deltas and outwash coasts
Retreating coasts
   Coasts of submergence
      Drowned glacial relief (fiords, drowned moraines, drumlins and outwash)
      Drowned river valleys
   Coasts of regression (eroded to cliff line)

Shephard (1973) outlined the following classification:[3]

I. Primary or youthful. Configuration dominated by nonmarine processes
   A. Subaerial erosion followed by submergence. Drowned river or glacial valleys (estuaries and fiords).
   B. Subaerial deposition. Deltas, alluvial plains, glacial deposits, dunes, landslides.
   C. Volcanic activity. Volcanic deposition, explosion.
   D. Earth movements. Faulting, folding.
II. Secondary or mature. Configuration dominated by marine processes
   A. Cliffed coasts made more or less regular by marine erosion.
   B. Marine deposition resulting in barrier coasts, cuspate forelands, beach plains, and mud flats or salt marshes.
   C. Organic deposition as reefs of various kinds.
   D. Organic entrapment of sediment by mangrove, marsh grasses, etc.

The southern Atlantic coast of the United States, which has been investigated by many workers and summarized and analyzed by Dolan and Ferm (1968), provides an excellent example of Type II B landforms (Shephard) on both micro and macro scales. The large-scale, crescentic coastline forms are attributable to reverse eddies from the main north-flowing Gulf Stream (fig. 7.26). These crescentic forms have been recorded with an amplitude of 1 to 10^8 m, ranging in hierarchical order from 1 to 8. Beach cusps of the order of 1 to 3 are believed related to shallow wave activity associated with in-shore current cells linked with larger rhythmic features of order 4. Larger order forms are represented by the Carolina capes.

1. C. A. M. King, Beaches and Coasts, 2d ed. © 1959 Edward Arnold, London. Reprinted by permission of Edward Arnold Publishers, Ltd.
2. After H. Valentin, Die Küste der Erde, Petermann's Geog. Mitt. Ergänzungheft, vol. 246, 1952.
3. Following F. P. Shephard, Submarine Geology, 3d ed. © 1973 Harper and Row, N.Y. Used with permission of Harper and Row Publishers, Inc.

Figure 7.26 *Hierarchical* landforms of the southern Atlantic coast of the United States

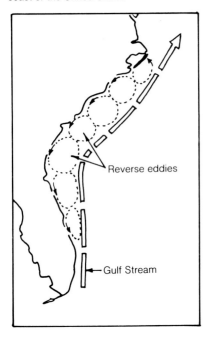

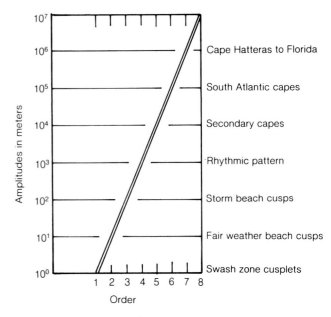

SUMMARY

The earth is predominantly a water world if only its surface area is considered. The salty water body of the world ocean is in continuous motion as it is stirred by currents, tides, and waves, and these motions are of great importance to those who live along its shores.

The interaction of the ocean (or large lakes) with its shores results in the formation of cliffs and beaches and the restless movement of sediment along the shore. Movement of this sediment may silt up harbors and close navigational channels or erode beaches of recreational importance.

Coastal engineering and protection is a major field of engineering endeavor today, and future ocean-related activities may be anticipated to greatly expand such work.

ADDITIONAL READING

Bascom, Willard D. 1964. *Waves and beaches: The dynamics of the ocean surface.* Garden City, New York: Anchor Books.

Bruun, P. 1978. *Stability of tidal inlets.* New York: Elsevier Science.

Dalrymple, R. A. ed. 1984. *Physical modelling in coastal engineering.* Rotterdam: A. A. Balkema Publishers.

Dyke, P. P. G.; Moscardini, A. O.; and Robson, E. H. eds. 1985. *Offshore and coastal modelling.* New York: Springer-Verlag.

Cates, D. R., ed. 1973. *Coastal geomorphology.* In *Publications in geomorphology.* Binghampton, New York: State University of New York.

Hails, J., and Carr, A., eds. 1975. *Nearshore sediment dynamics: An interdisciplinary review.* New York: John Wiley & Sons.

Ippen, A. T., ed. 1963. *Estuary and coastline hydrodynamics.* New York: McGraw-Hill.

Komar, P. D., ed. 1983. *CRC Handbook of coastal processes and erosion.* Boca Raton, Fl.: CRC Press.

MacMillan, D. H. 1966. *Tides.* New York: American Elsevier.

Muir Wood, A. M. 1969. *Coastal hydrodynamics.* New York: Gordon and Breach.

Schwartz, M. L., ed. 1982. *The encyclopedia of beaches and coastal environments.* Florence, Ky.: Van Nostrand Reinhold.

Shephard, F. P., and Wanless, H. R. 1970. *Our changing coastlines.* New York: McGraw-Hill.

Silvester, R. 1979. *Coastal engineering.* 2 vols. New York: Elsevier Science.

Stanley, D. J. and Swift, D. J. P., eds. 1976. *Marine sediment transport and environmental management.* New York: John Wiley & Sons.

Swift, D. J. P. and Palmer, H. D. 1978. *Coastal sedimentation.* Stroudsburg, Penn.: Hutchinson Ross.

Swift, D. J. P., Duane, D. B., and Pilkey, P. H., eds. 1972. *Shelf sediment transport: Process and pattern.* Stroudsburg, Penn.: Hutchinson Ross.

Tricker, R. A. R. 1965. *Bores, breakers, waves and wakes.* New York: American Elsevier.

U.S. Army Coastal Research Engineering Center. *Shore protection: planning and design.* 3d ed. Technical Report no. 4.

8

Mechanical Properties
of Rocks

INTRODUCTION

The rocks that comprise the earth's crust have been subjected throughout time to deforming forces and have yielded according to their nature and location. Rocks show the effects of *deformation* by the presence of such features as fractures, displaced blocks, flexures, and foliation. Whoever wishes to understand earth history must be able to unravel the deformational *geometry* and understand the deformational *mechanics*.

The resistance of rocks to deformation ultimately resides in the magnitude and orientation of the *forces* that must be overcome to move one mechanical unit with respect to another. At one extreme the mechanical units are atoms and the forces are interatomic bonds; at the other extreme the units may be cubic kilometers of rock or even continental-sized plates and the forces are frictional resistance and gravity. In the study of rock mechanics and the geologic structures that result from rock deformation, we are concerned with the interplay of external and internal forces tending to deform rocks, the mechanical behavior of rocks under these conditions, and the geometry of the resulting deformed rock masses.

MODELING

A hammer is symbolic of geologic work because the rocks studied are apparently hard, brittle substances. These properties, however, represent only the short-term view of the uninitiated because, on the large-size and long-time scales of geology, it is readily apparent that rocks have actually flowed, often into highly complex patterns. As a matter of fact, when *modeling* experiments are undertaken, which account for the scale strength as well as scale size, such substances as honey, paraffin, dough, tar, or wet clay are appropriate modeling materials; and analogy to such substances leads to a more realistic concept of the mechanical behavior of rocks. Unfortunately, both gravity and time are also important but cannot be readily modeled, although the effects of decreasing strain rate (increasing time) may be approximated by raising temperature.

Models are a convenient and sometimes necessary means for understanding geologic forms and phenomena (the basic principles of dimensional analysis [model theory] are set forth in appendix I). Scale models are familiar, but the *scaling of properties* other than size is generally not appreciated. For example, a model train faithfully scaled at 1:100 is seldom operated at scale speed (100 km/hr at scale speed is 1 km/hr or 0.28 m/s). In a scale-speed collision the damage is negligible; the engine would have to be constructed out of a material like tin foil to reproduce realistic crumpling. Likewise, though glacial ice may begin to flow when an accumulation of about 100 m is attained, to scale a flowing glacier whose model thickness is 1 m a model substance that will begin to flow under its own weight at this thickness is required. Unfortunately, the active force in both instances is gravity, so it might be necessary to perform the study in a centrifuge or rotating space station in order to achieve a true model.

In actual or mental modeling it is important to remember that linear properties scale linearly, areal properties as inverse squares, and volume properties as inverse cubes (i.e., a cube having 10 cm edges reduced to one whose edges are 1 cm undergoes a reduction in the area of each face from 100 to 1 cm² and a reduction in volume from 1000 to 1 cm³). Other properties are similarly reduced so that if the original cube is steel having a yield point under tension of 150 kg/mm², its model (having one-tenth the cross-sectional area) must yield at 150 / 10 kg/mm² = 15 kg/mm².

STRESS AND STRAIN

In order to discuss rock *strength* and deformation on a rational basis, it is necessary to have the meanings of several terms and concepts firmly in mind. A *force* is that which tends to change the state of rest or motion in matter and may be conveniently represented by an arrow or vector whose length and head respectively represent the amount of force and the direction in which it acts. When a force is applied to a body without counter action, the body is accelerated according to the relation $F = ma$, where F, m, and a are force, mass, and acceleration respectively. Obviously, the forces acting in the earth's crust are nearly or exactly in long-term balance since accelerating motions are seldom observed and the resultant of forces through any point in the earth's crust must be zero or nearly so.

Rock masses have extension in space and cannot be conveniently represented as single points. Also, the forces acting upon a given mass are not necessarily the same at all points within the mass, and forces acting at one point are transmitted through the mass and add or subtract to the forces at another point. In such a system of interacting and interdependent forces, it is most convenient to consider the system in terms of the *force per unit area* with which each part of the body acts upon the adjacent parts. This special kind of pressure need not be the same in all directions (as for hydrostatic pressure in a liquid); it varies according to the direction of its application. It is a **stress.**

For the purpose of analysis the system of active forces may be resolved into pairs of equal and opposite forces acting upon real or imaginary surfaces in or on the body. These forces per unit area (stresses) are further resolved into a *normal stress,* σ (sigma), acting perpendicular to the surface and

Figure 8.1 Resolution of forces and stresses on a unit cube

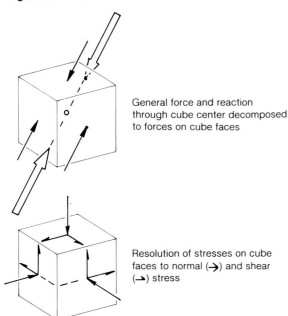

General force and reaction through cube center decomposed to forces on cube faces

Resolution of stresses on cube faces to normal (→) and shear (⇀) stress

stress planes. The normal stress is directed perpendicular to these planes along the principal stress axes, which are conventionally designated σ_1 (greatest), σ_2 (intermediate), and σ_3 (least); that is, $\sigma_1 \gg \sigma_2 \gg \sigma_3$.

It is common experience that if we push, pull, or twist various objects that they bend, break, or plastically deform. The relations between the forces acting and the deformations that result are expressed mechanically in the concepts of stress and strain; stress being the force per unit area acting upon a body and strain the resulting change (deformation) of size (dilation) or shape (distortion). Thus, *stress is a condition* and *strain a result, as shown by deformation of the original.* **Strain** is the result of applied stress and the *total* strain is the sum of the following three distinguishable forms of response:

Elastic Response   The applied stress goes to distort the body (say by stretching interatomic bonds). The body is, however, not broken and a relaxation of the stress allows the structure to return exactly to its original shape or size.

Viscous or Plastic Response   The applied stress either causes atoms or particles to slip by one another without loss of cohesion or it promotes solution and recrystallization of the constituents in the solid state. There is no loss of continuity of the body and the deformation is permanent.

Brittle or Clastic Response   The applied stress overcomes the cohesive properties of the substance and rupture takes place. Continuity is interrupted and the deformation is permanent.

The various kinds of strain response to applied stress may be visualized in mechanical analogy by considering the actions of a spring, a block sliding on a surface under applied stress, and a dashpot or liquid-filled cylinder with a loose-fitting piston (fig. 8.2).

a *shear stress,* τ (tau), acting parallel to the surface. In turn, shear stress may be resolved into two shear-stress components, τ_1 and τ_2, at right angles to each other. σ, τ_1, and τ_2 are thus mutually perpendicular stresses. Stress, being a force acting over an area, may be equated to pressure, and the units of stress are the pascal, Pa, the bar (10^5 Pa), and the kilobar, kb (10^3 bar).

Stress at any point may be illustrated. Figure 8.1, for example, shows the resolution of stresses (black arrows) resulting from the application of a single force (open arrows) on a vanishingly small cube. Since the system is closed, the opposing stresses must balance, and there then remain six, independent, stress components: three shear and three normal stresses. These stress components define the *stress tensor.* The geometry of the stress system may be simplified by choosing new axes such that the shear stress becomes zero on three mutually perpendicular planes. These are called the principal

Figure 8.2 Kinds of strain

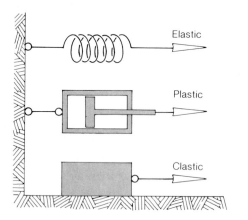

The spring, which returns to its initial position when a deforming force is removed, represents pure *elastic* behavior, the amount of deformation (stretch) being proportional to the deforming force. The block resting on a surface will move only when an applied force is greater than the frictional resistance. If this resistance is thought of as the internal friction of a solid body, this model represents pure *clastic* behavior with a rupture surface along the bottom of the block. Plastic behavior cannot be represented by such simple models but requires some combination of the spring and sliding block concepts with a dashpot. The piston in this liquid-filled cylinder moves at a rate proportional to the applied force (viscous flow) and stops when the force is removed. If the dashpot is hooked up in series or parallel with

Rupture in quarry floor caused by elastic expansion of rock, Louisville, Kentucky

a sliding body or spring, the motions that result when a force is applied then represent some simple cases of *plastic* deformation.

If the forces acting and the motions undergone by the linkages of figure 8.3 are equated respectively with displacement and time, then the resulting curves of strain response should and do have the general forms of some experimentally determined mechanical deformations. Obviously, the detailed shapes of the curves in general cases will vary with the restoring constants of the springs, the friction of the blocks, viscosity of the liquids in the dashpots and the rate of application of the stress.

CHANGES IN SHAPE AND VOLUME

The strain that arises in consequence of the stress applied to a body causes changes in its **volume,** its **shape,** or in both. Changes in volume but not in shape may be equated to ordinary hydrostatic pressure and the hydrostatic stress component P identified as $P = \sigma_1 + \sigma_2 + \sigma_3/3$. The remaining *deviatoric stress* tends to change the shape of a body and $\sigma_1 - P$, $\sigma_2 - P$, and $\sigma_3 - P$ measure the degree of this change on mutually perpendicular axes. Various kinds of strain may be distinguished (fig. 8.4) and measured as either the change in length or the position of an initial line. Any strain geometry can be found as some combination of these changes.

Figure 8.3 Examples of mechanical behavior

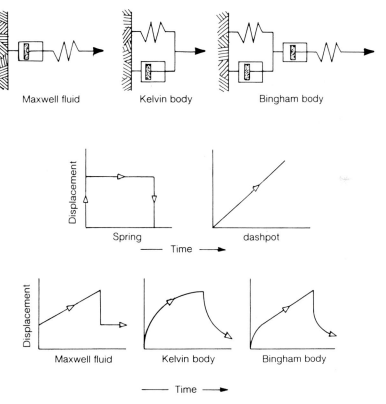

KEY

A. Model linkages
B. Graphs of the responses of the individual mechanical elements of the linkages
C. Graphs of the combined response of the linkages

Figure 8.4 Kinds of strain

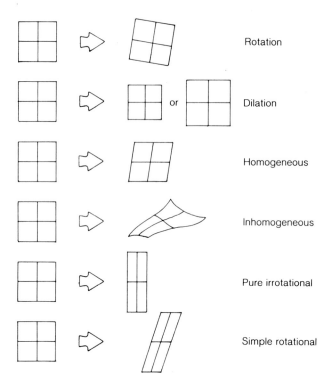

As a useful exercise to grasp the importance of the changes in volume and shape that occur in solid materials when they are deformed, consider the changes in area (volume) and perimeter (shape) of a square strained by compression or shear as shown in diagrams *a* and *b* of figure 8.5. The resultant changes in area at constant perimeter and in perimeter at constant area are shown by the graphs *c* and *d*. Further insight to the geometrical aspects of the deformation of solids may be gained by noting the change in length and orientation of line *a* to line *a'* in diagrams *a* and *b*. As shown in figure 8.6a, this line and its equivalents mark the boundary between regions of shortening and elongation, and thus the locus of maximum stress concentration and probable locus of failure. For solids undergoing shear deformation this locus may be found geometrically by replacing the square with a circle which, on deformation, is transformed into an ellipse (fig. 8.6b).

Lines connecting the intersections of the circle and the ellipse mark the locus of maximum stress concentration.

Strain Ellipsoid

Homogeneous strain may be usefully visualized by considering a tiny sphere of rock buried at some depth beneath the earth's surface where it is pulled downward by gravity and pressed in upon all sides by the surrounding rock. Should all of the stresses acting upon it be in balance, no change in shape will occur although a small reduction in volume representing elastic compression will, of course, be present. If, however, stresses are different in different directions, the sphere will tend to be deformed. Changes in the *stress field* could arise, for example, by loading or unloading the rock column of which the sphere is a part or by movement or volume changes in the surrounding material. The

Figure 8.5 Changes in area and perimeter as a consequence of deformation

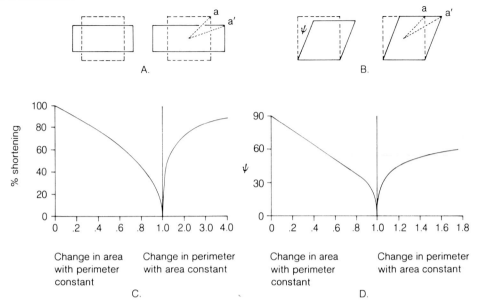

Change in area with perimeter constant

Change in perimeter with area constant

C.

Change in area with perimeter constant

Change in perimeter with area constant

D.

Figure 8.6 Locus of maximum stress concentration

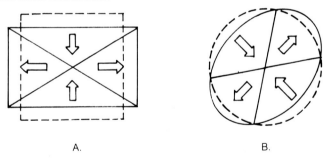

A.

B.

The rugged central portion of Tasmania is underlain by thick sills that retard groundwater infiltration and promote the accumulation of surface waters. The topography is such that hydroelectric power can be developed by carving penstock, power station, and tailrace within rock. The Poatina Power Station was under development in 1960 when rock spalling along the back (roof) of the access tunnel was noted; some roof bolts stood out as much as 0.3 m. Possible unequal tensioning of the roof bolts was investigated and shown not to be the cause of the spalling.

Drill holes were noted that had been deformed into elliptical cross sections with axial ratios of about 0.6, and their orientation suggested squeezing in the horizontal plane. This suggestion of laterally directed creep was followed up by measurements taken between fixed points on the tunnel walls. Steady inward creeping of the walls was documented at an average rate of .033 mm/h in tests run for over 100 hours.

A detailed in-situ measurement of rock stress coupled with theoretical calculations was initiated following the realization that rock stresses were both greater than anticipated and oriented in an unexpected way (i.e., horizontal greater than vertical). The results of the study showed horizontally directed shear stress to be 2.5 times greater than the vertically directed shear stress and the limiting value for nonspalling to be approximately 420 kg/cm^2. It was anticipated that this value would be exceeded by 100% in deeper excavations. The cause of the excess horizontal stress, although not determined, may most likely be ascribed to nearby faulting.

Following these studies, two modifications were made that resolved the difficulties and allowed the project to be successfully completed. The excavation schedule was changed so that tunnels at lower levels were opened before excavation of the main station in order to relieve some stress, and the tunnel walls were provided with a horizontal slot. This slot redirected the stress and reduced it to safe levels.

distinct conditions are those in which $\sigma_1 > \sigma_2 = \sigma_3$, $\sigma_1 = \sigma_2 > \sigma_3$, or $\sigma_1 > \sigma_2 > \sigma_3$ as shown in figure 8.7. In the first instance the sphere will be flattened into an *oblate ellipsoid* with σ_1 perpendicular to a circular section (shaded) in the plane defined by σ_2 and σ_3. In the second case the ellipsoid will be *prolate* with its longer axis parallel to σ_3 and the circular section in the $\sigma_1 \sigma_2$ plane. The third case (plane strain) generates a *triaxial ellipsoid* with axial lengths inverse to the applied stresses. Two circular sections exist in such an ellipsoid.

A means of visualizing the deformation that results from the application of a particular stress is at hand in a sponge rubber ball or a putty sphere. The deformation of spherical bodies in consequence

of differently oriented stress is illustrated in figure 8.8. Compressive stress (arrows) acting on a sphere (dashed) deforms it into an oblate ellipsoid with a circular section normal to the applied stress (fig. 8.8a). A similar situation arises when the sphere is placed under tensional stress generating a prolate ellipsoid (fig. 8.8b). Shearing stresses acting as shown in figure 8.8c result in a somewhat more complex situation; the rotational effects of the active couple (solid arrows) is offset by frictional and inertial forces in the rock represented by a passive couple (open arrows). The resultants of these couples are extensional and compressional forces acting upon the sphere. The orientation of the circular sections cannot be simultaneously perpendicular to

Figure 8.7 Deformation of a sphere by differential stress

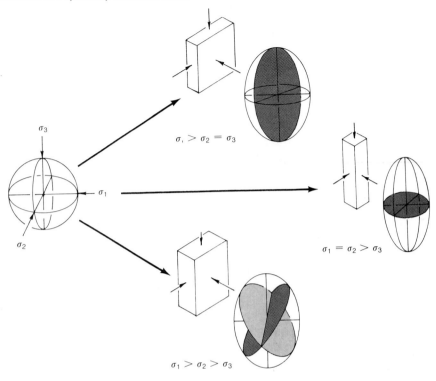

σ_3

σ_1

σ_2

$\sigma_1 > \sigma_2 = \sigma_3$

$\sigma_1 = \sigma_2 > \sigma_3$

$\sigma_1 > \sigma_2 > \sigma_3$

Figure 8.8 Strain products of stress
Kinds of strain

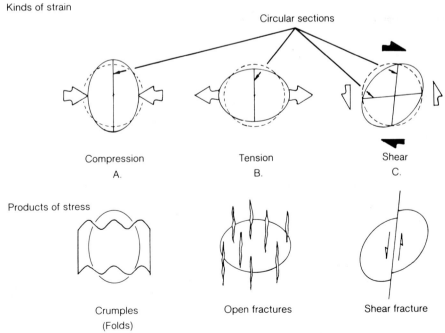

Circular sections

Compression
A.

Tension
B.

Shear
C.

Products of stress

Crumples
(Folds)

Open fractures

Shear fracture

Figure 8.9 Flinn diagram

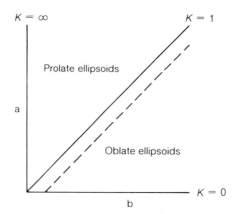

KEY

$X > Y = Z$	Axially symmetric extension, $K = \infty$ Constrictional strain, $1 < K < \infty$	Prolate ellipsoids
$X > Y > Z$	Plane strain (volume constant, $K = 1$)	Triaxial ellipsoids
$X = Y > Z$	Flattening strain, $0 < K < 1$ Axially symmetric flattening, $K = 0$	Oblate ellipsoids

these stresses, and the sphere is deformed into a triaxial ellipsoid having two circular sections that represent the *surfaces of maximum shear stress*. It is apparent that four points on the original circle are unchanged in length and position by the deformation and that the radii to them are the positions of maximum shear stress, which must arise between the alternatively elongated and shortened sectors. Since forces acting along a line are special cases and rather unlikely in most geologic situations, it is important to appreciate the geometry of shear deformation. From the preceding, it may be predicted that open fractures will lie parallel to the circular section of extensional stress, crumpling or flexure axes will parallel the circular section of compressional stress, and closed fractures showing wall movement parallel to the plane of fracture will lie in or near the circular sections of maximum shearing stress.

Changes in Volume

Shape changes of real materials are usually accompanied by volume changes and the amount of this change is equal to the final minus initial volume divided by the initial volume:

$$\Delta = \frac{V - V_o}{V_o}.$$

The volume of a unit sphere is $4/3\,\pi$ and that of the derived ellipsoid is $4/3\,\pi\,(XYZ)$ where X, Y, and Z are orthogonal axes of the ellipsoid. Thus, $\Delta = 4/3\,\pi\,(XYZ) - 4/3\,\pi/4/3\pi$; $\Delta = (XYZ) - 1$; or $\Delta + 1 = XYZ$. If e_x, e_y, and e_z are taken as representing the elongation along the X, Y, and Z axes and

$$e = \frac{l - l_o}{l_o}$$

(*Hooke's Law*), then $\Delta + 1 = (1 + e_x)\,(1 + e_y)\,(1 + e_z)$ where $+\,e$ represents elongation and $-\,e$ shortening.

Flinn (1962) devised a convenient geometrical means of representing these various strain states. If $1 + e_x\,/1 + e_y = a$ and $1 + e_y\,/\,1 + e_z = b$, then the ratio $a - 1/b - 1 = k$ distinguishes the kind and degree of ellipticity of the various strain ellipsoids (fig. 8.9). Axially symmetric extension with $1 < k < \infty$ describes prolate ellipsoids arising from constrictional stress, and $0 < k < 1$ describes oblate ellipsoids that result from flattening. The division of the two fields is that of a triaxial ellipsoid at constant volume for which $k = 1$. If volume changes are present then the boundary shifts, for example, to the dashed line in the figure.

Geometric Description of Strain

Strain may be measured as either the change in length of a line as in the discussion to this point, or as the change in the position of a line (fig. 8.10). In *pure shear* the position of the pre- and postdeformation axes are unchanged and the strained mass is indistinguishable from one deformed by changes in axial length consequent on the application of normal stresses. In simple shear, however, axes are rotated and the relation is given by shear strain, γ (gamma) = $\tan \psi$ (psi) where ψ is the deflection of an originally right angle. The relation between elongation and shear strain may be described by λ (quadratic elongation) = $(l/l_0)^2 = (l - e)^2$ where l and l_0 are respectively the new and initial lengths. In two dimensions a circle of unit radius deformed into an ellipse having λ_1 and λ_2 as its major and minor axis is called a **strain ellipse** and is a particularly useful means for establishing the relations of different mechanical products of deformation such as foliation, crumpling, or fracturing. This is especially true since one stress component must always be of intermediate value and a plane incorporating the least and greatest stresses will show the maximum strain results.

There is always a direct geometrical relationship between stress and strain, but the relationship may not be clear since the geometry of both stress fields and strain results may change in time and only the final strain state may be preserved. For example, in deformation by simple shear (see fig. 8.4) only the intermediate stress and strain axes correspond in the deformed body. The stress that has acted or is acting on rocks can seldom, therefore, be directly measured and in most instances must be a complex three-dimensional field. Some contributing factors to stress in rocks are (1) gravity forces that are always present, (2) internal stress arising from volume changes attendant on recrystallization with or without chemical change, (3) weight of the overlying rock, (4) pressure of contained pore fluids, (5) temperature changes, and (6) stresses arising from movement of the rock mass. Although it is generally impossible after the fact to completely resolve the stress field that acted on a body, the resultant permanent strain may sometimes be found

Figure 8.10 Change in length and position of a line as a result of strain

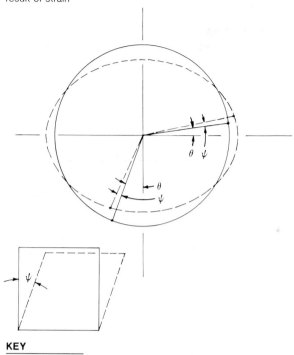

KEY

θ An original angle
ψ Angular increment due to strain

by examination of the deformed rock mass. A direct means is to determine the change in shape that has taken place in objects of known geometry such as tabular beds or dikes, planar cracks, spherules, pebbles, or shells.

MECHANICAL RESPONSE OF MATERIALS

Different materials respond differently to the same deforming stress and typically show different combinations of ideal strain responses. These may be identified as elastic strain and viscous strain. Perfectly elastic bodies obey Hooke's Law, $e = l - l_0/l_0 = \sigma/E$, or $\sigma/e = E$, where e is the elongation or extensional strain, σ is the stress, l_0 and l are respectively initial and new lengths, and E is a constant called *Young's Modulus of elasticity*. Also, $e = V - V_0 / V_0 = P / K$ where e is dilational

Figure 8.11 Ideal relations of strain response

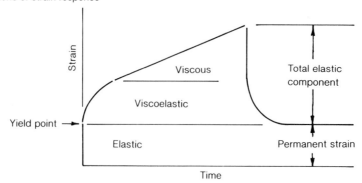

strain, V_o and V the initial and new volumes, P the hydrostatic pressure, and K a constant termed *compressibility*. Ideal viscous substances are those that exhibit permanent and nonrecoverable strain such as is shown by the deformation of a liquid. Such material exhibits "Newtonian" behavior described by the relation $\sigma = \eta \dot{e}$ where σ is applied stress, η (eta) is viscosity, and $\dot{e}$ the strain rate or rate of change of shape with time. For a constant stress the strain increases linearly with time: $e = \sigma t / \eta$. Rocks respond both elastically and viscously to deforming stress, and depending upon the dominant component, their response may be identified as being elastic, viscoelastic, elastoviscous, or plastic. An approximation to the total strain may be found by adding the elastic and viscous components: $e = \sigma / E + \sigma t / \eta$.

A material that exhibits basically elastic strain but requires time to reach its limiting value and time to return to its initial condition is *viscoelastic*. Earthquake aftershocks arise from this behavior. *Elastoviscous* materials basically obey the viscous laws but behave elastically for stresses of short duration. Asphalt is a good example. Materials that behave elastically at low values of applied stress but viscously above some critical yield-stress threshold are plastic. Most rocks, in fact, combine all three types of behavior as a function of the rock material, applied stress, and time. The relations for an ideal case are illustrated in figure 8.1.

Rocks are more or less porous mineral aggregates and their mechanical response to applied stress depends upon both the constituent minerals and the pore liquids. Pure elastic behavior is rooted in the structural distortion of minerals, a function of their atomic bonding. Permanent strain derives from slip and microfracture mechanisms often accompanied by recrystallization of mineral grains. Accelerated viscous flow arises from the spread and joining of slip and microfracture surfaces throughout the rock to generate a pervasive system of cracks that results in loss of cohesion and failure. In turn, these mechanisms are dependent upon mineral composition, temperature, fluid content, and strain rate. *Yield* and *failure stresses* of rocks, the points at which permanent deformation of a plastic or brittle character occurs, are thus also functions of these parameters. Their effects are given in table 8.1.

Increased *confining pressure* results in higher effective rock strength by raising both the yield and rupture stresses. Increased *temperature* (below the melting point) promotes the ability of atoms to interchange within a crystal and to migrate from points of maximum to points of minimum stress, and thus expands the *ductile range* of a rock by lowering its yield stress and raising the stress required for *brittle failure*. The pressure of *pore water* acts independently from the confining pressure due to rock load and may counter the pressure of the latter. The *effective pressure*, P_{eff}, is the pressure due to rock load, P_R, less the fluid pressure P_F:

$$P_{eff} = P_R - P_F.$$

Fluid pressure thus has opposite effects to confining pressure and effective pressure is the important parameter in engineering evaluation.

Table 8.1 Effects of External Parameters on Yield and Failure

Increased	Yield Stress, σ_Y	Rupture Stress, σ_R
Confining pressure	Raised	Raised
Temperature	Lowered	Raised
Pore pressure	Lowered	Lowered
Strain rate	Raised	Raised

Figure 8.12 Dependence of mechanical response on rate of stress application

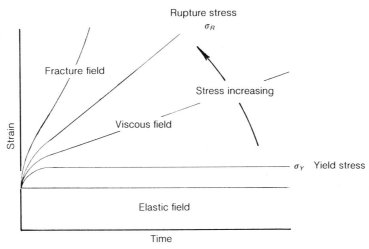

Strain Rate

Values for yield and rupture stress are much higher when measured over short periods of time than when determined on geologic time scales, and the long-term (creep) strength of rocks may be only 20–60% of the instantaneous strength of the same rock. The rate at which equilibration takes place is strongly dependent upon the internal resistance to change of the material and upon the time scale (fig. 8.12). Water, for example, readily deforms under minute stress differences but can be painfully hard when struck rapidly in a poorly executed dive; rock appears hard and brittle but can often flow significantly if subjected to small stress differences for millions of years. Even over geologic time, however, the internal resistance to change of a rock may be so high that the stress can build up to levels at which rupture rather than plastic flow takes place. Since rocks are usually constrained on all sides, this rupture may be anticipated to be conservative (i.e., to allow the maximum deformation consistent with a minimum increase in volume). A shear deformation in which the two parts of the rock mass move in opposite directions parallel to the rupture surface is such a deformation.

Experiments with materials such as paraffin or asphalt are especially instructive as to the effectiveness of temperature changes in modifying the strength of materials and even more instructive in evaluating the effects of varying strain rate. A sharp blow will cause rupture whereas a small stress applied for a long time will cause plastic deformation.

Figure 8.13 Maximum shearing stress

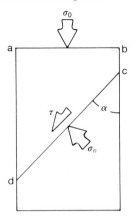

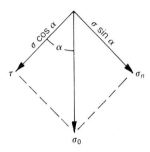

Let σ_0 ($= \sigma_1 - \sigma_3$) be the active deforming stress, τ the shear stress, and σ_n the stress normal to a slip or shearing plane. Since there is no acceleration all stresses are balanced by equal and opposite stresses.

taking $ab = 1$, $cd = 1/\sin \alpha$

$$\sigma_n = \text{force/area} = \sigma_0 \sin \alpha \, /cd$$
$$= \alpha_0 \sin \alpha \, /1/\sin$$
$$= \alpha_0 \sin^2 \alpha$$
$$= \alpha_0/2 \, (1\text{-}\cos 2\alpha)$$
$$\tau = \text{force/area} = \sigma_0 \cos \alpha \, /cd$$
$$= \sigma_0 \cos \alpha \, /1/\sin \alpha$$
$$= \sigma_0 \cos \alpha \, \sin \alpha$$
$$= \sigma_0/2 \, \sin 2 \, \alpha$$

The shearing stress, τ, will attain its maximum value when $\sin 2\alpha$ is maximum or when 2α is 90°. The maximum shearing stress will, therefore, lie in a plane 45° from the direction of the maximum stress σ_0.

MOHR DIAGRAM

The position of the plane of maximum shearing stress within the unit cube of figure 8.1 may be shown to be at 45° to the direction of greatest stress (fig. 8.13), which will be at 45° to the direction of σ_1 in the $\sigma_1\sigma_3$ plane. The shearing strength of a solid, however, is also a function of the stress normal to the shearing plane, and in general, the angle of shearing rupture to the maximum stress direction is always less than 45°, sometimes significantly so.

The characteristics of a state of stress necessary to cause shearing rupture are given by the relation (after Coulomb, 1773) $\tau = c + \mu \, \sigma_n$ where τ is shear stress, c a constant usually called cohesion, μ the coefficient of internal friction and $\mu = \tan \phi$ where ϕ is the internal friction, and σ_n is normal stress on the shear plane.

The stresses that can be developed at a given point within a given material are limited by the shear strength of the material and *ultimate strength* is determined by the shearing strength in a plane of slippage rupture. In turn, the shearing stress is a function of the normal stress acting on the same plane. Mechanical failure is thus related to the properties of the material and the interplay of shear and normal stress on a slippage plane.

The orientation of the stresses acting on any plane within a material is shown in figure 8.14a and the relations of the stress surface by figure 8.14b. Combining these diagrams yields figure 8.14c, which provides the general stress information for some point 0. For given values of the principal stresses, the ends of the arrows which represent the resultant shearing stresses lie on a circle, the *Mohr circle,* (Mohr, O., 1914) in figure 8.14d, which is a graphical solution of the interrelationship of stresses within a solid material.

Figure 8.14 Geometric relations of stress components

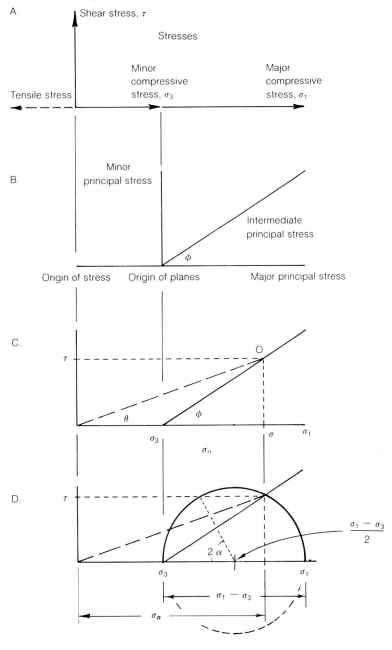

KEY

A. Orientation of stresses
B. Orientation of stress surfaces; ϕ is the angle of internal friction
C. *A* and *B* combined to give general stress information for point *O*; θ is the angle of obliquity
D. Mohr circle relations to *C*

Figure 8.15 Mohr diagram

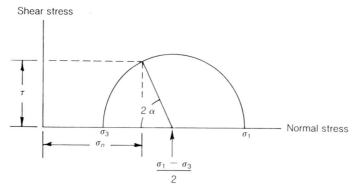

Shear stress

Normal stress

$2\,\alpha$

τ

σ_3

σ_n

σ_1

$$\frac{\sigma_1 - \sigma_3}{2}$$

KEY

$\sigma_1, \sigma_3, \sigma_n$ Maximum, minimum, and normal stress

τ Shear stress

α Angle between principal stress direction and surface of maximum shearing stress (see fig. 8.13)

Figure 8.16 Changes in stress leading to rupture

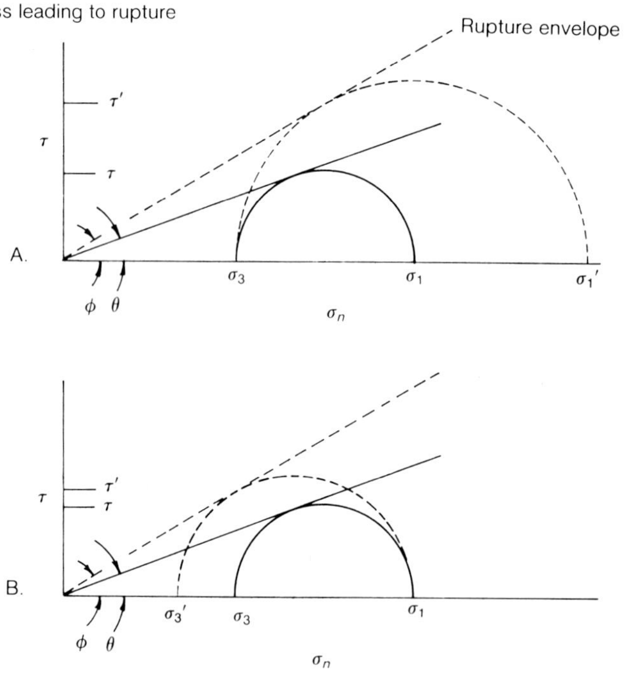

Rupture envelope

τ'

τ

τ

A.

σ_3

σ_1

σ_1'

σ_n

ϕ θ

τ

τ'

τ

B.

σ_3'

σ_3

σ_1

σ_n

ϕ θ

KEY

A.   Increase in angle of obliquity, θ, to angle of internal friction, ϕ, and rupture with changed maximum stress, σ_1 to σ_1'

B.   Rupture in consequence of decreased minimum stress, σ_3 to σ_3'

The relations of the principal stresses, σ_1 and σ_3, and their shear and normal stresses derivatives, respectively, τ and σ_n, may be represented on a nomographic plot called a *Mohr diagram* (fig. 8.15). Some of the important features of this diagram follow:

1. As 2α approaches $90°$, τ approaches its maximum value.
2. As the separation of σ_1 and σ_3 changes, both σ_n and τ change.
3. When σ_3 equals σ_1, the circle reduces to a point, vanishes, and the diagram represents purely hydrostatic conditions.

Substances such as dry sand have no cohesion and their resistance to deformation is entirely due to the friction between the constituent grains. At failure the *angle of obliquity, θ,* (fig. 8.16) reaches its maximum value and is termed the **angle of internal friction, ϕ.** For example in figure 8.16a, if σ_1 or τ are increased to the point of failure at σ_1 and τ^1 respectively, then θ is increased to equal ϕ. For any set of stress conditions this angle remains constant for the material; thus a decrease in σ_3 at constant σ_1 (fig. 8.16b) leads to the same value of ϕ. If ϕ is known the envelope of rupture, relating all combinations of σ_1 and σ_3 to the shear stress at rupture for a given material, may be drawn.

Many substances, for example solid rocks and damp clay, exhibit *cohesion,* the property of inherent strength that is independent of and exists in the absence of imposed stress. Since interparticle cohesion must be overcome by the applied shearing stress before failure can occur, the shear strength of the material is increased, although the ϕ angle of the Mohr rupture envelope is unchanged by the presence of cohesion (fig. 8.17). The angle of internal friction may be modified in a given substance, however, particularly by the presence of water or other fluids in its pores or by marked changes in temperature, and the rupture envelope becomes curved (fig. 8.18).

Figure 8.17   Effect of cohesion

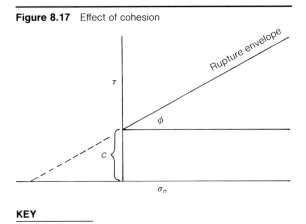

KEY

c   Cohension increases shear stress required for rupture

Figure 8.18   Effect of changing angle of internal friction

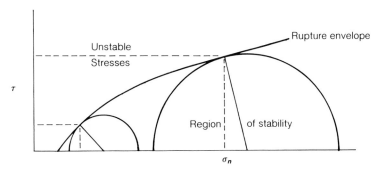

Mechanical Properties of Rocks   **167**

Figure 8.19 Effects of varying stress conditions

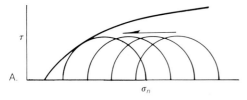

A.

B.

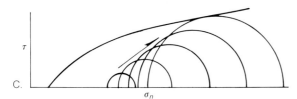

C.

Figure 8.20 Rupture envelopes for different materials

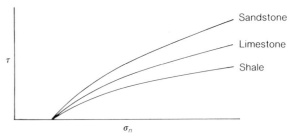

KEY

A. Rupture with constant stress differential and decreasing normal stress (unloading)
B. Rupture with increasing stress differential (loading)
C. Rupture with increases of both absolute and differential stress

The Mohr diagram may be used to anticipate the conditions leading to rupture under varying stress conditions. In figure 8.19a it may be seen that a lowering of the confining pressure with $\sigma_1 - \sigma_3 =$ constant, say by erosion or excavation, results in a decreased normal stress at constant shearing stress. Rupture occurs when a circle becomes tangent to the fracture line. In figure 8.19b, σ_3 is constant and σ_1 increases (as might occur in burial or loading). Both σ_n and τ increase, and rupture occurs when tangency of circle and ϕ curve takes place. In figure 8.19c, σ_1, σ_3, $\sigma_1 - \sigma_3$, and τ are all increasing; rupture again

occurs at the conditions of tangency. It is left to the student to suggest geologic and engineering situations involving changes in loading, volume, or fluid content that are consistent with such plots as represented in figure 8.19.

Different materials are characterized by different rupture envelopes. For example, shale, limestone, and sandstone are contrasted in figure 8.20. These differences become important to explain the different deformational responses observed in the same rock type at different locations, or when interlayered rocks have responded differently to the same stress field.

SUMMARY

The mechanics of rock deformation is complicated because the active stress is distributed in a solid continuum and variable strain responses arise in consequence of the differences in rocks and the effects of water content, strain rate, and temperature.

Modeling provides some insight as to the nature and consequences of stress application, changes in shape and volume of the continuous mass may be represented by strain ellipsoids, and the Mohr diagram is an invaluable nomograph to relate stress and strain.

ADDITIONAL READING

Brady, B. H. G. and Brown, E. T. 1985. *Rock mechanics*. Winchester, Mass.: Allen & Unwin.

Coates, D. F. 1981. *Rock mechanics principles*. 4th ed. Brookfield, Vermont: Renouf USA.

DeSitter, L. U. 1956. *Structural geology*. New York: McGraw-Hill.

Dreyer, W. 1972. *The science of rock mechanics*. Part I. Bay Village, Ohio: Trans Tech.

Farmer, I. W. 1983. *Engineering behaviour of rocks*. London: Chapman & Hall.

Goodman, R. E. 1975. *Geological engineering in discontinuous rocks*. St. Paul, Minnesota: West Publishing Co.

Habib, Pierre. 1983. *An outline of soil and rock mechanics*. New York: Cambridge University Press.

Hills, E. S. 1972. *Elements of structural geology*. 2d ed. New York: John Wiley & Sons.

Hobbs, B. E.; Means, W. D.; and Williams, P. E. 1976. *An outline of structural geology*. New York: John Wiley & Sons.

Jaeger, J. C., and Cook, N. G. W. 1979. *Fundamentals of rock mechanics*. London: Chapman & Hall.

Park, R. G. 1982. *Foundations of structural geology*. Glasgow: Blackie & Sons.

Price, N. J. 1966. *Fault and joint development in brittle and semibrittle rock*. New York: Pergamon Press.

Ramsay, J. G. 1967. *Folding and fracturing of rocks*. New York: McGraw-Hill.

Vutukuri, V. S., et al. 1974, 1976. Handbook on mechanical properties of rocks. v. 1 and v. 2. Bay Village, Ohio: Trans Tech.

9

Geologic Structures and the Geometrical Description of Deformed Rocks

INTRODUCTION

The response of rock to the deforming forces described in the preceding chapter results in a number of recognizable geometric features that are related to the stresses applied. Some rock units deform more or less plastically and are bent into sinusoidal forms termed folds; in other instances the rock units respond in a brittle manner—cracks form or blocks of rock are moved with respect to one another, sometimes for very long distances. The recognition and the description of these geologic structures involves three-dimensional insight, which should be cultivated as a geologic essential for the engineer.

ATTITUDE OF LINES AND PLANES

Deformed rock masses usually contain *planar* or *linear elements,* some of which are imaginary, whose *attitude* may be exactly specified by their relation to some coordinate system. The usual coordinate system for such geological purposes is simply the horizontal and a vertical plane located in space by geographical position and directon (*bearing, azimuth*).

Figure 9.1 Terminology for lines and planes

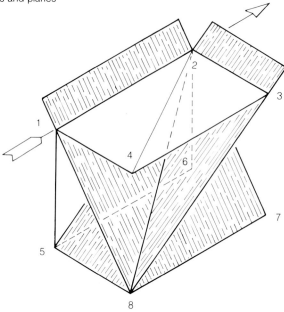

Figure 9.1 Terminology for lines and planes

KEY

Reference: horizontal plane 1234, vertical plane 1265, azimuth of line 12 taken here as north–south

Line	Strike	Plunge		Plane	Strike	Dip
12	N–S	0		1234		0
14	E–W	0		1265	N–S	90
24	NW–SE	0		1278	N–S	∡ 418
15		90		2385	E–W	∡ 438
18	E–W	— 418E				
38	N–S	∡ 438S				
28	NW–SE	∠ 428SE				

Maps and surveys provide locational data, horizontal and vertical planes are easily established in the field by use of levels or plumb lines, a compass provides a means for determining directions, and a clinometer may be used to measure vertical angles. An instrument such as a Brunton compass is commonly used to make these measurements. Figure 9.1 illustrates the use of *reference planes*. In the figure, the intersection of the horizontal and vertical planes is taken as north–south for convenience but as an exercise it is suggested that other directions be assumed and the accompanying key be rewritten.

The attitude of a plane is specified by its strike and dip. **Strike** is the direction of the line of intersection of an inclined plane with the horizontal and may be expressed either as the geographical direction (i.e., N 30° E or N 60° W, etc.) assuming a compass divided into quadrants. For a compass with a 360° dial these directions are respectively 30° and 120°. **Dip** is the dihedral angle between an inclined and horizontal plane. It is the largest angle between these planes and lies in a vertical plane that is perpendicular to the strike line. All other angles between these planes will be smaller and are called *angles of apparent dip* (e.g., the apparent dip angle 428 is less than the true dip angle 418 of the inclined surface 1278).

The attitude of lines is specified by their bearing and plunge. The **bearing** of a line is the direction of its projection on the horizontal plane and is taken as the "downhill" direction; thus lines 18 and 28 in figure 9.1 project into the horizontal as lines 14 and 24 with strikes of east and southeast respectively. The **plunge** of a line is its inclination to the horizontal measured in a vertical plane that contains the line. The plunge of lines 18 and 28 are the angles 418 and 428 respectively.

When a linear feature lies in an inclined plane it is sometimes convenient to describe the line with reference to this plane. The angle between the strike of an inclined plane and a line measured in the inclined plane is called *rake*. The rake of line 28 in figure 9.1 is the angle 128 in the easterly dipping plane or the angle 328 in the southerly dipping surface.

Measurable *planar* or *subplanar features* in rocks may arise from a number of circumstances; all have implications as to the history of the rock, the more commonly noted follow:

Mineral orientation   Dimensional orientation or segregation of blocky or platy mineral grains leading to a *banded* or *foliated* (schistose) appearance; common in metamorphic rocks.

Bedding   More or less thick and parallel layers of sediments and sedimentary rocks with the layers separated by bedding surfaces or planes.

Contacts   Interfaces between rocks of distinctly different kind or age such as an unconformity, or boundary between an intrusive igneous mass and its country rock.

Fractures   More or less uniformly spaced and oriented surfaces of discontinuity caused by brittle fracturing; *joints* having no sensible movement in the plane of rupture and faults having significant movement.

Imaginary   Surfaces of symmetry as between the limbs of a folded sheet or the midplane between planar fractures.

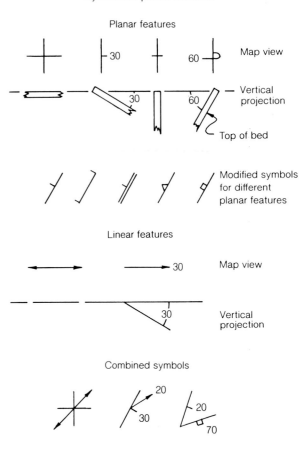

Figure 9.2   Common map symbols
Symbols for planes and lines

Linear elements are less common in rocks but are seen in mineral orientation as features related to erosion and deposition and as imaginary intersections of curved and planar surfaces.

For purposes of plotting on maps, it is usual to adopt a set of conventional symbols such as are shown in figure 9.2 in order to show the location and attitude of various lines and planes.

GEOMETRY OF A SINGLE FOLDED SURFACE

The deformation of rocks sometimes generates wavelike forms known as **folds.** Students may find it convenient to have a simple model for reference, and one may be easily made by bending a sheet of paper. (Note that the bending may be accomplished by pushing either parallel or perpendicular to the sheet). This bent surface represents a single bed or a bedding surface within a stack of deformed layers.

Flexures or folds bent upward (against gravity) are called **anticlines,** or antiforms, and those bent downward (with gravity) are called **synclines,** or synforms. The degree of bending is indicated by the terms *open, tight* (or *closed*), and isoclinal (fig. 9.3).

Figure 9.3 Axial planes and tightness of folding

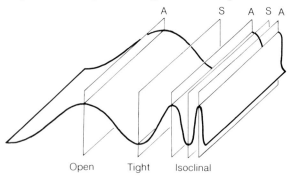

Open   Tight   Isoclinal

KEY

A. Anticline
S. Syncline

Contorted bedding, Andes Mountains

Asymmetric anticline

Outcrop of plunging folds, Oklahoma

Each fold is bent symmetrically about an imaginary surface, usually roughly planar, that is termed the **axial plane** of a fold as shown in the figure. A linear element representing the intersection of this plane with the folded surface is an **axial line,** or fold axis (called hinge line by some), as shown in figure 9.4 where *ab* and *a'b'* are axial lines in the anticline and syncline respectively.

The various positions that folds may have in space with respect to the horizontal and vertical may be visualized by the rotation of the fold around an axial line, a line perpendicular to an axial line, and a combination of these motions. Rotation around an axial line causes the fold to change from an upright through an overturned to a recumbent position (fig. 9.5). The distinction between upright and overturned folds is made upon whether or not one limb or flank has been rotated through the vertical and has its originally upper surface underneath. Recumbent folds are those whose axial plane is essentially horizontal. It should be noted that, with the exception of upright folds, the axial line does not coincide with the uppermost points on an anticlinal surface or with the lowest points on a syncline. Since it is often the locus of highest or lowest points that is determined by mapping in the field, it is important to distinguish this locus, the **crest line,** from the axial lines.

Rotation of folds about a horizontal line normal to an axial line causes the axial lines to be at some angle to the horizontal and the folds to plunge. Most folds plunge at relatively small angles but some are steeply inclined, and in rare instances

Figure 9.4 Axial lines

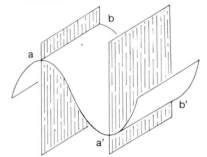

Figure 9.5 Fold positions resulting from the rotation of a fold around an axial line

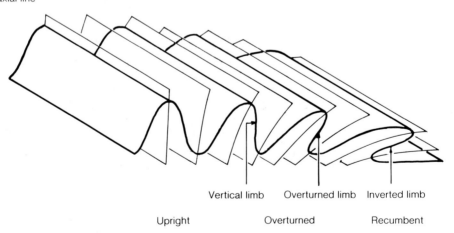

Vertical limb   Overturned limb   Inverted limb

Upright   Overturned   Recumbent

folds can have their axial lines vertical or even over-turned. Since rocks deform plastically it is also possible, and as a matter of fact probable, that a fold may plunge by different amounts or in opposite directions in different portions of the structure (fig. 9.6).

Combination of the possible rotations about an axial line and a line normal to the axial plane leads to fold attitudes whose position in space seem difficult to describe. Their geometrical description, however, is readily accomplished by specification of the attitude of the axial plane (strike and dip) and an axial line (bearing and plunge or rake). Figure 9.7 shows how these elements combine in the description of a fold in a generalized attitude.

Figure 9.6 Longitudinal views of axial lines of plunging folds

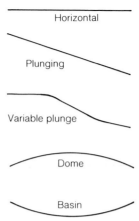

Horizontal

Plunging

Variable plunge

Dome

Basin

Figure 9.7 Eroded folds in general position

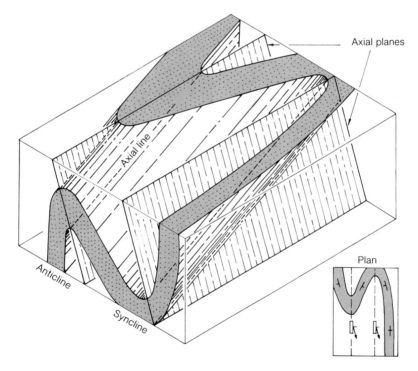

GEOMETRY OF A FOLDED STACK OF BEDS

The beds comprising a fold may be unchanged in *stratigraphic thickness* (perpendicular distance between bedding surfaces) or approximately constant in vertical separation at all points depending upon the mechanics of deformation. The first case corresponds to *concentric folds* and the second to *similar folds* (fig. 9.8). It should be noted that concentrically folded anticlines become smaller with depth while synclines increase in size. In similar folds no changes of this kind occur. In many sequences of folded rocks the different beds have differing abilities to transmit stress and yield in different ways so that the resultant package of folded beds contains interbedded layers of concentrically folded *competent* beds and similar-folded *incompetent* beds.

It is mechanically necessary that layers that fold concentrically without change in thickness do so by sliding past one another (as shown by the small couples in figure 9.9a). Layers that are changed in thickness undergo plastic flow from the limbs into the crests and troughs of similar folds (fig. 9.9b). These internal adjustments of the rock layers may generate small crumples, or **drag folds;** fracture cleavage and slaty cleavage; boudinage (sausage structure) or **slickensides** (striated polished surfaces) whose orientation holds a fixed relationship to the major fold.

In circumstances in which rocks can be shown to have folded without change in bed thickness, the graphical reconstruction of the structure becomes quite simple since any plane tangent to a bed must be perpendicular to a radius of curvature. Consider, for example, the reconstruction of the folds obtained from the measured dips of beds at the surface, which are represented by the small open rectangles in figure 9.10 following the method of Busk (1929). Normals to adjacent beds are drawn and intersect at the center of concentricity. Their extension defines the concentric zone from each center. Arcs swung from each center are carried through each zone and locate the particular bed in its other intersections with the sur-

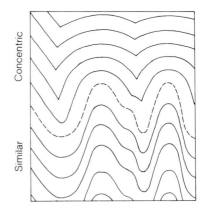

Figure 9.8 Similar and concentric folds

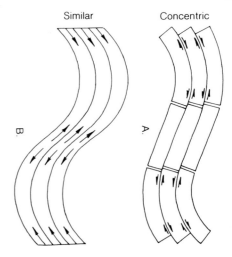

Figure 9.9 Mechanics of concentric and similar folding

face and in the subsurface. Such constructions are useful not only to represent the shapes of folds but also to predict their form at depth, to determine stratigraphic thickness and to estimate the position of buried surfaces. Bed *A,* for example, should also be found at *A'* and a well drilled at *x* will intersect bed *A* at depth *d.* If concentric folding is continued to depth the size of synclines increases while anticlines shrink, and a point is eventually reached, represented by 0, beyond which a lack of space precludes

Figure 9.10 Reconstruction of folds

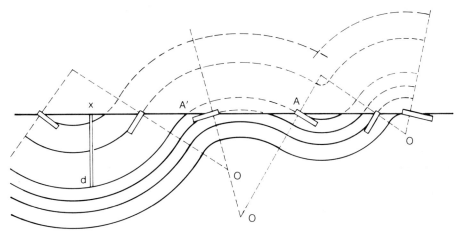

Figure 9.11 Décollement

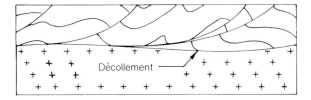

further concentric folding. At this depth the geometry of the folding must change or rupture (*décollement, detachment*) must occur. Décollement is also, and more commonly, the result of differential resistance to deformation by the upper and lower portions of crustal rocks and often takes the form of crumpling of overlying sedimentary layers above a relatively undisturbed crystalline basement (fig. 9.11).

GEOMETRY OF FOLD GROUPS

Folds seldom if ever occur alone. Microscopic-size crumples are arched over small folds that in turn are part of a still larger flexure. Each anticline is paralleled by a syncline and vice versa. These small and great folds arise from the deformation of rocks in the same stress field and are therefore essentially parallel crumples. The resultant parallelism of axial lines and planes thus allows the extrapolation of measurements on one fold to a general attitude for a large folded region and, for example, observations at a single outcrop may serve to characterize the general attitude of folded rocks over an area of several square kilometers. Some features of groups of folds are illustrated in figure 9.12.

Single folding episodes characterize many deformed regions of the earth's crust, but multiple periods of folding under differently oriented stress fields

Figure 9.12 Folded rocks in Pennsylvania

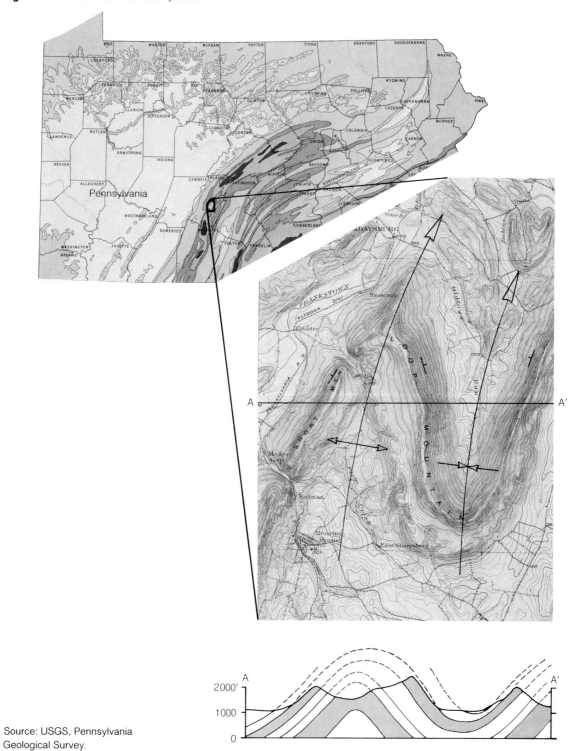

Source: USGS, Pennsylvania
Geological Survey.

are also common. Refolding will preserve some features of the earlier deformational event and add others, often resulting in particularly complex geometrical relationships.

Geometry of Folds: Field Observations

The field observations that enter into the working out of the structural relations of folded rocks are the correlation of beds, the attitude of beds (dip, strike and original top surface), the topographic corrections, and the determination of the location and attitude of axial planes and crest lines.

Crumples and drag folds have their axial lines and planes parallel to those of the main fold. Fracture cleavage and slaty cleavage are surfaces of parting that approximately parallel the axial plane and whose intersection with bedding is parallel to the axial lines. Further, the intersection of bedding and cleavage may be used to find the direction to anticlinal and synclinal crests from the point of observation. Boudinage (sausage structure), drawn pebbles (flattened or ribbonlike cross section), rolled pebbles (circular cross section) and striae formed on the surface between plates of moving rock (slickensides) are all oriented either parallel or perpendicular to the relative motion between the beds. If perpendicular to the direction of this motion, their elongation is parallel and if parallel their elongation is perpendicular to an axial line. Some of these relationships are diagrammatically shown in figure 9.13.

GEOMETRY OF FAULTS

Faults are rock ruptures in which significant and often very large movement has taken place parallel to the rupture surface. They arise in consequence of the same forces that generate folds and may be intimately related to them. Generally faults either die out through flexures into undeformed rocks (fig. 9.14) or blend into or terminate against another fault. A fundamental difference between the kinds of fault lies in whether the fault rupture has occurred in consequence of a shortening or an extension of the rock mass (i.e., as a result of compression

Figure 9.13 Minor features associated with folds

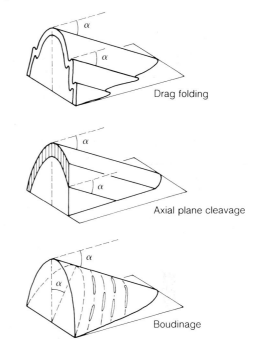

Drag folding

Axial plane cleavage

Boudinage

Figure 9.14 Transformation of folding to faulting

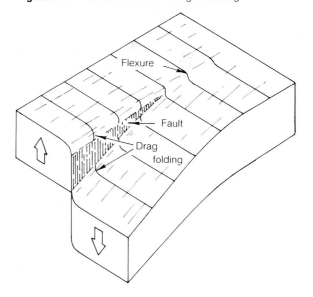

Flexure

Fault

Drag folding

Slickensides, Nahant, Massachusetts

Fault breccia, Maine

Figure 9.15 Geometrical terminology for faults

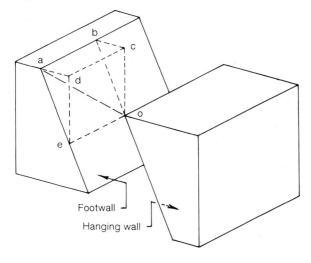

KEY

aboe = Fault surface
ab = eo = Strike of fault plane
∡ *dae* = ∡ *cbo* = Dip of fault plane
ab = eo = Strike-slip component of fault motion
ae = bo = Dip-slip component of fault motion
ao = Net slip

or tension). It is traditional to term the tensional product a **normal fault** and dislocation resulting from compression as a reverse or **thrust fault.**

The geometrical description of a fault entails identification of the attitude of the rupture or fault surface and the relative movement of the adjacent rock masses. Figure 9.15 represents a generalized fault between two blocks, which have moved in such a way that initially adjacent points are separated to *a* and *o*. The inclined faces of the two blocks are identified as the **footwall** and **hanging wall** from their position respectively below and above, for instance, an observer in a tunnel along the strike of the fault surface. In figure 9.15 the hanging wall moved downward with respect to the footwall.

It is nearly always impossible to tell in absolute terms which block has moved up or which down, or whether both have moved in the same direction

Figure 9.16 Fault surface details

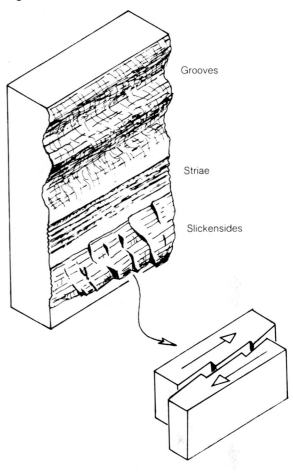

Grooves

Striae

Slickensides

but at different amounts. Only the relative movement may be detected. Clues to the direction of the motion may be found by the presence of grooves, striae, or slickensides on the fault surface; in the dragging of beds into fault plane; in associated fractures; in the separation of initially continuous features (e.g., beds, dikes, other faults, etc.); in the presence of topographic scarps; and in many other ways (fig. 9.16).

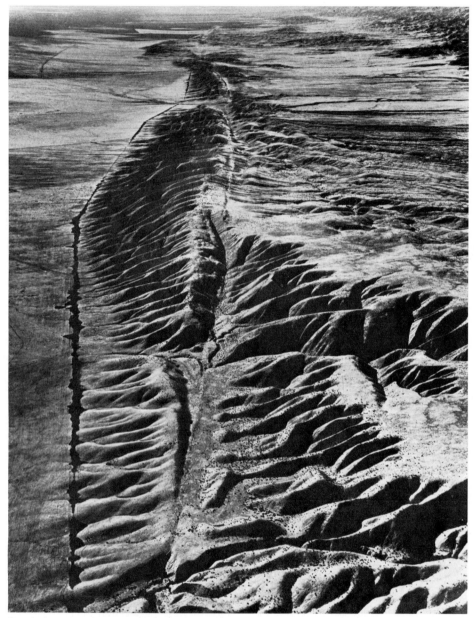

San Andreas fault (strike-slip movement), California

Normal fault, Virginia

Small-scale faulting (actual size)

Kinds of Faults

A large and complex group of terms are in general use for the description of particular types of faults and their genetic relationships. A sampling of these follows, and some are illustrated in figure 9.17.

Dip of the Fault Surface

High- or **low-angle faults** Dipping more or less than 45°

Apparent Motion in the Fault Surface

Dip, strike, or **oblique (diagonal) faults** The dominant motion parallels the dip or strike of the fault surface or is between these directions.

Right- and **left-lateral faults** Strike-slip motion with a clockwise (dextral, right-handed) or counterclockwise (sinistral, left-handed) sense.

Normal or **reverse faults** Dip-slip motion in which the hanging wall moves respectively down or up with respect to the footwall.

Stress Orientation

Tension fault Rift
Gravity fault Tensional stress
Thrust fault Compressional upthrust, overthrust, underthrust or ramp (steeply-inclined fault surface)
Tear fault Strike-slip fault striking parallel to thrust fault movement

Relationship of the Fault to Regional Structure or Bedding

Strike, dip, bedding, or **diagonal fault** Fault motion in relation to the attitude of the beds that it displaces.

Transverse and **longitudinal faults** Steeply inclined faults respectively crossing or parallel to the regional structure or grain.

Fault Sets

Intersecting, parallel, en echelon (staggered), **peripheral,** or **radial faults**
Distributive faults Multiple parallel fault surfaces within a zone of faulting.
Imbricate structure Successive overthrust slices

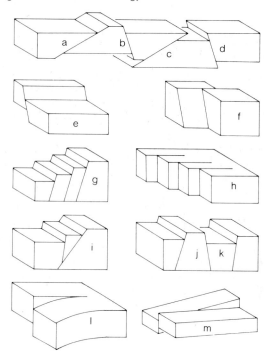

Figure 9.17 Fault terminology

KEY

a. Low-angle normal fault
b. High-angle normal fault
c. Low-angle thrust fault
d. High-angle thrust fault
e. Dip-slip motion on normal fault
f. Strike-slip motion
g. Step faults
h. En echelon strike-slip faulting
i. Fault wedge
j. Horst
k. Graben
l. Scissors fault
m. Pivot fault

Fault blocks

Fault wedge
Horst Uplifted block, upthrown block
Graben Down-dropped block, downthrown block
Rotational (hinge or scissors fault)

Jointing in granite, Cape Ann, Massachusetts

Columnar jointing in basalt, island of Skye, Scotland

Geologic Structures and the Geometrical Description of Deformed Rocks 187

Recognition of Faulting

Direct evidence of faulting may sometimes be seen in grooved or slickensided fault surfaces; gouge and breccia zones; or in the displacement of beds, dikes, or geologic boundaries across a fault trace. **Slickensides** are fault surfaces showing polish, small-scale grooving, and minute ramps and steps; the grooving orients the fault motion and the steps its direction. **Gouge** is rock macerated to claylike consistency, and **breccia** or cataclased zones are angular rubble of broken rock, which may or may not be cemented. It should be noted that the presence of gouge and breccia does not provide any indication of the amount of movement on the fault and that the apparent displacement on a fault may be misleading unless coupled with other information. In figure 9.18, for example, the apparent strike-slip movement of x meters is actually due to a dip-slip displacement of y meters followed by erosion of the upthrown block; the key is the direction of the slickensides.

Postfaulting erosion and deposition are usually effective in removing the direct evidences of faulting. In consequence, most faults are obscured and may be recognized only by careful interpretation of the geology, which usually requires considerable geologic insight. It is important in this context to realize that geological discontinuities may result from erosional unconformity and intrusion of igneous rocks as well as from faulting. Geologic evidence of faulting is usually seen in such features as the bending or dragging of beds or other two-dimensional bodies into the fault surface; the omission or repetition of geologic units; the offset of units with or without overlap; or the abrupt termination of structures along their trend. The presence of faulting is often also suggested by linear topographic lows, topographic features such as a **scarp,** a straight stream course or lake edge, or by anomalous patterns in magnetic or gravity data obtained in geophysical surveys.

Figure 9.18 Apparent and true movement

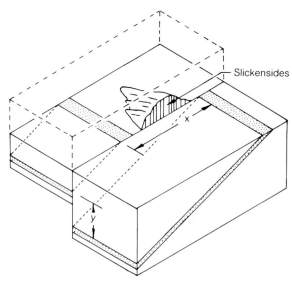

Slickensides

JOINTS

When observed in the field most rocks show cracking that separates the rock mass into discrete blocks of more or less regular shapes. The term *joint* is applied to these rock ruptures in which there is no sensible movement parallel to the rupture surface. The joints in rock have important consequences to bulk mechanical properties, ease of excavation, transmission of fluids or seismic waves, and uses for aggregate and dimension stone. Note of their *spacing* and *orientation* is an essential part of any engineering geology site description. Joints are more abundant in the near surface, decreasing with depth in number and from open to closed to invisible but potential cracks. Incipient joints are often exploited in the quarrying of dimension stone since their presence allows the stone to be more easily split.

The origin of joints is often unclear and is certainly not a mechanically simple process. Extensional stress coupled with a shearing component appears to be the principal mechanism. Generally speaking, it would appear that joints are the brittle fractures that arise in rocks in consequence of a change in the volume or the shape of a rock mass.

Changes in volume, which place the rock mass under tensional stress, can arise from (1) shrinkage due to dehydration or cooling that results in polygonal cracks perpendicular to the active surface, (2) expansion of rock when overlying material is removed (unloading) that results in a sheeting or spalling (chipping) parallel to the surface, or (3) failure due to the passage of a compressional-extensional shock wave from an explosion that results in radial cracking around the shot point, or blast area.

Changes in shape, which place the rock under shearing stress, occur as a result of settlement or of folding or faulting, and the orientation of the resultant joints are a reflection of the active stress field. Changes in both volume and shape can be anticipated when elastically compressed rocks are unloaded by erosion or excavation and expand in an anisotropic way. Not only may differently oriented stress fields be set up in a rock mass, but it must be remembered that the strength of most rocks is different in different directions. The interaction of changes in shape, volume, and strength may thus lead to complicated patterns of jointing.

Although the mechanics of formation of many kinds of joints is obscure, the origin of shrinkage joints is obvious when a mud layer dehydrates forming a network of mudcracks or when a lava cools yielding columnar basalt. When a material such as lava or mud cools or dehydrates all portions of the mass are placed under tensional stress by the resultant decrease in volume. Any slight imperfection in the mass will then localize a tensional rupture, which relieves the stress locally and temporarily. Further stresses may affect the now mechanically isolated portions and rupture again occurs within

them. Eventually the mass is broken up into a series of polygonal plates or columns. Many joints in undeformed sedimentary rocks are probably primary shrinkage joints, closed under the pressure of burial and reopened when the rock is unloaded.

All buried rock is elastically compressed by the weight of superincumbent material and this compression may be taken to be essentially hydrostatic. When erosion or excavation removes weight the rock expands nonuniformly because of the differences in lateral and vertical constraints. Sheeting roughly parallel to the earth's surface identifies the upward direction of easiest relief. These sheeting joints typically have undulatory surfaces and are probably pure extensional phenomena. Elastic response to unloading, however, cannot be restricted to adjustments to vertical stress only. Response to modification of lateral stress must also take place and is probably controlled to a significant degree by directional strength properties within the rock mass. A granite, for example, is known by quarriers to have different strengths in different directions, which they call the "rift," "grain," and "hardway." If lateral extension is anisotropic, then unbalanced shear stress must be present and the conservative failure will be a planar, not an undulatory, joint. A particularly compelling argument for the origin by shear of many joints is their clean passage through cobbles in a conglomerate rather than a plucking of the cobble in a boss and socket manner. Other features include a plumose (feather) pattern on the joint surface or the presence of a ramp-and-step feature reminiscent of slickensiding.

The expansion of rock following the removal of compressive stress may be noted in many ways. "Rock bursts" or "bumps" in mines are caused by the explosive spalling of rock fragments from a tunnel roof or wall. Upward expansion of the floor or the binding of drills or channel-cutting machines may be experienced in quarry operations. Drill holes used for the blasting of a quarry face or road cut may be seen to have a slight offset. Joints crossing geologic boundaries or one another may show minute displacements.

Figure 9.19 Gash joints opened by fault movement

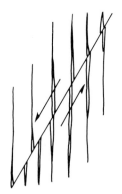

Figure 9.20 Axial plane cleavage

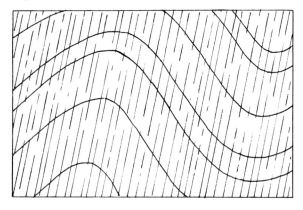

Joints often have some geometrical relationship to larger structural features. Gash joints representing the brittle fracturing of walls are sometimes associated with faults (fig. 9.19), and rocks near a fault may be reduced to a granular rubble—brecciation or cataclasis. Jointing approximately parallel to the axial plane is often developed in folded rocks (fig. 9.20). This latter jointing is sometimes very closely spaced and is variously termed *axial plane, slaty* or *shear cleavage,* or *foliation.* These terms imply an origin different than that for extensional jointing and this is probably correct. However, the mechanical result is rock broken by closely spaced parallel cracks whose properties are nowise different from one jointed by other means.

A regional view of jointing includes the whole assemblage of joints present—the *joint system*—and will often show the presence of two or more *joint sets* (groups of joints of common origin) that are usually more or less parallel. Not uncommonly, jointing in one or more direction is particularly well developed and persistent. These *master joints* may have a profound effect on topography by their control of surface erosion and subsurface drainage patterns and their presence may strongly influence the ease of excavation or tunneling.

SUMMARY

Real bodies of rock exist in varied and complex shapes in consequence of the manner in which they formed and the deformation they have undergone. Large-scale, long-term forces have caused rocks to bend and break. These forces have been mentioned in chapter 3 (plate tectonics) and discussed in some detail in the previous chapter.

The results of rock deformation as seen in folded, faulted, and jointed rocks may be described by the attitude of the real and imaginary lines and planes that they incorporate. This allows geometrical information to be presented by the plotting of measured information on maps or by simple written description.

Joints markedly affect the ease of excavation and permeability of rocks. Faults may be the locus of earthquake movement and even when inactive (as most are) they may juxtapose rocks with different properties separated by a permeable zone of brecciation. Folds, when eroded, also may result in rock types of different character being present in the foundation of a project.

ADDITIONAL READING

Dennis, J. G. 1967. *International tectonic directory.* Tulsa, Okla.: American Association of Petroleum Geologists.

Dennison, J. M. 1968. *Analysis of geologic structures.* New York: W. W. Norton.

DeSitter, L. U. 1956. *Structural geology.* New York: McGraw-Hill.

Hills, E. E. 1972. *Elements of structural geology.* 2d ed. New York: John Wiley & Sons.

Hobbs, B. E.; Means, W. D.; and Williams, P. E. 1976. *An outline of structural geology.* New York: John Wiley & Sons.

Park, R. G. 1982. *Foundations of structural geology.* Glasgow: Blackie & Sons.

Platt, J., and Challinor, J., 1974. *Simple Geological Structures.* Winchester, Mass: Allen & Unwin.

Roberts, J. L. 1982. *Introduction to geological maps and structures.* Elmsford, New York: Pergamon Press.

Suppe, J., 1985. *Principles of structural geology.* Prentice-Hall.

Thomas, J. A. G. 1977. *An introduction to geological maps.* Winchester, Mass.: Allen & Unwin.

PART II
Engineering Considerations

The division of this text into two parts is arbitrary and done on the basis of emphasis rather than material. Part I has established the geologic framework of concepts, processes, and materials within which engineering activities are carried on; Part II addresses the properties of earth materials, natural hazards, and some areas of applied engineering geology.

To paraphrase Karl Terzaghi, an important contributor to the foundations of both engineering geology and soil mechanics, at one time the theoretical methods promised to eliminate the need to depend on geologic information. However, the nature and importance of the uncertainties associated with the results of even the most conscientious explorations depend entirely on the geologic characteristics of the site. Engineering geology cannot be separated from geology. It must be based on a firm understanding of geologic fundamentals but go beyond them to their impact on the engineering characteristics of a site. What sets engineering geology apart from other geoscience fields is its close relationship with engineering disciplines. Indeed, it is the engineering consequences of a particular geologic setting that are of paramount importance from a functional viewpoint.

Engineering geology has evolved over nearly a century through the efforts of geologists serving as consultants on engineering projects, as regular employees in federal agencies (Corps of Engineers, Bureau of Reclamation, Tennessee Valley Authority), and today as members of state agencies and private engineering and environmental firms. The Association of Engineering Geologists (AEG), founded in 1963, grew out of an earlier association in California. Other important organizations are the Engineering Geology Division of the Geological Society of America and the American Society of Civil Engineers, which maintains a Permanent Joint Committee on Engineering Geology.

10

Laboratory and Field
Description of Soil
and Sediment

INTRODUCTION

All geologic materials, whether massive rocks or unconsolidated earth materials, have certain intrinsic properties that are important to their various engineering uses as construction material, in excavation, or as foundations. The kind of data that is commonly measured varies, of course, with the nature of the engineering work. An indication of the importance of material properties is given by table 10.1 in which some of the more commonly measured properties of soils are tabulated. Detailed descriptions of these tests may be found in the references given at the end of the table. It should be noted that not all of the tests listed need be performed at any particular site; also that other information may be required and determined. It is also important to note that the engineering classifications derived from tests intended for the prediction of the field performance of temperate-zone soils will usually fail if applied indiscriminately to soils of arctic regions or to lateritic soils developed in the humid tropics.

Table 10.1 References for Standard Test Procedures

Test	References
Moisture content	ASTM D2216, Ref.1*
Moisture and ash of peat	ASTM D2974, Ref.1
Specific gravity	
< 4 mesh	ASTM D854, Ref.1
> 4 mesh	ASTM C127, Ref.1
Atterberg limits	
Liquid limit	ASTM D423, Ref.1
Plastic limit	ASTM D424, Ref.1
Shrinkage limit	Ref.3
Gradation	ASTM D422, Ref.1
Corrosivity	
Sulfate and chloride content	Ref. 4
pH	ASTM D1293, Ref.1
Resitivity (field)	Ref.5
Permeability	ASTM D2434, Refs. 1,2,3
Consolidation	Ref.2
Swell	AASHTO T258, Ref.6
Shear strength	
Direct shear	ASTM D3080, Refs. 1,2
Unconfined compression	ASTM D2166, Refs. 1,2
Triaxial compression	ASTM D2850, Refs. 1,2,3
Expansion pressure	AASHTO T174, Ref. 6

*ASTM, American Society for Testing and Materials
†ASSHTO, American Association of State Highway and Transportation Officials
Reference Sources:
1. American Society for Testing and Materials. *Annual book of ASTM standards.* Part 19: Natural building stone, soil and rock, peat, mosses, and humus; Part 14: Concrete and mineral aggregates. ASTM Philadelphia, PA.
2. Lambe, T. W., 1951, Soil Testing for Engineers, John Wiley and Sons, New York.
3. Office of the Chief of Engineers, 1970. Laboratory Soils Testing. Dept. of the Army, Engineering Manual EM 1110-2-1906. Washington, D.C.
4. American Society of Agronomy and the American Society for Testing and Materials, 1965. Methods Soil Analysis, Chemical and Micrological Properties, Part 2. Black, C. A., ed.
5. National Bureau of Standards. Underground Corrosion. Circular C450. U.S. Gov't. Printing Office.
6. American Association of State Highway and Transportation Officials, 1978. Standard Specifications for Transportation Materials and Methods of Sampling and Testing, Part II. AASHTO, Washington, D.C.

Material abstracted from *Soil Mechanics Design Manual 7.1.*
Department of the Navy, Naval Facilities Engineering Command, 1982.

SIZE, SORTING, AND GRADING

Both the *size* and the *distribution* of size of grains are important considerations in the properties of granular materials. Because of variations in particle *shape* and because the usual means of measurement is passage through a sieve, the size of a particle is customarily expressed as the size of the smallest square hole through which it will pass. In practice a weighed sample is placed on the top of a stack of standard sieves with successively smaller apertures, is shaken thoroughly, and the fraction retained on each sieve is weighed. The information may then be presented as a *histogram* (fig. 10.1) in which amounts retained on a particular sieve are identified as + (plus) and those which pass as − (minus). In the figure all of the sample is −6 +60 mesh.

The relation of mesh size of the standard sieves to particle size is given in table 10.2.

A more usual and useful method of data presentation than a histogram is a *cumulative curve* in which the portion that passes successively finer sieves is plotted (fig. 10.2) from the data (fig. 10.1).

Table 10.2 Standard Sieves

Mesh	Particle Size in mm (approx)
4	4.8
6	3.2
8	2.4
14	1.5
20	.8
30	.6
40	.41
60	.45
100	.15
150	.11
200	.051 (approximate limit of visibility to the naked eye)

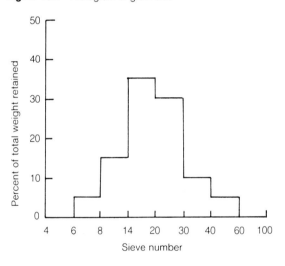

Figure 10.1 Histogram of grain size

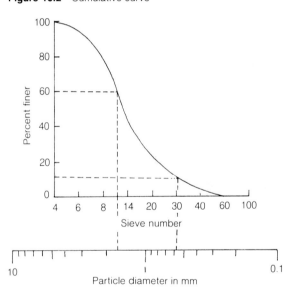

Figure 10.2 Cumulative curve

Geologists are particularly interested in the **sorting** of granular materials because this parameter is a key to the efficiency of the winnowing processes by the agents of transportation. For geologists, a *well-sorted* material is one with all particles of about the same size. Engineers, on the other hand, look at the distribution of grain sizes and consider a *well-graded* sediment to be one in which the amounts of successively smaller grains are present in just the amounts needed to fill the interstices between larger grains. It should be appreciated that well-sorted sediments are poorly graded and vice versa.

Apparatus for measuring settling rate of fine particles (elutriation)

The cumulative curve of figure 10.2 allows the ready determination of gradation of the sample to be found by means of the *Hazen uniformity coefficient*. This coefficient, C_u, is defined as the ratio of particle diameter at the 60% finer point to that at the 10% finer point on the cumulative curve. From the figure,

$$C_u = \frac{D_{60}}{D_{10}} = \frac{1.7}{.58} = 2.9.$$

The *coefficient of curvature, C_z,* is also used to measure gradation:

$$C_z = \frac{(D_{30})^2}{D_{10} \times D_{60}} = \frac{(1.1)^2}{.58 \times 1.7} = 1.23.$$

As a rule of thumb D_{10} may be taken as a measure of the size of the void spaces in coarse-grained soils and sediments (see chapter 6).

The sizes of particles passing a 200-mesh sieve can be determined by their settling rate in water or by the use of *hydrometer analysis* (ASTM D422). A hydrometer is a device to measure fluid density as in an automobile battery. The density of a fluid containing suspended particles will decrease as the particles settle out, and a calibrated hydrometer read at predetermined intervals indicates the amount and size of suspended material remaining. A cumulative curve may be drawn directly from the data.

The size of particles is expressed in accepted size terms that, however, have been somewhat differently defined by various engineering groups and other groups. Figure 10.3 provides some index information for granular materials and soils. Figure 10.4 provides appropriate descriptive terminology.

DENSITY AND SPECIFIC GRAVITY

Density, D or ρ, is the weight per unit volume of a substance and **specific gravity,** SG or G, is the weight of a substance compared with the weight of an equal

Figure 10.3 Grain size terminology

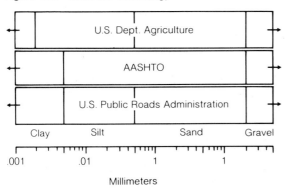

Figure 10.4 Triangular classification chart for sediments

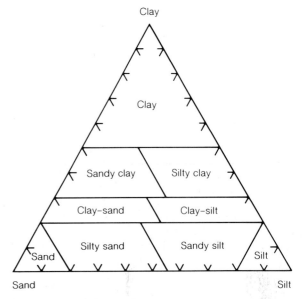

Source: U.S. Corps of Engineers, Lower Mississippi Valley Division.

volume of pure water at 4°C. In metric units density and specific gravity are numerically equal since one cubic centimeter of water weighs one gram (or 10^3 kg/m³ or 1 tonne/m³).

In all determinations of density or specific gravity it is important to recognize that except for massive (poreless) rocks or minerals the material in its natural condition contains some pores occupied by air, water, or other fluids. Samples must thus be prepared in such a way as to allow measurements meaningful to their purpose. For example, the dry density of material to be excavated seriously under-estimates the tonnage to be removed if pore water is present.

Specific gravity is usually determined by weighing the sample first in air and then suspended in water. A solid immersed in water displaces a volume of the water equal to its volume; that is, a unit volume of solid is substituted for a unit volume of water. The weight of the immersed solid relative to that of the displaced water is specific gravity:

$$G = \frac{\text{weight in air}}{\text{weight in air} - \text{weight in water}}.$$

An alternate means of determining the specific gravity of small amounts of granular or powdered material is by the use of a *pyncnometer* (see chapter 1).

The specific gravity measured on bulk samples is the weighted average of the particles and their pore fillings or matrix. If the interstitial material is water the apparent *bulk specific gravity* found, which is frequently used, will be less than that of the particles alone:

$$G = \frac{\text{dry weight in air}}{\text{saturated weight} - \text{saturated weight in water}}.$$

The amount of water in the pores, W_p, is equal to the weight saturated less the weight dry and the true specific gravity of the particles comprising the aggregate is thus

$$G = \frac{\text{dry weight in air}}{\text{saturated weight} - W_p - \text{saturated weight in water}}.$$

Table 10.3 Specific Gravity of Rocks and Minerals

Mineral	SG	Rock	Approx. SG
Montmorillonite	2.0–2.7	Ice	0.9
Graphite, halite	2.2	Peat	1.1
Gypsum	2.3	Coal	1.3
Kaolinite, K-feldspar	2.6	Soil, dry	1.6–1.9
Plagioclase	2.6–2.7	Soil, wet	1.7–2.2
Quartz	2.65	Clay	1.8–2.2
Calcite, chlorite, talc	2.7	Sand and gravel, dry	1.9–2.2
Biotite	2.7–3.1	Sand and gravel, wet	1.9–2.3
Muscovite	2.8–3.0	Shale	2.0–2.8
Dolomite	2.9	Sandstone, conglomerate	2.1–2.7
Hornblende	2.9–3.3	Limestone, dolomite	2.4–2.9
Anhydrite	3.0	Schist	2.4–3.0
Fluorite	3.2	Granite	2.5–2.9
Augite	3.2–3.6	Gypsum	2.3
Olivine	3.3–4.4	Marble, gneiss	2.6–3.0
Diamond	3.5	Slate	2.7–3.0
Garnet	3.5–4.3	Basalt	2.8–3.0
Kyanite	3.6	Gabbro	3.0–3.2
Limonite	3.6–4.0		
Sphalerite	3.9–4.2		
Corundum	4.0		
Chalcopyrite, rutile	4.2		
Barite	4.5		
Ilmenite	4.5–5.0		
Zircon	4.7		
Pyrite, pyrolusite	5.0		
Magnetite	5.2		
Hematite	5.3		
Galena	7.6		
Native copper	8.9		
Native gold	19.3		

Bulk density may be readily measured in the field by removing a sample with a spade, weighing the material removed, and backfilling the hole with dry sand from a calibrated container to obtain the volume.

The specific gravity of some common minerals, rocks, and aggregates is given in table 10.3.

The unit weight of porous substances must be specified as dry, partly saturated, saturated, or submerged to be meaningful for calculation purposes. When the porosity is small the difference in dry and saturated weight may be negligible, however. Additionally, it is important to remember that submerged rock displaces an equal volume of water and loses weight due to buoyancy. This may be an important factor in the analysis of such problems as the bearing properties of submerged foundation material or the integrity of an earth-fill dam. As an example of the magnitude of this effect consider a rock with 20% of its volume occupied by pores and whose dry weight is 2 185 kg/m³. When submerged its loss in weight due to buoyancy would be 0.8 × 1 000 kg/m³ (0.8 times the weight of a cubic meter of water) or 800 kg/m³ and the submerged weight = 2 185 − 800 or 1 385 kg/m³, about two-thirds of its dry weight. Typical values for unit weights of granular materials are given in table 10.4

Table 10.4 Soil Index Properties

	Particle Size & Gradation			Voids				Unit Weight (t/m³)†					
	Size range (mm.)	D_{10}^* (mm.)	Uniformity coef., C_u^*	Porosity (%)		Void ratio		Dry wt.		Wet wt.		Submerged wt.	
	D_{max} D_{min}			n_{max}	n_{min}	c_{max}	e_{min}	Min.	Max.	Min.	Max.	Min.	Max.
				(loose)	(dense)	(loose)	(dense)	(loose)	(dense)	(loose)	(dense)	(loose)	(dense)
Granular materials													
Uniform materials													
Equal spheres (theoretical values)	— —	—	1.0	47.6	26.0	0.92	0.35	—	—	—	—	—	—
Clean, uniform sand (fine or medium)	— —	—	1.2 to 2.0	50	29	1.0	0.40	1.33	1.89	1.35	2.18	.83	1.17
Uniform, inorganic silt	0.5 0.005	0.012	1.2 to 2.0	52	29	1.1	0.40	1.28	1.89	1.30	2.18	.82	1.17
Well-graded materials													
Silty sand	2.0 0.005	0.02	5 to 10	47	23	0.90	0.30	1.39	2.03	1.41	2.27	.86	1.27
Clean, fine to coarse sand	2.0 0.05	0.09	4 to 6	49	17	0.95	0.20	1.36	2.21	1.38	2.37	.85	1.38
Micaceous sand	— —	—	—	55	29	1.2	0.40	1.22	1.92	1.23	2.21	.77	1.22
Silty sand and gravel	100 0.005	0.02	15 to 300	46	12	0.85	0.14	1.43	2.34	1.44	2.48	.90	1.47
Mixed soils													
Sandy or silty clay	2.0 0.001	0.003	10 to 30	64	20	1.8	0.25	.96	2.16	1.60	2.35	.61	1.36
Well-graded gravel, sand, silt and clay mixture	250 0.001	0.002	25 to 1000	41	11	0.70	0.13	1.60	2.37	2.00	2.50	.99	1.51
Clay soils													
Clay (30 to 50% clay sizes)	0.05 0.5μ	0.001	—	71	33	2.4	0.50	.80	1.79	1.51	2.13	.50	1.14
Colloidal clay (−0.002 mm. ⩾ 50%)	0.01 10	—	—	92	37.	12	0.60	.21	1.70	1.14	2.05	.73	1.06
Organic soils													
Organic silt	— —	—	—	75	35	3.0	0.55	.64	1.76	1.39	2.10	.40	1.11
Organic clay (30 to 50% clay size)	— —	—	—	81	41	4.4	0.70	.48	1.60	1.30	2.00	.29	.99

$$^*C_u = \frac{D_{60}}{D_{10}}$$

† Taking G = 2.65 for granular soil, 2.70 for clays, and 2.60 for organic soils

From *U.S. Navy Soil Mechanics Design Manual* NAVFAC DM−7.1, 1982.

POROSITY, SATURATION, AND PERMEABILITY

Porosity is defined as the ratio of the volume of a continuous fluid matrix, more usually called *pore space* or *voids,* to the overall volume of the material, usually expressed in percent:

% porosity, *n,*

$$= \frac{\text{volume of voids}}{\text{volume (voids + particles)}} \times 100.$$

It may be anticipated that there will be a general relationship between the porosity, the bulk specific gravity, and the mode of origin of various geologic materials. Igneous and metamorphic rocks formed by crystal growth at elevated pressures have very low porosities and high specific gravities; coarse granular sediments initially have high porosities and correspondingly low specific gravities, which approach crystalline rock values as the pores are filled with solid matrix material; very fine-grained clayey sediments have high porosity when first deposited that is markedly reduced by compaction. Their specific gravity is low.

All rock and sediment pores are filled with water below the water table, but filling is incomplete above this level. This partially saturated or vadose zone is a very important environment to the civil engineer and others since its properties may be variable in time due to changing *water content.*

The water content of granular materials may be determined by a two-step procedure in which the material is first thoroughly dried and weighed and then saturated and weighed. Drying should be done in an oven at 105°C for 24 hours and saturation by immersion in water for 48 hours for more usual materials. For relatively impermeable materials it may be necessary to force water into the sample by pressure or vacuum.

The *degree of saturation* in a porous system is the ratio of the volume of water present to the total pore volume; its determination, however, is difficult. Water in the vadose zone is divided between that which is strongly attracted or adsorbed on particle surfaces, *pellicular water,* and that which is unattracted or free. Measurement of the total water content thus requires long heating to evaporate adsorbed water and to obtain a completely dry sample.

The determination of total *pore volume* is also difficult. Obtaining an *undisturbed* sample is no simple matter, and water uptake on immersion is apt to be incomplete due to trapped air bubbles. The usual procedure is to disregard this trapped air by establishing a relative scale of values based on immersion in water for a specified period of time at a specified temperature. The ratio between sample volume and the volume of sorbed water is the *percent sorption* by volume under the prescribed conditions. Percent sorption may also be expressed by weight:

$$\frac{\text{weight saturated} - \text{weight dry}}{\text{weight dry}} \times 100.$$

Since there are *n* cubic meters of voids and $1 - n$ cubic meters of solid matter in porous material, the **void ratio, *e,*** is

$$e = \frac{n}{1 - n}.$$

For coarse dry sand, $n \approx 0.33$ or 33% voids and *e* is, therefore, about 0.5. In clayey aggregates the pores, although small, are very numerous and *n* may have values of 0.5 or more giving void ratios greater than unity. Values for porosity and void ratio in some common granular materials and soils are given in table 10.4.

Pores in geologic materials are usually connected by more or less open channels, thus providing a tortuous network of pathways through which fluids may migrate in response to a pressure gradient. Obviously, the rate of flow and quantity of fluid transmitted through a given cross section will be related to fluid viscosity; the size and tortuosity of the channels, which are directly related to the frictional resistance to flow; and the pressure differential. The basic relation is that of Darcy and was discussed in chapter 6.

Frictional resistance to the transport of a fluid should increase as pore dimensions are decreased, and pore dimensions decrease as particle size decreases or perfection of grading increases. The effect on permeability of the filling of pores in a granular quartz aggregate with different amounts of finer materials is shown in figure 10.5; permeability decreases as pores become smaller and surface attraction for water by clay surfaces increases.

There are many types of field permeability tests that can be performed of which equilibrium tests are the most common. These include constant and variable head gravity tests and pressure (packer) tests conducted in single borings and percolation tests. Constant and variable head gravity tests, as their names imply, involve the rate of supply of added water needed to maintain a constant level in the hole or measure the rate of water-level drop. Packer tests involve the isolation of segments of the borehole with an expandible balloonlike packer to test the permeability of selected zones.

Percolation tests are used to ascertain the acceptability of a site for septic tank systems or the shallow disposal of other liquid wastes. The method most generally used is to dig six holes 10–30 cm in diameter to the depth of the proposed absorption trench and backfill them with 10–15 cm of sand or gravel. The holes are prepared for testing by wetting the soil for at least 4 hours after which 15 cm of water above the sand layer is added. The rate of head loss in 30 minutes is then used to assess the suitability of the soil to meet local regulations. Percolation rates below 2.5 cm/hr are unsuitable for any type of soil absorption system. Figure 10.6 illustrates how the infiltration rate changes as a function of soil type.

The object of all field permeability tests is to determine the *coefficient of permeability* (hydraulic conductivity) of in-place materials. The coefficient is the ratio of fluid discharge per unit of cross-sectional area, Q/A, to the hydraulic gradient (head) inducing the flow $Q/A = {}^h/_l (K)$ where $K =$ coefficient of permeability; $Q/A =$ volume per unit time, Q, through cross-sectional area A; and ${}^h/_l =$ hydraulic gradient.

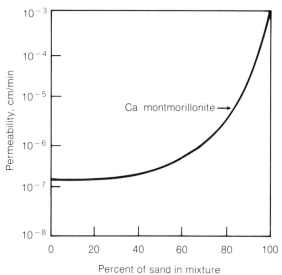

Figure 10.5 Permeability of sand–clay mixtures

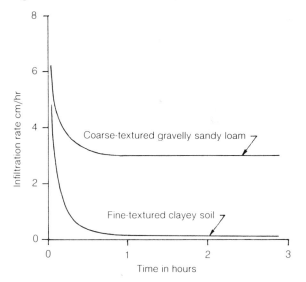

Figure 10.6 Infiltration in different soils

Generally speaking, the applicability of permeability tests in the field will be influenced by the position of the water table; permeability and depth of the material tested; and the type and homogeneity of the material.

Permeability may be measured in the laboratory using permeameters to determine the quantity of water that will flow through a sample of known cross-sectional area and length in a unit time. A constant head is maintained for testing cohesionless soils of high permeability and a variable head observed in a standpipe is employed for impermeable materials.

AGGREGATION AND COHESION

Rocks, sediments, and soil are typically composed of two or more mechanically different components in an intimate mixture. One component is the solid particles of various shapes, sizes, and kinds and the other is the continuous solid or fluid material that occupies the interstices between the grains.

In minerals, the constituent atoms are fixed in place by interatomic bonds to form extended coherent solids, and many crystalline rocks of igneous or metamorphic origin and some sedimentary rocks have a fabric made up of interlocking mineral crystals. Deformation of such materials is opposed by the need to break strong atomic bonds. At the other extreme the constituent particles in fluids are individual molecules having negligible mutual attraction and hence able to rearrange themselves with little hindrance. Between these two extremes lie a number of important *incoherent, semisolid,* and *plastic materials* such as soil, loose sand, and clay aggregates of considerable engineering interest whose physical properties may be qualitatively related to the internal details of their makeup.

The mechanical properties of an unconsolidated granular material does not depend on the relative proportions of particles and matrix but on whether the particles are in contact, how strongly they cohere, and the direction of applied stress. When solid particles are touching they act as an intricate linkage that transmits compressive stress without affecting the matrix. The strength of a material in compression is thus that of the particles assuming that they cannot move relative to one another. If the particles do not touch, the compressive strength is that of the matrix as it is for all tensional stress. Shearing involves both compressional and tensional components, and the shear strength of an aggregate will be intermediate to its compressive and tensile strengths. It may thus be anticipated that the strength of geologic materials will be a function of the kind of applied stress, and further, that of compressive > shear > tensile strength in general.

Solids may be defined as extended masses whose internal integrity is such that they maintain their shape, whereas fluids deform to the shape of their container. Unconfined aggregates of particles lacking intergrain bonding (such as dry sand) deform as do fluids and are termed *cohesionless solids.*

The degree of cohesion of various materials or aggregates is suggested by the ease with which they flow or are deformed and may be visualized as indicating their strength or viscosity. Fluids such as air or water become heavier and more viscous when solids are suspended in them; unconfined dry sand flows readily because the grains are not "sticky"; clay deforms readily because the grains are weakly bonded; and granite, with its interlocking crystalline fabric, will flow only at temperatures far above those of the surficial environment and under considerable differential stress.

A mass of individual mineral grains or rock fragments in a matrix of air may be likened to a liquid composed of individual molecules, each particle being internally coherent but the mass able to flow if it is unconfined. In other matrices, however, the individual particles may be more or less strongly bonded together depending upon the nature of the cementing material, which may be water, plastic clay, or rigid mineral matter.

The surface tension of water causes it to cover grain surfaces and to bridge intergrain gaps as shown in figure 10.7. It should be noted that this sandy mass is a three-component system of solid particles, pellicular water, and interstitial air. If the latter is replaced by water (the assemblage becomes water

Figure 10.7 Pellicular water

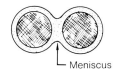

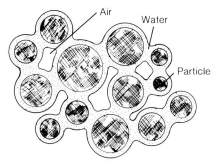

Meniscus

Air

Water

Particle

Figure 10.8 Strength of quartz sand aggregates

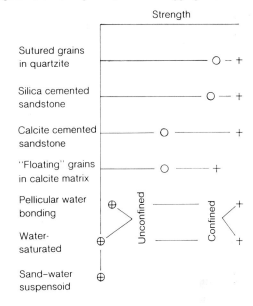

Strength

Sutured grains
in quartzite

Silica cemented
sandstone

Calcite cemented
sandstone

"Floating" grains
in calcite matrix

Pellicular water
bonding

Water-
saturated

Sand–water
suspensoid

KEY

+   Compressive
O   Tensile

saturated), cohesion arising from surface tension is lost and the mass is fluidal. It should also be noted in figure 10.7 that the grains are not in actual contact; the presence of water causes a small increase in volume or *bulking* of the sand. Alternately, removal of the fluid films causes a reduction in volume and may result in surface subsidence if underground fluids are withdrawn.

The actual attractive force between the grains is related to the pulling force of the meniscus (curve) as the surface tension of water against air attempts to reduce its curvature and amounts to approximately $.7 \times 10^{-3}$ N/cm² (newtons per square centimeter). The smaller the grains, the finer the pores, and the greater the overall cohesion. There is thus an extended range of cohesive properties related to water bonding from negligible effects on coarse fragments through moldable sands and silts to plastic clays. The strength of this bonding is attested by the vertical faces to be found in sand pits.

Water not only acts itself as a bonding agent but provides a medium through which dissolved matter may migrate. Either percolation of ground water or upward migration of near-surface water to replace that lost by evaporation provides the means of transporting dissolved material. Since chemical conditions change from point to point, solution and

deposition will take place and deposited material will *cement* grains together. First, solid films are deposited on grain surfaces, these coalesce, and gradually all interstices are filled. The ultimate strength of such a cemented mass of grains will lie in the strength of the cement. In nature cements are typically fine-grained carbonates, iron oxides, clays, or silica. Only silica cementation yields a consolidated product equivalent in strength to a crystalline igneous or metamorphic rock.

As specific examples of cohesion, consider the range of properties exhibited by grains of quartz sand in various modes of aggregation and cohesion as illustrated by figure 10.8. In quartzite the grains interdigitate and no intergranular matrix is present; in cemented sandstones the matrix more or less completely fills the interstitial voids and provides a glue whose strength is that of the matrix material. (Reduction of the numbers of particles may leave them "floating" in the matrix and the rock is no longer a carbonate-cemented sandstone, say, but

Table 10.5 Mechanical Response of Clay

State	Limits	Indices
Liquid		
	Liquid Limit, LL	
Plastic		Plastic Index, $PI = LL - PL$
	Plastic Limit, PL	
Semisolid		Shrinkage Index, $SI = PL - SL$
	Shrinkage Limit, SL	
Solid		

rather a sandy limestone). Sands containing pellicular water have sufficient cohesion to stand in vertical banks but when saturated are cohesionless and slump to their angle of repose. When sand grains are suspended in water and are not in continuous contact, the resultant mixture is a suspensoid having the properties of a heavy liquid and occasionally encountered as **quicksand.**

MECHANICAL PROPERTIES OF SOILS AND SEDIMENTS

The more important *mechanical properties* of soils and sediments in civil engineering practise relate to their plasticity, strength, volume changes, and compactness or consistency.

Plasticity

The mechanical response of a clay or clay-bearing aggregate will change from brittle to plastic to liquid as a function of the increasing amount of water contained in the mass. Of particular interest is the range within which the aggregate is plastic, and this determination is readily made by use of the **Atterberg Limit Tests.**[1] These tests are described by Casagrande (1932) and, in brief, are as follows:

1. **Liquid Limit** (LL) = percent by weight moisture content of oven-dried clay that will just begin to flow when slightly jarred. This is a measure of the maximum amount of

[1]Atterberg, A., 1911, Über die physikalische Bodenuntersuchung und Über die plastizität der Tone. *Internationale Mitteilungen für Bodenkunde,* V. I.

water that can be held on particle surfaces with substantial rigidity; more water and the clay-water aggregate is a liquid.

2. **Plastic Limit** (PL) = lowest moisture content at which a clay sample can be rolled into 1/8″ diameter cylinders or threads without breaking. This measures the water content slightly in excess of that which is rigidly fixed, thus allowing easy relative movement between the clay particles without destroying their mutual attraction. With less than this amount of water the aggregate behaves as a brittle solid.

3. **Plasticity Index** (PI) = $LL - PL$, is the range of moisture over which a given clay sample is plastic.

These relations are summarized in table 10.5

The Atterberg limits, particularly the liquid limit, LL, and the plasticity index, PI, are an essential part of the classification of fine-grained soils (see table 4.4) and a number of important soil properties may be represented on a plot of LL vs. PI (fig. 10.9). With increasing LL, materials plotting above the **"A" line** show increasing plasticity and those plotting below it show increasing compressibility. Changes in other properties of fine-grained soils with increasing values of LL are a decrease in dry strength and toughness at the plastic limit. Dry strength and toughness increase with an increase in the plastic index while compressibility is essentially unchanged.

Figure 10.9 Utilization of Atterberg plasticity limits

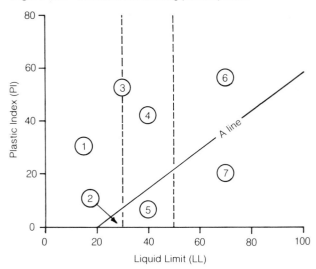

Figure 10.10 Triaxial compression testing

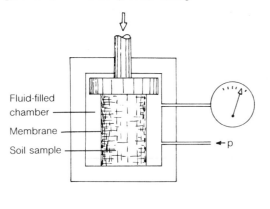

Figure 10.11 Direct shear testing. Sample-filled split box with movable upper portion. The shearing force needed to slide the upper part under different confining pressures is measured.

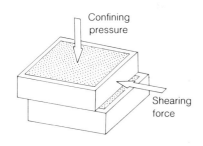

KEY

1	Cohesionless soils
2	Inorganic silts of low compressibility
3	Inorganic clays of low plasticity
4	Inorganic clays of medium plasticity
5	Inorganic and organic silts of medium compressibility
6	Inorganic clays of high plasticity
7	Inorganic silts and organic clays of high compressibility

Strength

The more common *tests of strength* applied to soil and sediment are those for unconfined compressive strength, triaxial compressive strength, and resistance to direct shear.

Unconfined compression tests measure the *cohesion* and *consistency* of soil samples. In the test a cylindrical sample is loaded to failure, which typically occurs along diagonal planes, and cohesion, *c,* is taken as one-half of the unconfined compressive strength.

Triaxial compressive strength is determined by placing a cylindrical soil sample encased in a thin membrane in a fluid-filled chamber where it can be subjected to both fluid confining pressure and axial compressive load (fig. 10.10). The load is increased until the sample shears, and the total principal stress is equal to the fluid pressure plus the principal stress difference at failure.

Direct shear testing may be done in the laboratory by placing the soil sample in a box split horizontally, applying a constant vertical load, and determining the horizontal force required for displacement (fig. 10.11). Testing in the field is accomplished by the use of such devices as penetrometers or vaned shearing rods by which the resistance to displacement of the probe is measured much as a torque wrench is employed.

Table 10.6 Soil Strength by the Standard Penetration Test

Material*	Penetration Resistance, Blows/Foot	Angle of Internal Friction, ϕ
Very dense sands	50	41°
Dense sands	30–50	36–41
Medium dense sands	10–30	30–36
Loose sands	4–10	28–30
Very loose sands	10	28
Compact, well-graded sands and gravels	30	40–45
Loose sands and gravels	10	35–40

*Strength and ϕ typically greater for mixtures
Source: U.S. Corps of Engineers

Table 10.7 Consistency of Fine-grained Soils

Approximate Compressive Strength, kg/cm²	Consistency	SPT Resistance, Blows/Foot	Description
.1			
	Very soft	< 2	extruded between fingers when squeezed
.25			
	Soft	2–4	molded by light finger pressure
.5			
	Medium	4–8	molded by strong finger pressure
1			
	Stiff	8–15	readily indented by thumb but penetrated with great effort
2			
	Very stiff	15–30	readily indented by thumbnail
4			
	Hard	> 30	indented with difficulty by thumbnail
10			

Redrawn from U.S. Navy Soil Design Mechanics Design Manual NAVFAC CM 7.1, 1982.

Compactness and Consistency

Compactness in the field may be determined by any of several standardized tests of which the Standard Penetration Test, usually abbreviated to SPT, (ASTM D1586) is the most common. In the test, the number of blows of a 140 lb hammer falling 30″ that are required to drive a 2″ OD, 1 3/8″ ID split-barrel sampler one foot are counted and the number of blows is the standard penetration resistance. Table 10.6 provides some quantitative information for sands and gravels in different states of packing. The consistency of fine-grained soils is described and measured as shown in table 10.7.

Another test, the California Bearing Ratio Test, is used to evaluate subgrade, subbase, and base materials for pavement design purposes. In the test the load required to gradually drive a 3-inch-square piston head 0.1 or 0.2″ is compared with that required to penetrate a standard compacted stone layer. This testing is similar to cone penetrometer testing in which the rate of application of pressure for each increment of penetration is measured.

FIELD DESCRIPTION OF SOILS

The *description* of soil for engineering purposes should include the identification of its constituents, its appearance and structural characteristics, and the determination of its in-place compactness or consistency.

For *coarse-grained soils,* which are defined as having more than one-half of their constituent particles below 3″ in size distinguishable by the naked eye (i.e., coarser than 200 mesh), it will ordinarily suffice to report the color, grain size, grading, structure, compactness, and such features of the grains as hardness, angularity, coatings, and cementation. *Fine-grained soils,* those that are more than half comprised of −200 mesh particles, are to be similarly described with special emphasis on their structure and, of course, only such information regarding the constituent grains as can be indirectly inferred. For example, fine-grained material that grits between the teeth contains silt-size grains.

For completeness the descriptive material gained by observations in the field is combined with laboratory determined *soil index properties* of particle size and gradation, porosity and void ratio, and unit weight information. Table 10.4 provides some typical index values for various soils and sediments.

SUMMARY

Laboratory testing of soil and sediment is an essential early step in the planning of an engineering work. Many of the tests that have been described are routinely conducted and others should be made as called for by the particular situation. The object of testing, of course, is to accurately predict the conditions that will obtain as the work progresses. It is thus necessary that the material tested be representative (i.e., be a sample in the true sense). The discussion in chapter 16 about accuracy and precision may be useful in this context.

ADDITIONAL READING

American Association of State Highway Officials. Standard recommended practise for the classification of soils and soil aggregate mixtures for highway construction purposes. *Standard specifications for highway materials and methods of sampling and testing.* Washington, D.C.

Attewell, P. B., and Farmer, I. W. 1976. *Principles of engineering geology.* London: Chapman & Hall.

Department of the Navy, Naval Facilities Engineering Command. 1982. *Soil mechanics design manual 7.1.*

Foth, H. D., and Turk, L. M. 1972. *Fundamentals of soil science.* 5th ed. New York: John Wiley & Sons.

Hough, B. K. 1969. *Basic soils engineering.* 2d ed. New York: The Ronald Press.

Kézdi, A. 1980. Handbook of soil mechanics. V. 2, *Soil testing.* New York: Elsevier Science.

Krynine, P. D., and Judd, W. R. 1957. *Principles of engineering geology and geotechnics.* New York: McGraw-Hill.

Lambe, T. W., and Whitman, R. V. 1979. *Soil mechanics.* New York: John Wiley & Sons.

Mitchell, J. K. 1976. *Fundamentals of soil behavior.* New York: John Wiley & Sons.

U.S. Waterway Experiment Station. 1953. Unified soil classification systems. Technical Memo 3–357.3 Vicksburg, Miss.

11

Mechanics of Unconsolidated Materials

INTRODUCTION

A very large proportion of the material to be dealt with in engineering practise consists of unconsolidated aggregates of varying degrees of complexity. Typical are materials containing different kinds and sizes of solid particles with a liquid, usually water, either partially or completely filling the interstices between the grains. This, for example, is a description of soil. Obviously, the mechanical properties of such aggregates are of prime engineering importance.

Most soils are dominated by mixtures of sand, clay, and water. In this chapter features of the sand–water system and the clay–water system are first considered separately and then combined for an examination of the mechanics of soils.

SAND–WATER SYSTEM

The mechanical properties of sand grains in a fluid matrix have been previously discussed in a general way in chapters 4 and 10. Some important aspects of this system, however, deserve more emphasis because sand and water mixtures are so commonly encountered in engineering works and each is an important constituent in soils.

The **void spaces** in a mass of granular material are a function of the arrangement, or **packing,** of the constituent grains. If the grains are imagined to be uniform spheres in closest packed array (on the corners of a cubo-octahedron), the percentage of void space is minimal and all other arrangements of uniform spheres have a greater volume of voids. If, however, the grains are either nonuniform in size or nonspherical, the void volume may be significantly changed (fig. 11.1). The changes in void space as a function of changes in packing may be described under the term **dilatancy.** Closest-packed arrangements are found in nature, but some degree of openness of packing is more common, especially in recently deposited material.

The *packing style* of unconfined, in-place deposits may be changed by deformation or vibration either related to earthquakes or to human–induced shocks. As the packing changes so does the void volume. If the voids are filled with incompressible water, it appears as a suddenly wetted surface on the sample if void space is decreased; if internal void space is increased the sample will have a dried surface. Examples are the appearance of water on the surface of concrete being laid as the aggregate is shaken or troweled into close packing, or the appearance of an area of dry sand around one's foot when walking on a wetted-sand beach because the pressure from the foot distorts the sand into more open packing which sucks in the water. A change in packing obviously requires a change in volume of the granular mass: Total volume = volume of particles + volume of matrix or voids (see fig. 11.1). This is easily accomplished if the mass is unconfined but an increase in volume from a close- to open-packed arrangement in a confined mass is opposed by the strength of the container. For example, a thin-walled steel cylinder filled with close-packed sand will have a compressive strength equal to the tensile strength of the container; that is, the sand acts as a solid body because of its inability to deform without expanding.

Figure 11.1 Packing and void space

Close-packed arrangements

Open-packed arrangements

This phenomenon of **rigidity** of granular materials when confined occasionally has useful engineering applications, especially in foundation work. Unconsolidated sand or gravel lenses enclosed in plastic clays are encountered in river alluvium, glacial deposits, or seacoast areas. Because of the constraint of the plastic clay envelope these lenses act as rigid members and may be satisfactory footings for structural foundations. Similarly, stable road beds, levees, or similar structures may be built in areas of muck or clay by excavating a trench that is back-filled with sand or gravel vibrated into close-packing and held in this arrangement by its plastic envelope.

Another important consequence of the dilatant principle is that unconfined granular materials have a large **angle of repose;** that is, when poured into a pile their slopes make a large angle with the horizontal. These angles are in the order of 30° for cohesionless sand or gravel as compared with angles of only a few degrees for such other cohesionless materials as muds or mucks. The theoretical limit for the angle of repose in close-packed uniform spheres is 60°, but variations in size, shape, and packing tend to make the particles mechanically unstable and the pile to fail in shear. The angle is thus reduced to about 30° (fig. 11.2). Many natural slopes, especially those mantled with soil or other granular debris, tend toward stable slope angles of about 30°. Care must be taken in engineering work not to oversteepen or remove support at the toe of such slopes without appropriate precautions against landsliding.

Table 11.1   Typical Values of Friction for Dry Granular Materials

Materials	Angles in Degrees
Clay	0*
Silt, uniform fine to medium sand	26 to 30
Well-sorted sand	30 to 34
Sand and gravel	32 to 36

*May be greater, but usually taken as 0 for conservative calculations.

Abstracted from B. K. Hough, *Basic Soils Engineering*, 2d ed., The Ronald Press Co., N.Y., 1969.

Figure 11.2   Angle of repose

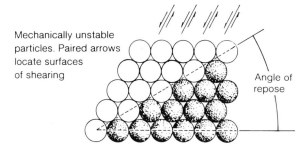

Mechanically unstable particles. Paired arrows locate surfaces of shearing

Angle of repose

Figure 11.3   Angle of internal friction

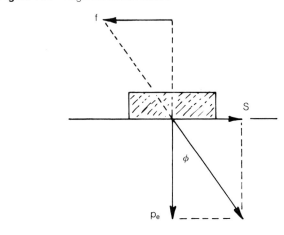

KEY

f   Coefficient of friction
S   Shearing strength
p_e Effective pressure
ϕ Angle of internal friction

The angle of repose is a limiting case of a more general mechanical property of materials called the *angle of internal friction*. The shearing strength of materials is related to the coefficient of friction and the effective pressure normal to the sliding surface: $S = fp_e$ where S is the unit shearing strength; f is the coefficient of friction, a constant principally dependent on the physical character of the sliding surface; and p_e the solid-to-solid contact pressure on the sliding surface (fig. 11.3). Since the coefficient of friction is equal to tangent ϕ at maximum resistance to sliding, the relation is usually expressed as $S = p_e \tan \phi$. Some typical values of friction angles for natural materials are given in table 11.1.

The properties of the *sand–water system* so far discussed are based on the continuous contact of the individual grains. Under conditions in which the grains are separated or in discontinuous contact a set of fluid properties of interest arises. The point at which sand grains in an aggregate are free to move may be called the *liquid limit* for this system and

the change from fluid to solid mechanical response is abrupt.

A sand grain will be suspended in water or air when the lifting force imparted to it by the surrounding fluid equals the acceleration due to gravity. It should be remembered that the submerged weight of a particle is its weight in air less the weight of the water it displaces. The weight of a quartz grain of density 2.65 g/cc in water of density 1.00 g/cc is only 1.65 times that of the water. The movement of a particle by running water in a stream may require rather high water velocity because only the vertical component of the applied force is effective. Should,

Figure 11.4 Suspensoid relations

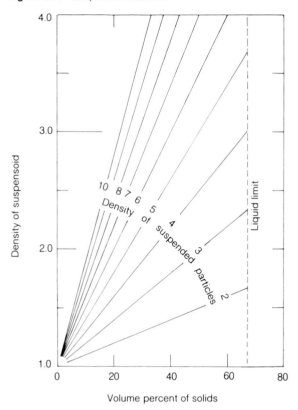

Figure 11.5 Liquefaction potential

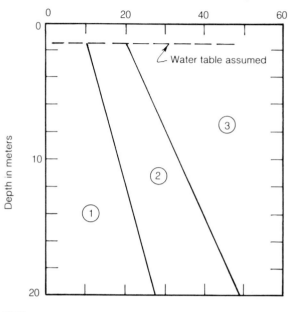

KEY

1   Liquefaction very likely
2   Liquefaction a function of soil type and earthquake
    magnitude
3   Liquefaction very unlikely

however, a mass of sand be subjected to a relatively small upward flow of water the grains will be rather easily suspended and the mass will act as a fluid. This is **quicksand.**

Quicksands or other suspensoids (solid–fluid mixtures whose grains are free to move) have an effective specific gravity that is the weighted average of the components present. Thus for a suspensoid composed of 30% quartz particles (SG = 2.65) and 70% water by volume, the specific gravity is $0.3(2.65) + 0.7(1.00) = 1.50$. Greater specific gravities may be attained by a larger proportion of solids or by use of heavier particles. Figure 11.4

shows graphically the specific gravity of suspensoids as a function of the amount and specific gravity of immersed solids.

Suspensoids find many applications in industry where heavy liquids are used to separate materials of different densities. For example, coal with a specific gravity of 1.3 can be floated on a sand–water suspensoid of proper consistency while the shaley waste (SG about 2.3) sinks.

When increased water pressure results in the loss of friction between grains in a sandy deposit the whole mass may become suspensoid. Soil *liquefaction* from this cause is an ever-present earthquake hazard (see chapter 12) but may also occur if porewater pressure builds up to the point at which it

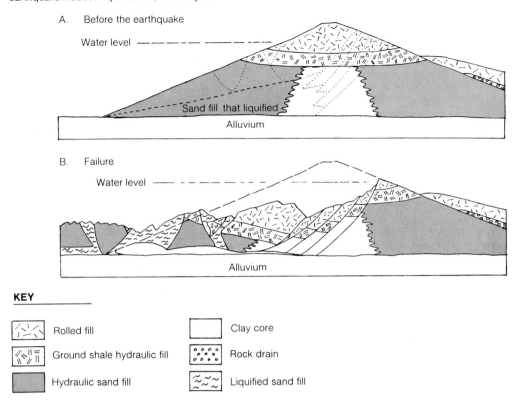

Figure 11.6 Failure of the San Fernando Dam resulting from earthquake-induced liquefaction, February 1971

A. Before the earthquake

Water level

Sand fill that liquified

Alluvium

B. Failure

Water level

Alluvium

KEY

Rolled fill

Ground shale hydraulic fill

Hydraulic sand fill

Clay core

Rock drain

Liquified sand fill

equals the overburden pressure. Should this occur the effective stress becomes zero and the sand loses its strength and behaves like a thick fluid.

Liquefaction appears to be restricted to geologic and hydrologic environments in which layers of loose to medium-dense sands, silts, and sandy soils are found over a high water table (i.e., one within 10 m of the surface). Typical are recently deposited deltaic, river channel, flood plain, or wind-borne deposits and poorly compacted fills.

Most of the documented cases of liquefaction are related to the passage of seismic waves through the material where the pulsations cause a decoupling of the grains. The liquefaction potential of a deposit may be assessed by modeling this phenomenon. Figure 11.5 shows the results of such testing using a pulsating load of about 5 kb and a period of about 0.5 s resulting in accelerations of about 0.25 g.

As examples of liquefaction the San Fernando, California, earthquake of 1971 caused the partial collapse of an earthen dam (fig. 11.6); 6 000 km² of soil liquified in the 1977 Argentina earthquake forming numerous sand boils and destroying many foundations; and the pounding of waves during Hurricane Camille in 1969 liquified the foundation of a drilling platform 25 km off the Louisiana shore.

San Fernando Dam, earthquake damage, 1971

CLAY–WATER SYSTEM

Clay minerals, in a strict sense, are hydrated, aluminum, silicate-sheet structures having a very small size, typically less than four micrometers. These particles are active participants in a sensitive clay–water system that results in a number of important engineering properties.

All clay minerals are built up of three basic parts—a *silica sheet,* an *octahedral layer,* and a *water layer*—which, in different combinations, provide for the large variety of individual species and bulk properties. The silica sheet is comprised of linked silicon-oxygen tetrahedra, all of whose apices point in one direction (see chapter 1); and limited to moderate solid solution of other ions, typically aluminum for silicon and possibly hydroxyl radicals

for oxygen can occur. The octahedral layer is made up of edge-sharing octahedra of oxygen and hydroxyl radicals surrounding trivalent aluminum, or less commonly, bivalent ions such as magnesium or ferrous iron. Two apical oxygens from the silica sheet and one coplanar hydroxyl form one face of an octahedron while the opposite face may be either the same or composed of three hydroxyls. These tetrahedral and octahedral sheets are combined as an open-faced sandwich in kaolinite and halloysite or as an octahedral layer between two facing silica sheets in montmorillonite (see fig. 1.7).

The faces of the silica sheet, the octahedral layer, and the water layer are hexagonal arrays of atoms that may be fitted together in various orientations to yield different repeat sequences along the

Figure 1   San Fernando Valley earthquake, 1971

The San Fernando Valley area of California lies within the major fault and earthquake zones of the west coast of the United States. At dawn on February 9, 1971, a major earthquake struck the San Fernando Valley area causing widespread damage to buildings, roads, and overpasses. Although five interstate road overpasses fell and were completely destroyed, the absence of heavy traffic helped to account for the relatively few fatalities. Only 64 died while 3,200 buildings were either severely damaged or destroyed. A major earthfill dam (fig. 11.6) almost failed and a catastrophic loss of life could have resulted.

The street view is typical of the type of damage inflicted. Pavements and parking lots are buckled, utility lines are broken, and buildings demolished. Some engineering design features can minimize such damage and should always be used in known earthquake areas.

Figure 11.7 Liquid limit and plasticity

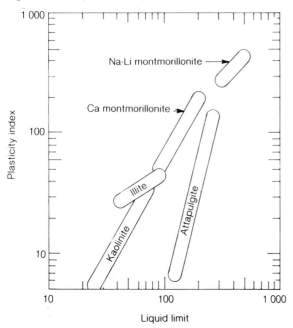

stacking axis as layers are added by crystal growth (see fig. 1.7). Regular or occasional shifts in orientation may occur or random stacking may dominate. These possible geometric variations thus lead to numerous differences in crystallographic detail between clay minerals of identical composition that, in turn, affect their mechanical properties.

The third component of all clay minerals is a structural water layer more or less contaminated with various foreign ions. There is disagreement as to the details of the arrangement of this layer, but some sort of rigidly held nonliquid layer of H_2O attached to the exposed surfaces of the sheet must be present to explain many of the properties exhibited by clays. This layer is not of fixed thickness but may be several, perhaps 5 to 10, molecules thick before its rigidity is lost. The attachment of water molecules to the basal surfaces of clay mineral particles provides both an interparticle bond and an interparticle lubricant depending on the number of water layers involved. Changes in its thickness due to wetting or drying can cause significant *swelling* or *shrinkage* in a clay mass; overthickening leads to

dispersal of the particles, overthinning to increased *rigidity* and optimum thickness to *plasticity*.

The basal surfaces of clay mineral particles carry a net negative charge, and because the water molecule is polar (has positively and negatively charged ends), the first water layer is rather firmly attached and assumes an icelike configuration. Succeeding layers show less and less organization and at 5 to 10 water layers the molecular ordering is lost. Water layers on two facing clay mineral surfaces may interact through a zone of disordered molecules whose thickness increases as the separation of particles increases.

The interparticle water serves to bind the individual particles together and provides modest resistance to deformation of the mass. The clay–water mass will, however, yield to relatively small differential forces without rupture (i.e., be plastic). The degree of plasticity is thus a function of the water content; as the amount of water is decreased, its ability to lubricate the system lessens, and eventually, at about the water content represented by icelike water molecule organization, failure will be brittle. Increasing the amount of interparticle water eventually separates the particles until there is no cohesion between them and the system is liquid. The Atterberg limits, which describe these mechanical differences, are dependent upon many factors, among them being the clay mineral species; kind, amount, and ionizability of ions adsorbed or included in the water layers; particle size; ionic substitution within the clay mineral structure; and dessication history. The range of plasticities as a function of liquid limit is shown in figure 11.7 and represents changes with both mineral species and the effect for montmorillonite of different adsorbed ions. Plastic and liquid limits and plasticity index all increase with decreasing particle size. The determination of the limits and range within which plastic properties prevail in the clay–water system is readily made by the Atterberg Limit Tests described in chapter 10.

The minute size and platy nature of most clay mineral grains means that there is a very large surface presented by a small volume of material. In addition, the particles are so small that significant

Table 11.2 Exchange Capacities, meg/100g

Mineral	Cations	Anions
Kaolinite	3–15	7–20
Halloysite	5–40	—
Montmorillonite	80–150	20–30
Illite	10–40	5–15
Vermiculite	100–150	5
Chlorite	10–40	—

Note: Measure of the number of moles of singly charged negative sites in 100 grams of exchanger determined at pH 7 as milliequivalents of sodium. For example, one equivalent of sodium expressed as Na_2O would be a combining weight of 31 (molecular weight of Na_2O = 62, and for each atom of sodium the combining weight is thus 31), 1 meq/g = 3.1%, and 1 meq/100g = .031% Na_2O.

From D. Carroll, Cation exchange capacities in clay minerals and soils. *Bulletin of the Geological Society of America*, vol. 70, 1959.

uncompensated electrical charges exist on their surfaces, and in consequence, relatively large numbers of dissolved ions in the surrounding water may be fixed on the particle surfaces by sorption. **Exchange capacities** vary markedly within and between clay and other platy mineral species; table 11.2 provides some general ranges. Sorption also occurs on nonclays, increasing with decreasing particle size, and some natural and synthetic products such as zeolites, activated charcoal, and various resins are particularly active. The nature of the exchangeable cation is suggested by the pH of a slurry; pH 9 as sodium, 7.5 as calcium, and below 7 as hydrogen.

In engineering considerations, the most important clay minerals are allophane, kaolinite, montmorillonite, and illite (see table 1.9). The term *allophane* is given to those constituents that are amorphous to X-ray diffraction.

Kaolinite minerals (kandites), Al_2 (Si_2O_5) $(OH)_4$, are generally crystallized into well-formed hexagonal plates perhaps 0.05 to 2 μm thick and 0.3 to 4 μm across. There is little or no solid solution, but geometrical shifts and crystalline disorder occur. The most important variant is halloysite, which exists in two forms—one hydrated (4 H_2O) and the other not. Hydrated halloysite tends to occur in tubular units with an average diameter of 0.07 μm and a length of several micrometers. Irreversible dehydration takes place at about 50°C and causes the tubes to collapse, split, or unroll.

Montmorillonite minerals (smectites), Al_4 (Si_4O_{10}) $(OH)_8 \cdot$ n H_2O, occur in extremely small platelets with widths 10 to 100 times their thickness. The double sheets are aggregated by weak bonds due to the presence of exchangeable ions, and water or other polar molecules can readily enter and expand the mass. The composition of montmorillonite always differs from the theoretical formula in consequence of solid solution. Typically, aluminum and possibly phosphorus substitute for silicon in the silica sheet. Aluminum in the octahedral layers may be replaced by magnesium (nontronite), ferrous iron (saponite), or rarely zinc or nickel. The hydroxyl ion is commonly partially replaced by fluorine. The structure of a smectite is always electrically unbalanced by these solid solutions, and the net charge per unit cell is usually about −0.65. Balance of this excess charge is accomplished by exchangeable cations between the layers.

Illites represent an intermediate mineral type between montmorillonitic clays and true micas. They are structurally identical to montmorillonite except that some silicon is replaced by aluminum and charge compensation is accomplished by interlayer potassium ions. Illites appear to differ from true micas in having less substitution of aluminum for silicon, smaller amounts of potassium (essential for most micas) and having a relatively small amount of interlayer water.

When clay particles are deposited from suspension in quiet water they may settle very gradually, particle by particle, or may flocculate (clump) and settle out rapidly. The difference is due to the concentration of positive ions in solution; if high these ions act to form a weak intergrain bond, and if low the repulsion of negatively charged grains is controlling. A common locus for **flocculation** is a near-shore marine environment where dispersed suspended clay particles in a fresh water stream are delivered to salty water. Even in fresh water, however, there is usually a sufficient concentration of cations to cause flocculation, and most sedimented clay deposits were probably so formed.

The particle arrangement of dispersed and flocculent clays is probably as shown in figure 11.8. The flocculent state is characterized by a weakly adhering, continuous, open framework and large-diameter void passages. This arrangement is very susceptible to mechanical distortion that will be enhanced by intergrain osmotic pressure. Shearing the flocculent framework reduces the intergrain angles and increases the osmotic pressure in the contact area. Since the pressure is inversely proportional to the particle spacing, the system moves toward a new equilibrium arrangement of equally spaced parallel platelets (fig. 11.9). This arrangement of particles is termed a *dispersed structure*. Its distinctive features are the approximate parallelism of the clay platelets and their existence as solid inclusions in a continuous liquid matrix. The strength and behavior of such a system depends largely on the characteristics of the clay mineral and upon osmotic pressure in particular.

The groundwork for an understanding of the mechanical properties of aggregates of clay particles and interstitial water was developed in the previous pages and earlier in chapter 1. Generally speaking, such systems have low strength, are impermeable, and exhibit a high degree of plastic behavior: all properties of considerable importance in engineering works.

Strength of Clay Mineral Masses

The unconfined compressive strength of a clay mass is determined by the compressive stress required to cause failure of an undisturbed sample tested under unconfined conditions. It depends upon a number of factors including the composition, size, shape, and arrangement of the clay particles; the amount of water; the presence of nonclay minerals; and particularly upon the prior stress history (geologic history) of the mass.

The shear strength of a clay aggregate is basically determined by interparticle cohesion and by the coefficient of internal friction, which defines the resistance to the movement of particles past one another. Both of these are affected by the compressive loading, so the shear strength within a clay mass is also a function of depth of burial. Shear strength, the shear stress at maximum displacement before

---

Figure 11.9 Mechanics of dispersion. The thickness of the arrows indicates the relative amount of repulsive force.

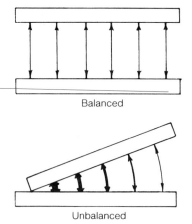

Balanced

Unbalanced

Figure 11.8 Dispersed and flocculent clay aggregates

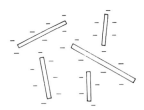 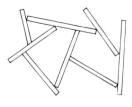

Dispersed state, repulsion by negatively charged grains

Flocculent state, attraction due to presence of cations

rupture, is usually measured under conditions of increasing load. It should be noted that the resistance to shearing of an undisturbed clay mass may be considerably greater than its strength when remolded, all other factors being equal.

Cohesion is a form of shear strength that is clearly independent of compressive loading since it exists in the absence of external load. It is often termed "no-load shear strength" and serves to distinguish clayey cohesive aggregates from sandy cohesionless masses. Obviously, the amount of cohesion is related to the proportions of sand and clay in mixed aggregates.

The *sensitivity* of a clay mass is defined as the ratio of strength of an undisturbed clay or soil mass to that of the remolded material at the same water content. A log scale is used to distinguish the varying degrees of sensitivity (fig. 11.10). Insensitive soils and clays are typically those that have been heavily overconsolidated during their geologic history, usually by having had to support a thick ice mass during the Pleistocene epoch of continental glaciation north of about latitude 40°. Sensitive clays are characterized by high water contents, frequently having a liquid limit greater than unity. **Quick clays,** like quick sands, are or may be readily transformed into suspensoids and act as fluids.

The ability of clay aggregates to regain strength when dispersed and then allowed to stand at constant moisture content is termed **thixotropy.**

Figure 11.11 illustrates this property as a function of different clay minerals. It should be noted that the strength of one clay, montmorillonite, is increased by this procedure and that kaolinite, the principal component of quick clays, has no appreciable regain of strength. Under compression the void ratio decreases as a function of increasing sensitivity (fig 11.12). Since the load increases with depth in natural deposits, the void ratio may be expected to decrease downward also; figure 11.13 shows this decrease for several different types of clay.

Figure 11.10   Terms for the sensitivity of clay aggregates

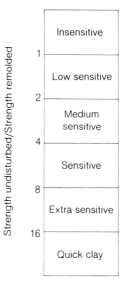

Figure 11.11   Strength recovery in clays

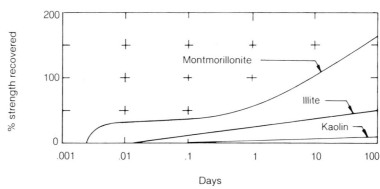

Figure 11.12 Change of void ratio of clays with loading

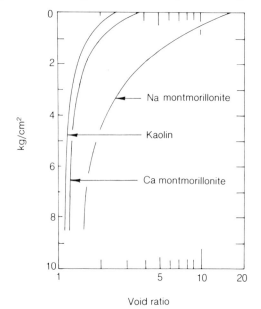

Modified after S. G. Samuels, *The Effect of Base Exchange on the Engineering Properties of Sediments,* Note c176, Building Research Station, Great Britain, 1950.

Figure 11.13 Void ratio and sensitivity in clays

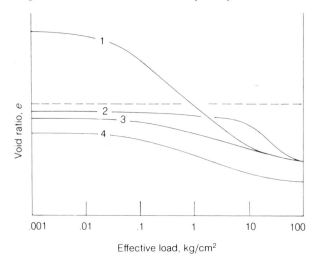

KEY

1 Undisturbed, extra sensitive
2 Undisturbed, sensitive
3 Undisturbed, low sensitive
4 Remolded

Relaxation of compressive stress in the presence of water allows for the enlargement of intergrain capillary water films, and additional water may be imbibed by clay minerals such as montmorillonite that have an expanding structure; the result is a swelling of the clay mass. For dried clays the indrawing of capillary water may cause the swelling mass to exert a considerable pressure, and *swelling* clays thus make excellent sealants when emplaced as dikes to control underground seepage or to seal rock pores in oil well work. On the other hand, the same phenomenon may disrupt foundations or underground utility lines. As examples of the scale of this swelling, the expansion of dry kaolin when wetted is usually between 5 and 20%, of illite 10 to 120%, of Ca-montmorillonite 90 to 150%, and of Na-montmorillonite as much as 1 500%.

The rate of **consolidation** of a clay or soil mass is dependent upon both the applied load and time. Consolidation rates for several clay minerals in aggregate are illustrated in figure 11.14.

SOIL MECHANICS

For engineering purposes, soil may be considered to be a complex and variable mixture of more or less water-saturated silty or sandy particulate matter and clay-sized particles, more or less admixed with organic material. Soils are mechanically thus related to both the sand–water and clay–water systems.

It is not the purpose here to discuss **soil mechanics (geotechnics)** at length since this is a recognized subdiscipline of Civil Engineering. On the other hand, it is not possible to disregard some mechanical aspects of the material that mantles most of the land surface of the earth and must be dealt with in nearly all civil engineering jobs. Some of the more important properties of soils are (1) their strength, usually measured as unconfined compressive strength but sometimes as shear strength; (2) their sensitivity, which describes changes in their strength as a function of disturbance; and (3) such features as their compressibility, permeability, erodability, susceptibility to corrosion, and ease of excavation (see chapter 10).

Figure 11.14 Consolidation rates for clays

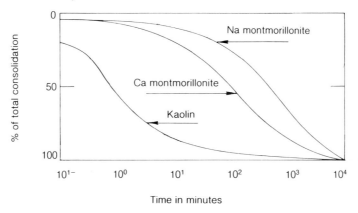

Modified after S. G. Samuels, *The Effect of Base Exchange on the Engineering Properties of Sediments,* Note c176, Building Research Station, Great Britain, 1950.

Limit Equilibrium

A cohesionless mass of clean sand or gravel is in a state of *limit equilibrium* if $S_s = \sigma_n \tan \phi$ where S_s is the shearing strength of the material, σ_n is the normal pressure, and ϕ is the angle of internal friction. The value of ϕ depends upon the type, shape, size distribution, and packing of the constituent particles with larger values being associated with angular, well-sorted, close-packed, and undisturbed materials. The value of ϕ in natural subaerial sand and gravel deposits varies between 30 and 40° or slightly more, and under water ϕ is one or two degrees less. It is common to use values of 30° ($\tan \phi = 0.577$) for subaerial and 25° ($\tan \phi = 0.466$) for submerged masses. The angle of repose, that angle the side of a pile of unconsolidated material makes with the horizontal (i.e., its natural slope) is slightly smaller than the angle of internal friction so that in calculations the angle of repose is often used to introduce a safety factor.

In cohesive (clayey) soils the shear strength is given by the Coulomb formula, which may be expressed as $S_s = c + \sigma_n \tan \phi$ and is identical to the relation for cohesionless masses but with an added term c representing unit cohesion. The effect of cohesion may be seen in figure 11.15 where shearing strength, S_s, is plotted as a function of normal stress, σ_n.

Figure 11.15 Effect of cohesion on shear strength, S_s

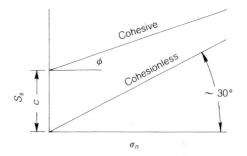

Pressures in a Loaded Earth Mass

There are vertical and horizontal *pressures* on an earth mass in equilibrium prior to its loading (or unloading), and horizontal and vertical shearing stresses are in balance. Vertical pressure may be readily found if bulk density and depth are known. If, for example, a cubic meter of soil weighs 2 tonnes the pressure on the base of a 1×1 m soil column at a depth of 5 m within the mass would be $2 \times 5 = 10$ t/m². Horizontal pressures also exist everywhere within the earth mass and generally have lower values than those of the vertical pressure at the same point. The ratio of horizontal to vertical pressure is termed the *coefficient of earth pressure*

Mechanics of Unconsolidated Materials 223

Figure 11.16 Stress in layered sediments

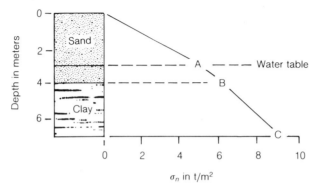

at rest and is approximately 0.4 in sands and somewhat greater in clayey sediments. The horizontal pressure in a sand mass at the point discussed above would thus be $0.4 \times 10 = 4$ t/m².

Stress, which tends to cause deformation of the solid portion of the soil, is referred to in engineering practice as *effective stress*. In calculation it is necessary to consider both the contributions of the solid portion of the soil and the contained groundwater, particularly whether the ground water acts to increase or decrease the weight of the solids. Above the water table, interstitial water, which is supported by the grains, adds to the total weight and hence to the vertical stress, so *unit wet weight* is used in vertical stress calculations. Below the water table, water is independent of the grains for support and reduces the weight of saturated solids by buoyancy; the *unit submerged weight* is thus used in calculations. Consider, for example, that the layered soil shown in figure 11.16 has the following properties:

	Sand Layer	Clay Layer
Porosity	40%	60%
Specific gravity of solids	2.65	2.72
Water content above water table	8%	—
Water content below water table	saturated	saturated

The effective vertical stress at the depth of the water table is the unit weight of the material present times the depth of 3 m:

unit weight of dry sand	$= (2.65)(1 - 0.4)$	$= 1.59$ t/m³
unit weight of water	$= (1.59)(0.08)$	$= \underline{0.13}$
	Total weight	$= 1.72$ t/m³

$\sigma_n = (1.72)(3) = 5.16$ t/m² (point A).

At the base of the sand the weight of 1 m of submerged sand must be added:

unit weight of submerged sand $= (2.65 - 1)$ $(1 - 0.4) = 0.99$ t/m³; $\sigma_n = 5.16 + 0.99$ $= 6.15$ t/m² (point B).

At the base of the clay, the added weight of 3 m of saturated clay is present:

unit weight of submerged clay $= (2.72 - 1)$ $(1 - 0.6) = 0.69$ t/m³; $\sigma_n = (6.15) + 3$ $(0.69) = 9.22$ t/m² (point C).

Structures are generally built upon earth masses initially in equilibrium and only those pressures caused by the structure are usually computed. It must always be remembered, however, that these are a pressure increment and do not give the total pressure at a point at depth.

Loading causes a soil mass to be compressed leading to a tendency of vertical downward movement, or *settlement,* of the structure. Except for a very small or very rigid structure, settlement may

Figure 11.17 Pressure P_v within an earth mass at depth and lateral distance r due to a surface load P

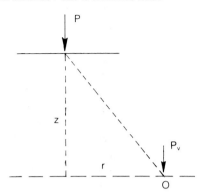

Figure 11.18 Boussinesq and Westergaard coefficients

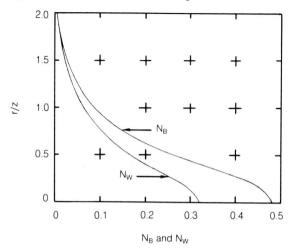

N_B and N_W

From D. W. Taylor, *Fundamentals of Soil Mechanics*, John Wiley & Sons, N.Y., 1948.

also be expected to be differential in consequence of both nonuniform structural loading and inhomogeneity of the earth mass. Large uniform settlement is generally to be preferred to a small amount of differential settling—as attested by diagonally cracked walls, uneven floors, and misfitting doors in differentially settled structures.

The foundation surface for the majority of structures is horizontal and the earth mass below, bounded at the top by the foundation surface, may be considered to be semi-infinite. In its unloaded condition every point in the earth mass is subjected to pressure on all sides that increases linearly with depth, and the magnitude of these pressures should be determined to assess possible settlement or failure. Exact measurement of the pressures below a loaded surface is very difficult to obtain, but can be estimated with reasonable accuracy using formulae for idealized bodies.

For vertical pressures the basic geometry is given in figure 11.17 where P represents the vertical pressure, O the test point at depth z and offset r below the applied load, and P_v the pressure at the test point.

If the earth mass is inherently plastic as, for example, clays or peats, P_v may be approximately determined by use of the **Boussinesq formula:**[1]

$$P_v = N_B \frac{P}{z^2}$$

1. J. Boussinesq. 1885. *Application des potentiels à l'étude de l'équilibre et du mouvement des solides élastiques.* Paris: Gauthier-Villars.

where P is the load assumed concentrated under the center of the mass of the structure, z the depth, and N_B a coefficient related to r/z given in figure 11.18.[2] As an example of the use of this relation, determine the additional vertical pressure at a depth of 50 m and an offset of 15 m under a load of 850 tonnes. From figure 11.18, $r/z = 0.3$ and $N_B = 0.38$ and thus

$$P_v = 0.38 \frac{8\ 500}{15^2} = 14.4 \text{ t/m}^2.$$

If the foundation material is sedimentary rock or sandy soil, a better approximation may be obtained with a simplified **Westergaard formula,** which assumes that the mass is prevented from being strained laterally. The formula is the same as quoted earlier but with a different coefficient, N_w (see fig. 11.18); for the previous example,

$$P_v = N_W \frac{P}{z^2} = 0.24 \frac{8\ 500}{15^2} = 9.1 \text{ t/m}^2.$$

2. Note that when $r/z > 2$ the value of N_B becomes negligible regardless of the point load and that for all points on the vertical axis of loading $r/z = 0$, $N = .477$, and the stress increment due to a given load varies inversely with the square of the depth.

CASE HISTORY 11.2 Foundation Failure, Duffus Castle

Figure 1 Foundation failure

Duffus Castle, near Elgin, Moray, Scotland, existed by 1151 as a wooden fortification surmounting an artificial hill, or motte. At some time during the next two centuries the wooden structure was replaced by a mortared stone castle. The weight of the castle tower exceeded the bearing strength of the foundation soil and has slid and sunk as a unit to its present position. Medieval castle builders were obviously not competent foundation engineers, but the quality of their mortar was superb.

The total pressure at points beneath the load is the summation of the pressures due to the weight of the superincumbent material and that induced by loading. For a simplified analysis of this pressure field consider the results of placing a standpipe containing 1 000 m³ of water (= 1 000 t) upon a thick, uniform earth mass whose bulk specific gravity is 2.0. Before loading, pressure increases at 2 t/m² per meter of depth. Overpressures, calculated by using the simplified Westergaard formula and added at each measured point in the mass, give the pressure distribution shown by the solid lines in figure 11.19. Overpressures may be seen to be concentrated under the load and to die out both laterally and vertically.

If it is desired to determine the overload pressure on a horizontal surface at some depth beneath a load the "2:1 method" provides a practical means of rough estimation. It is assumed that pressures under a loaded area whose dimensions are a and b spread in the form of a truncated pyramid whose sides slope in the ratio of 2 vertical units for 1 horizontal unit and that at a depth z the load is spread over an area of $a + z/2 + z/2 = a + z$ and $b + z/2 + z/2 = b + z$ (fig. 11.20). If the loaded area is irregular or consists of several structures the loading on overlapping areas is not doubled. As an example, let a and b in figure 11.20 be 10 and 40 m respectively and the area to be loaded at 100 kg/m² for a total load of 40 000 kg. At a depth of 8 m the load will be spread over an area of $18 \times 48 = 864$ m². The average added pressure at this depth is thus $40\,000/864 = 46.3$ kg/m². Usually, 50% is added to the average pressure so determined to obtain the maximum vertical pressure.

In design practice it is generally assumed that the pressure distribution under a uniformly loaded structure is also uniform. This assumption is apparently satisfactory because no failures traceable to its practice have been recorded. It is well, however, to remember that uniform soil pressures should exist only under uniformly loaded, perfectly flexible structures as loads on a perfectly uniform earth mass. Allowable bearing capacities of earth materials are given in table 11.3.

Figure 11.19   Pressure distribution in an earth mass due to a point load

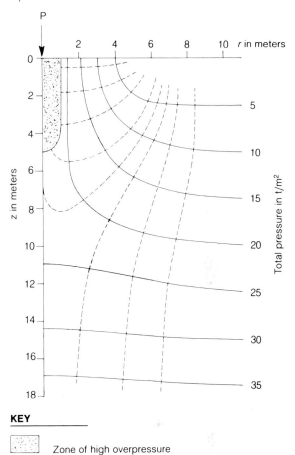

KEY

Zone of high overpressure

Figure 11.20   2:1 method

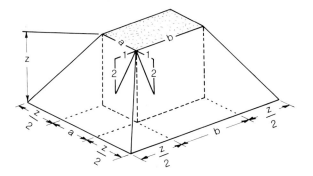

Table 11.3 Allowable Bearing Capacities

Material	Allowable Bearing Capacity, t/m²
Medium-soft clay	16
Medium-stiff clay	27
Sand, fine, loose	21.5
Sand, coarse loose; compact fine sand; loose sand-gravel mixture	32
Gravel, loose; compact coarse sand	43
Sand-gravel mixture, compact	65
Hardpan and exceptionally compacted or partially cemented gravels or sands	108
Sedimentary rocks such as hard shales, sandstones, limestones, and siltstones in sound condition	161
Foliated rocks such as schist or slate in sound condition	430
Massive bedrock such as granite, diorite, gneiss, and trap rock in sound condition	1 076

Based on Section 906.2 of the 1955 edition of the National Building. Code of the American Insurance Group. Used with permission of the American Insurance Group, Inc.

Stresses in a Loaded Earth Mass

Forces acting on an earth mass produce stresses within it whose orientation may be visualized by considering a small cube within the loaded mass. Vertically and horizontally oriented stresses, σ_v and σ_h, act on the horizontal and vertical cube faces respectively (fig. 11.21). The vertical stress, σ_v, tends to flatten the cubelet and this deformation is resisted by the horizontal stress, σ_h. The interaction of these stresses generates *tangential* and *horizontal shear stress* by the means shown in the figure.

The tendency to move an upper cubelet horizontally with respect to a lower cubelet is resisted by the shearing strength of the material. Should this be overcome, *lateral flow* on a spoon-shaped surface will occur, resulting in settlement of the structure and bulging of the ground surface beside it.

The stress field beneath a surface load may be roughly determined if the pressure field introduced earlier is known. Referring to figure 11.19, normals to the pressure contours (shown as dashed lines) give the direction of maximum pressure change and the perpendicular and horizontal components of the normal pressure and horizontal stress respectively. Thus, from figure 11.19, at a point on the 15 t/m²

contour at $r = 3.5$ and $z = 5$ m the slope perpendicular to the contour is about 57°. From figure 11.22 $\sigma_h = P \sin 57° = (15)(.8387) = 12.6$ t/m². Referring again to the pressure distribution with depth under a load shown in figure 11.19, it may be seen that lines drawn perpendicular to the pressure contours connect points of greatest pressure difference and hence delimit surfaces of maximum *differential stress*. Near the point of loading, their shapes, except for the upward central spike, approximate a nested sequence of spherical shells of equal radius (fig. 11.23a) and are potential surfaces of shear rupture. Should the shear stress induced by loading be greater than the shear strength of the foundation material, failure will occur. The direction of easiest relief is upward and lateral stress will cause the mass to flow on one or more upward-curving surfaces creating a surface bulge paralleling the load (fig. 11.23b).

Prandtl (1920) described a theory of rupture that gives a generally applicable geometry for the rupture surface within loaded uniform materials as a function of the angle of internal friction ϕ and incorporates a realistic distributed load rather than a

Figure 11.21 Development of shear stress

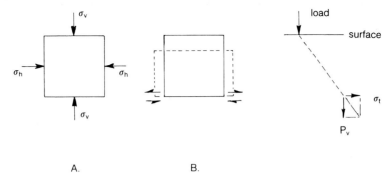

A.   B.

Figure 11.22 Pressure distribution at a buried point (refer to fig. 11.19)

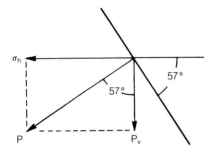

Figure 11.23 Potential and actual surfaces of shear failure

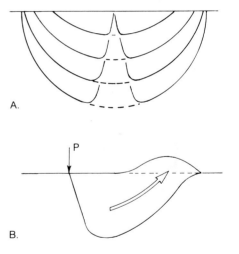

A.

B.

point load (fig. 11.24). Region I under the load can be seen to act as a wedge driving regions II and III laterally and upward.

If a slope rather than a horizontal plane is loaded, the surfaces of potential shear rupture still exist with the nested shells successively farther from the slope surface and the spike is still vertical. Because easiest relief is outward, failure will follow a surface shown in cross section in figure 11.25a and in contoured plan view in figure 11.25b. Failure of this kind is called **slump** and is discussed at greater length in chapter 13.

Figure 11.24 Prandtl construction

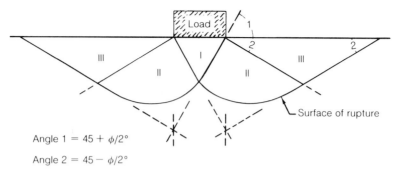

Angle 1 = 45 + $\phi/2°$

Angle 2 = 45 − $\phi/2°$

Figure 11.25 Shear failure in a slope

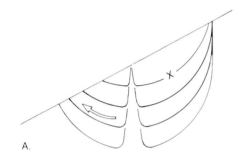

A.

B.

KEY

A. Cross section showing potential rupture surfaces
B. Contour plot of surface X

Nonvertical Loading

The design of many engineering structures requires that considerations be given to the loading of a structure or to the passive resistance of fill material in other than a vertical direction. A few examples follow.

Cohesionless fill behind a retaining wall exerts a *horizontal pressure* (active pressure or thrust) on the wall that increases linearly with depth. The value of this horizontal pressure may be estimated by

$$K_a = \frac{1 - \sin \phi}{1 + \sin \phi}$$

where K_a is the coefficient of active pressure and ϕ the angle of internal friction of the material. Using $\phi = 30°$, a commonly used value, $K_a = 0.33$. The horizontal pressure or thrust, E_a, is then $E_a = \frac{1}{2}$

$K_a \rho H^2$ where ρ is the density and H is the fill thickness. This thrust may be considered to be concentrated at the "third point" or at a point one-third of the height of the wall (fig. 11.26). The thrust is often determined in practice by considering the fill to be a fluid of equivalent density.

The *turning moment* of walls made of rigid members (sheet piling, for example) is opposed by the passive resistance of the earth mass in which they are embedded. For cohesionless material this passive resistance, K_p, is the reciprocal of K_a; thus if K_a is 0.33, K_p is 3.0 and the resistance is assumed to be applied one-third of the distance below ground level. Since the turning moment is the force (or resistance) times the lever arm the conditions for safe practice of failure can be represented as in figure 11.27 in which the turning moments on walls due to

Figure 11.26 Thrust at "third point"

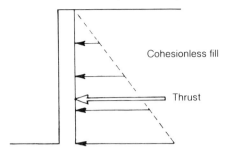

different thrust combinations are shown. Both the amount and position of the thrust vectors are important considerations.

The foregoing examples treat the soil mass as though it were a heavy liquid but in real material the vertical and horizontal pressures may be redistributed and relieved by the action of shearing stresses that arch within the soil mass.

COMPRESSIBILITY

Soil masses will ordinarily be significantly decreased in volume when loaded; the volume of the particles is unchanged but the amount of pore space is reduced. In addition, some shortening in a loaded soil mass may result from the elastic bending of bridges formed by chains of fortuitously oriented particles. It may be noted that in most engineering practice it is assumed that the volume change is entirely due to a shortening in the vertical direction. Should the load be removed from a compressed soil some increase in its volume usually occurs, which is termed *rebound*. Soil testing for *compressibility* is usually done using devices called consolidometers, which essentially model the loading and response of a particular situation.

Within the range of ordinary loading of virgin soils the change in void ratio, Δe, as a function of the change in applied pressure, Δp, has been found to be approximately linear when plotted in a semilog diagram. The curve slope may, therefore, be expressed as a constant called the **compression index,** C_c

$$C_c = \frac{\Delta e}{\Delta \log p}$$

Figure 11.27 Turning moments of retaining walls

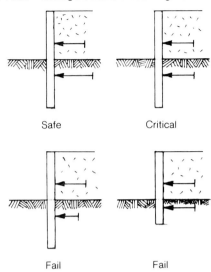

Safe   Critical

Fail   Fail

The relationship between the compression index and the *virgin void ratio, e_o,* has been established by testing for many types of material and may be expressed as $C_c = a(e_o - b)$ where e_o = no load void ratio; a = slope constant for the particular soil type dependent upon particle size, shape, and sorting; and b = x-axis intercept (constant for particular soil and a close approximation of the minimum void ratio of the soil).

For saturated soil, volume change cannot take place unless water is either expelled or drawn into the soil mass. The rate at which water can migrate through soil thus has an important influence on the rate at which compression can occur. Usually this parameter is not considered for permeable coarse-grained soils but assumes importance for fine-grained soils or coarse-grained material enclosed by an impervious envelope.

The **rate of compression** for earth masses closely approximates a parabola represented by the equation $T = \pi/4\ (U^2)$ where T = time factor (dimensionless) and U = average percent consolidation.

Figure 11.28 Consolidation of materials of different permeability

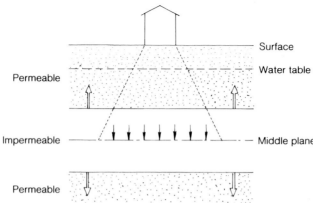

Permeable
Impermeable
Permeable

Surface
Water table
Middle plane

If the total amount of settlement to be experienced has been determined by laboratory measurement the rate is readily estimated. For example, assuming the total settlement is 50 cm, settling will be half completed (25 cm) in 0.2 of the total time of settlement: $T = \pi/4 \; (U^2) = \pi/4 \; (.5)^2 \cong 0.2$.

Consolidation

Loads applied to a water-saturated earth mass containing flexible particles such as clay minerals or organic debris and remaining thereon for a long period of time will lead to consolidation of the mass. The rate of loading may be very slow, as in normal burial by geologic processes; geologically sudden, as in an overriding ice sheet or rapid sedimentation; or quickly applied as in the erection of a structure or the dumping of fill.

Three stages of consolidation may be recognized: instantaneous *initial compression,* on loading related to elastic compression of the mass; followed by *primary compression,* as a result of the squeezing out of pore water, which leads to a decrease in porosity and hence volume; and a final stage of *secondary compression* related to internal rearrangements of particles by their deformation or recrystallization, which eventually yields a sedimentary rock. Engineering and geologic terminology differ here since the engineer considers consolidated clay, for example, as a material for which the consolidation process is complete for his

purpose whereas the geologist would require the clay to be lithified into a shale or claystone.

To examine a specific situation consider a structure built on layered foundation material consisting of soil, soft clay, and permeable sandstone (fig. 11.28). Under normal loading, the unit weight of the overburden is the preconsolidation unit load, which is computed from the ground surface to the middle plane of the clay layer if water may escape both upward and downward, or to the top of the layer if water release is upward only. Some of the parameters needed to determine the amount and rate of consolidation must be determined at the site while others must be determined in the laboratory using a consolidometer. The unit vertical pressure on the middle plane may be calculated by the 2:1 method knowing depths and bulk densities and the void ratio, e_o, of the saturated clay before construction is found by measurement, (see chapter 10).

Consolidometer measurements yield information regarding the changes in thickness, ΔH, as a function of the applied load, initial and final void ratios, e_o and e_f, and time

$$\Delta H = H \frac{e_o - e_f}{1 + e_o}$$

An example of the change in the void ratio as a function of applied pressure is shown in figure 11.29. The void ratio decreases with increased pressure (*AB*), increases with decreasing load (*BC*), and decreases with subsequent reloading (*CD*).

Figure 11.29 Change in void ratio with loading history

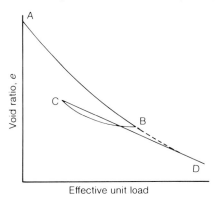

KEY

AB	Loading
BC	Unloading
CD	Reloading

SWAMPS, MUSKEGS, AND PEAT BOGS

Swamps are typically closed or poorly drained areas over relatively impermeable bedrocks or clay. They are filled with water-saturated, decaying vegetable solid matter of a density ranging from 1.2 to 1.7, having a very high porosity, and whose natural moisture content may be several hundred (exceptionally 1 000) percent by dry weight. In consequence, this material will show large shrinkage on drying and respond plastically to loading.

The bearing capacity of swamp material is so low that only light structures may be safely supported. Alternate means of support for heavier structures are by the use of piling driven through the swamp fill or by removal of the muck and its replacement with an adequate foundation material. Some less usual, but occasionally applicable, means are to float the structure or to freeze the swamp material.

SUMMARY

Soil is surely the most important geologic material to be dealt with by a majority of civil engineers. The physical properties of soil cannot, therefore, receive too much emphasis.

Most soil is a mixture of more or less water-saturated granular fragments, clay-sized material, and vegetable matter. The mechanical response of each of these materials is different, and in various proportions they give soil a wide range of physical properties. The more important properties of soil are its plasticity, strength, volume changes, and consistency, which may be measured by the Atterberg limits, angle of internal friction, compressive and shear strength, penetration resistance, and index properties (particle size, gradation, void ratio, unit weight).

ADDITIONAL READING

Cassie, W. F. 1976. *The mechanics of engineering soils.* London: Chapman & Hall.

Craig, R. F. 1983. *Soil mechanics.* 3d ed. Florence, Ky.: Van Nostrand Reinhold.

Farmer, I. 1983. *Engineering behaviour of rocks.* 2d ed. New York: Chapman & Hall.

Jumkis, A. R. 1983. *Soil mechanics.* Melbourne: Robert E. Kreiger.

Keedwell, M. J. 1984. *Rheology and soil mechanics.* New York: Elsevier.

Krynine, P. D., and Judd, W. R. 1957. *Principles of engineering geology and geotechnics.* New York: McGraw-Hill.

Lambe, T. W., and Whitman, R. V. 1979. *Soil mechanics.* New York: John Wiley & Sons.

Sanglerat, G.; Olivari, G.; and Cambon, B., eds. *Practical problems in soil mechanics and foundation engineering.* New York: Elsevier Science.

Terzaghi, K. 1952. *Soil mechanics for engineers.* London: H. M. Stationery Office.

Williams, P. J. 1982. *The surface of the earth: An introduction to geotechnical science.* New York: Longmans.

Zaruba, Q., and Mencl, V. 1976. *Engineering geology.* New York: Elsevier Science.

12

Geologic Hazards: Earthquakes and Volcanoes

INTRODUCTION

The earth is a restless planet and its surface is in continuing readjustment to varying chemical and mechanical forces. Usually these forces are tiny and the adjustments correspondingly small, but occasionally they are additive over time and the sudden adjustment of earth materials may be correspondingly very large. Some hazards such as floods, earthquakes, volcanic eruptions, excessive tides, and seismic sea waves are sporadic while others such as clayey foundation soils, failure of slopes, and erosion are omnipresent.

It is essential that civil engineers anticipate the possibility of natural catastrophe in the siting and design of their works. The damage and loss of life accompanying a geologic catastrophe is very often the direct consequence of a structural failure or dangerous location and not of the geologic phenomenon itself—it is the failed dam, falling building, or seismically triggered landslide and not the earthquake vibration that kills.

The siting of an engineering work may be as important an aspect of its safe design as its construction. In some instances, of course, the siting is

Figure 12.1 Synthetic record for an aperiodic event

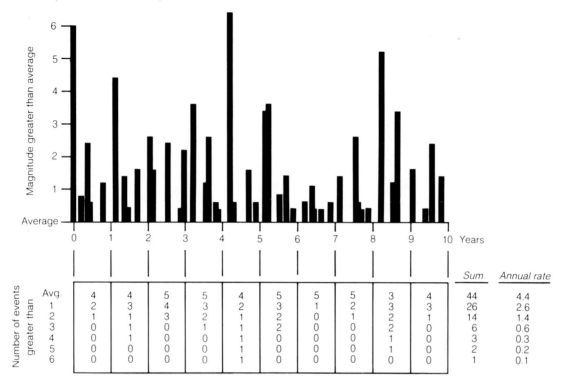

greater than	Avg.										Sum	Annual rate
Avg.	4	4	5	5	4	5	5	5	3	4	44	4.4
1	2	3	4	3	2	3	1	2	3	3	26	2.6
2	1	1	3	2	1	2	0	1	2	1	14	1.4
3	0	1	0	1	1	2	0	0	2	0	6	0.6
4	0	1	0	0	1	0	0	0	1	0	3	0.3
5	0	0	0	0	1	0	0	0	1	0	2	0.2
6	0	0	0	0	1	0	0	0	0	0	1	0.1

dictated by the purpose, but in many there will be alternative sites having different inherent risks. *Risk assessment* is thus a serious aspect of engineering design, and risks should be thoroughly investigated for all potential sites prior to development or construction.

It is probably within the competency of engineers to locate and construct any work in such a manner as to be safe in the face of almost any natural hazard. It is not, however, realistic to overdesign beyond a reasonable margin of safety because of the additional costs involved. Reasonable safety factors must be incorporated in any planning and construction, but excessive factors are both costly and unnecessary.

GEOLOGIC RISKS AND PREDICTION

Prediction, a fundamental characteristic of science, is the extrapolation from concepts or known facts to specific future conditions (i.e., when, where, and magnitude). Prediction implies accurate foreknowledge of the place, time, and magnitude of an occurrence and for many phenomena has rarely been within our capabilities. Considerable strides in prediction have been made, however, for a number of natural hazards. Risk assessment, on the other hand, leads to a statistical or probabilistic statement of the location, magnitude, and time bracket but not of the specific timing of a future event. A statement of risk would thus be that "three tides greater than 5 meters above average may be expected at a given point within the next century".

Determination of future risk is based upon an extrapolation from past experience and hence upon available records of the phenomenon at a particular place; and the better the records, the better the assessment of the risk of future occurrences.

The determination of the probability of occurrence of aperiodic natural events (e.g., the number and intensity of earthquake shocks, amount

Figure 12.2 Analysis and extrapolation of event record of figure 12.1

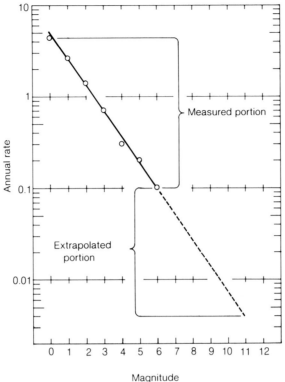

of rainfall, flood stages, or tidal levels) for the purpose of engineering design may be accomplished by the following procedure: (1) adequately document records of the local phenomenon, (2) find the probability of reoccurrences, (3) determine the magnitude and risk factor, (4) establish acceptable risks and design costs to find optimum level design.

Document records An adequate local record of the magnitude, level, or intensity of the phenomenon of interest over a period of years should be significantly longer than the planned lifetime of the engineering work. Figure 12.1 presents a synthetic record for a period of 10 years as a series of bars indicating the timing and magnitude of events exceeding the average value for the documented period. A numerical summary of the record is provided below the bar diagram. This numerical data is then plotted on semilog paper (fig. 12.2). The best-fit

straight line joining the plotted points is extrapolated to predict the annual rate and magnitude of future events. In the example, an event of magnitude 5 can be anticipated at an annual rate of 0.2 per year or once every 5 years, whereas an event of magnitude 10 will occur only once in about 150 years (recurrence interval = 1/annual rate). If the lifetime of a structure designed to withstand events whose magnitude will probably be encountered is taken as 50 years, then 0.02 events per year or a total of one event of magnitude 8.4 may be anticipated. A design to handle events up to 9.4 should provide an adequate safety margin.

Figures 12.3 and 12.4 are plots constructed from local flood and rainfall records. Note that in figure 12.4 the ordinate has been folded at the average value (annual rate = 1.0) to facilitate prediction for below-normal precipitation.

Figure 12.3 Record of flooding at Louisville, Kentucky

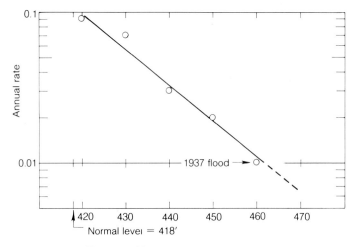

Elevation of flood crest at Louisville, Kentucky

Figure 12.4 Distribution of above and below average rainfall, Bowling Green, Kentucky

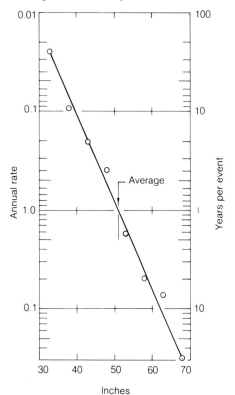

Probability of reoccurrences   It is often desirable to know the probability that some number of occurrences can be expected in order that risks may be assessed. If N is taken as the average number of occurrences per year (annual rate) and T the length of the observation period in years, then $NT = m$, the average number of occurrences during the period T. The probability, P, that a number of x occurrences can be expected is given by Poisson's law:

$$P(x) = \frac{m^x}{x!} \, e^{-m}$$

where e is the base of the natural logarithm. For $m = 1$ (one x occurrence during the time period considered) the probability is 37% and the corresponding probability that this magnitude will be exceeded is thus 63%.

Risk factors   Since the problem of assessing risks involves interdependence of the magnitude of the event, the level of associated risk, and the duration of exposure to risk, diagrams such as figure 12.5, which was developed from the data of figure 12.2, may be usefully employed. A selected magnitude may be interpreted in terms of its associated risk factor for different durations (e.g., a magnitude 4 event has approximately a 40% probability of occurring in 1 year and 80% in 5 years). Similarly,

Figure 12.5 Period and level of risk

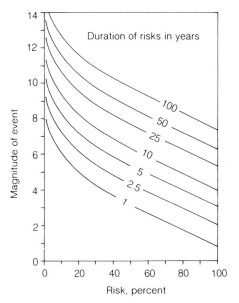

Figure 12.6 Optimization of cost and risk

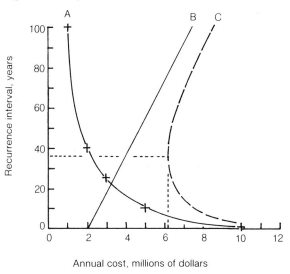

KEY

A. Annualized cost of damage
B. Cost of protective works
C. Summation of A and B; the inflection point identifies the optimum situation.

there is a fifty-fifty chance that there will be an event of magnitude 5 in 5 years and one of magnitude 9 in 50 years. If a basic frequency of say, $N = 0.5 \times 10^{-2}$, or if 50 years per event of given magnitude is chosen for design purposes, then the risk factor for various lifetimes is readily obtained. The present example indicates that the chances are 1 in 10 (10%) of an event of magnitude 11.5 occurring in 50 years, 1 in 4 of a magnitude 10 event and a certainty of an event in excess of magnitude 7.

Acceptable costs Decisions regarding appropriate design level versus acceptable risks must include economic factors whose treatment is beyond the scope of this text. It is apparent, however, that costs will rise as safety margins are increased and that some optimum level design must exist beyond which increases are no longer justified because of excessive costs. If costs of land acquisition, materials, construction, and maintenance for successively safer structures are known and if costs of estimated damage for events of different magnitude can be established, an optimum design level can be determined. Construction costs will follow some curve reflecting scale, time, and materials (taken to

be linear in fig. 12.6). The cost of expected damage is obviously higher for events of larger magnitude, but their longer recurrence-intervals generate lower probable annual costs of damage than the annual costs for more common events of lesser magnitude. Adding the two curves for costs of construction and for expected damage gives the total cost curve whose inflection point identifies the best economic balance of preventative costs and risk.

Assume, for example, that the land acquisition cost is $2 million and anticipated annual costs of damage for various recurrence intervals (curve A, fig. 12.6) are as follows:

Recurrence Interval	Total Damage Cost	Annual Cost
100 years	$100 million	$1 million
40	80	2
25	75	3
10	50	5
1	10	10

Curve B (fig. 12.6), which appears as a straight line for illustration purposes, shows the anticipated costs of land acquisition, construction, and maintenance for works adequate to protect against events of successively larger recurrence interval. The optimum cost–benefit choice will be at the inflection point of curve *C,* which is the sum of the two anticipated cost curves.

EARTHQUAKES AND SEISMICITY

Seismic waves are generated whenever energy is suddenly released within an elastic medium. These shock waves travel outward at the *speed of sound* in the material with the energy being attenuated as it is spread over an increasing area. Rocks are elastic bodies and seismic waves may be generated in them by explosion, meteor impact, volcanic eruption, or clastic (brittle) failure; the result in the latter instance is an **earthquake.** In an isotropic medium the waves that characterize an earthquake travel outward from their point of generation as spherical shells, but physical discontinuities and differing elastic properties of rock masses may be expected to modify the shape of the wave front.

Seismic waves may be propagated in an elastic medium both as **body waves** and as waves along its surface. Body waves are comprised of **P waves** (dilational, compressional, primary) and **S waves** (transverse, secondary, shear). **Surface waves** are either a special kind of S wave or a wave with a combined S-P motion.

P waves are transmitted as a series of compressions and rarefactions of the medium in a manner analogous to the propagation of sound waves in air. Their velocity is given by the relation

$$V_p = \sqrt{\frac{k + 4/3\,\mu}{\rho}}$$

where k is the compressibility, μ the modulus of rigidity (shear modulus) and ρ the density of the medium.

S waves travel as vibrations normal to the direction of wave propagation in a manner analogous to that of light waves. Their **velocity** is

$$V_s = \sqrt{\frac{\mu}{\rho}}.$$

P waves travel at a higher velocity than S waves so the time difference in the arrival of these two kinds of waves at an observation point is a measure of the distance they have traveled from the earthquake origin or focus. Intersecting circles, whose radii are the distance from a surface point above the **focus** drawn from three or more observation points, will locate the **epicenter,** or surface point, directly above the point of energy release.

Some idea of the velocity of P and S waves in different geologic materials is given in table 12.1 and figure 12.7. It may be noted that measurement of V_P and V_s allows the determination of Poisson's ratio, δ, and thus the determination of this important physical property for otherwise inaccessible rocks:

$$\frac{V_p}{V_s} = \sqrt{\frac{k}{\mu} + \frac{4}{3}} = \sqrt{\frac{1 - \delta}{\frac{1}{2} - \delta}}$$

$$\delta = \frac{(V_p/V_s)^2 - 2}{2(V_p/V_s)^2 - 2}.$$

Body waves generated by earthquakes are of interest to the seismologist who utilizes them to interpret the internal makeup of the earth and of concern to the engineer when they affect structures in or on bed rock. Probably of greatest engineering interest, however, are the complex surface waves that develop at and travel along the earth's surface. Surface waves travel along bounding surfaces between two physically different media, typically rock and air or water. **Love waves** (fig. 12.8) are essentially polarized S waves vibrating transverse to their direction of propagation but restricted to vibrating parallel to the surface. **Rayleigh waves** (fig. 12.8) vibrate normal to the surface with each particle following an elliptical orbit in retrograde motion reminiscent of the particle motion in water waves. In fact, Rayleigh and water waves are geometrically the same and differ in that the restoring force for a Rayleigh wave is elasticity whereas it is gravity for a water wave.

The effects of an earthquake at points distant from the epicenter are qualitatively described by the use of the **Modified Mercalli scale** (table 12.2) in which the seismic **intensity** is judged by the sensations and damage that are produced. This scale has been particularly useful in working out the historical seismicity of areas since it may be applied to observers' reports in contemporaneous journals and newspapers; useful insight as to seismic risk is thus provided.

Table 12.1 Representative Seismic Wave Velocities in m/s

	P waves	S waves	Rayleigh waves
Sands, shales, marls	850–3 000		200–550
Sandstone, shale	1 100–4 200		
Limestone, gypsum,			
anhydrite,	1 100–5 100	2 800–3 300	2 200
chalk,	3 000–4 200		
salt	4 500–7 700		
Igneous rocks	3 600–5 400	2 400–3 300	2 200–2 800
Metamorphic rocks	3 000–5 000		

Figure 12.7 P wave velocity of various materials in m/s

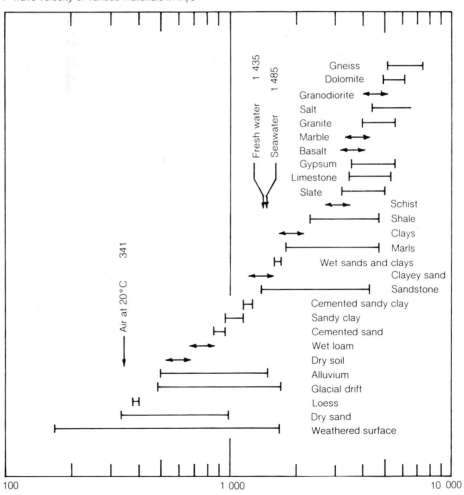

Source: Plotted from tabulated data in Heiland, C.A. 1946.
Geophysical Exploration. Prentice-Hall, N.J.

Figure 12.8   Surface waves

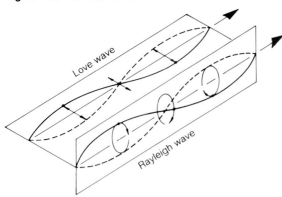

Love wave

Rayleigh wave

Modern earthquake reports use the quantitative **Richter scale.**

Earthquake magnitude [is] a measure of the strength of an earthquake, or the strain energy released by it, as determined by seismographic observations. C. F. Richter first defined *local magnitude* as the logarithm, to the base 10, of the amplitude in microns of the largest trace deflection that would be observed on a standard torsion seismograph at a distance of 100 km from the epicenter.[1]

1. Robert L. Bates and Julia A. Jackson, eds. *Glossary of geology,* 2d ed. (Falls Church, Va.: American Geological Institute, 1980).

Table 12.2   The Modified Mercalli Scale of Seismic Intensity, Abridged

I. Not felt except by a very few under especially favorable circumstances.

II. Felt only by a few persons at rest, especially on upper floors of buildings. Delicately suspended objects may swing.

III. Felt quite noticeably indoors, especially on upper floors of buildings, but many people do not recognize it as an earthquake. Standing motorcars may rock slightly. Vibration like passing truck.

IV. During the day felt indoors by many, outdoors by few. At night some awakened. Dishes, windows, and doors disturbed; walls make creaking sound. Sensation like heavy truck striking building. Standing motorcars rocked noticeably.

V. Felt by nearly everyone; many awakened. Some dishes, windows, etc., broken; a few instances of cracked plaster; unstable objects overturned. Disturbances of tree, poles, and other tall objects sometimes noticed. Pendulum clocks may stop.

VI. Felt by all; many frightened and run outdoors. Some heavy furniture moved; a few instances of fallen plaster or damaged chimneys. Damage slight.

VII. Everybody runs outdoors. Damage negligible in buildings of good design and construction; slight to moderate in well-built ordinary structures; considerable in poorly built or badly designed structures. Some chimneys broken. Noticed by persons driving motorcars.

VIII. Damage slight in specially designed structures; considerable in ordinary substantial buildings, with partial collapse; great in poorly built structures. Panel walls thrown out of frame structures. Fall of chimneys, factory stacks, columns, monuments, walls. Heavy furniture overturned. Sand and mud ejected in small amounts. Changes in well water. Persons driving motorcars disturbed.

IX. Damage considerable in specially designed structures; well-designed frame structures thrown out of plumb; great in substantial buildings, with partial collapse. Buildings shifted off foundations. Ground cracked conspicuously. Underground pipes broken.

X. Some well-built wooden structures destroyed; most masonry and frame structures destroyed with foundations; ground badly cracked. Rails bent. Landslides considerable from river banks and steep slopes. Shifted sand and mud. Water splashed (slopped) over banks.

XI. Few, if any (masonry), structures remain standing. Bridges destroyed. Broad fissures in ground. Underground pipelines completely out of service. Earth slumps and land slips in soft ground. Rails bent greatly.

XII. Damage total. Waves seen on ground surfaces. Lines of sight and level distorted. Objects thrown upward into the air.

U.S. Coast and Geodetic Survey

Figure 12.9 Epicentral seismic energy, magnitude, and intensity

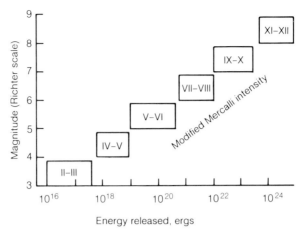

Drawn from data in C. F. Richter, *Elementary Seismology.* © 1958 W. H. Freeman and Co., San Francisco. Used with permission of W. H. Freeman and Co.

In practice, seismographic data is corrected both to the "standard instrument" and for the actual distance of the observation from the earthquake epicenter. Since magnitude using the Richter scale is expressed as exponents to the base 10: a magnitude of 6 is 10 times greater than magnitude 5; 100 times greater than magnitude 4; etc. The approximate relations between energy release, magnitude, and intensity are given in figure 12.9.

To design earthquake-resistant structures or to assess the effects of earthquake-induced accelerations on slopes requires that the engineer evaluate the degree of strong ground motion associated with earthquakes of different intensities. The maximum horizontal acceleration is usually employed, and although precise results are difficult to obtain, results of the kind shown in figure 12.10 give the orders of magnitude to be expected.

Earthquakes and Seismic Risk

Earthquakes are natural phenomena usually related to the sudden release of stress within the earth with the resultant strain seen as a relative motion of two portions of the earth's crust separated by a surface of shearing rupture (a fault). The energy released, which may be very large, propagates outward from the point of failure as complex seismic waves within

Figure 12.10 Strong ground motion as a function of earthquake intensity

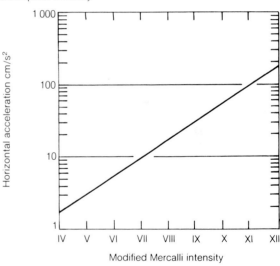

Figure 12.11 Principal areas of high seismic risk in relation
to tectonic plate edges

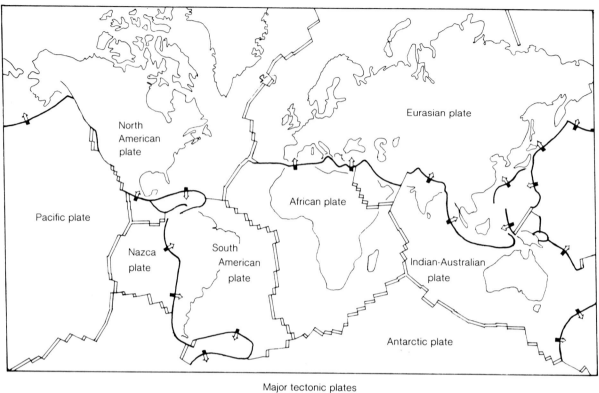

North
American
plate

Eurasian plate

Pacific plate

African plate

Nazca
plate

South
American
plate

Indian-Australian
plate

Antarctic plate

Major tectonic plates

KEY

Spreading center

Subduction zone

the rock and along the surface that shake and twist the rocks as they pass. A conceptual model of an earthquake can be simply made: slightly overlap a fingertip of each hand, palms facing in opposite directions, and push palmward. The buildup of stress is felt and then failure occurs with an audible snap, representing seismic wave production, as the fingers separate. Finally, the hands are separated in an unstressed condition having moved parallel to an imaginary fault surface.

The earth's surface is crisscrossed with faults marking loci at which crustal stresses were relieved. In the majority of instances these faults are fossilized evidence of past events and are not seismically active today. They are of concern to the engineer because of associated broken ground and subsurface water flow, or to mining engineers as mineralized zones, but not as potentially destructive earthquake locations. The seismically active areas of the world are moderately to very well known, and active research is proceeding on the cause, prediction, and even control of earthquakes.

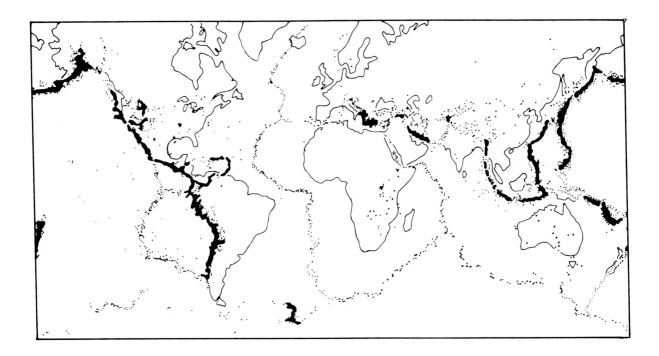

A fundamental cause of earthquakes on the world scale is the relative motion of huge plates of lithospheric material over the underlying asthenosphere (see chapter 3). At their boundaries, these plates may separate; collide head-on; plunge beneath, over-ride, or slide past one another, and resistance to these relative motions provides the stresses that cause earthquakes. The phenomena of generation, motion, and eventual dissipation of these crustal plates is called **plate tectonics.**

The plates, 70 to 100 km thick and moving at 2 to 5 cm/year, are probably moved along their courses by the drag of slowly convecting material in the earth's mantle. New material, rising as magma along *extensional zones* more or less centered in all of the world's oceans, is plastered on the trailing edges of the plates, and the leading edges are shortened by collisional crumpling or consumed by melting after sinking or being pushed to depth (subduction). Figure 12.11 shows in a general way the

plan of these plates together with the distribution of earthquakes. It is immediately apparent that plate edges are principal areas of high *seismic risk* in the world.

The point of energy release, or focus, of an earthquake may be at any depth so long as brittle deformation can occur, and the range of depths within the earth is from a few to about 700 km. *Shallow earthquakes* caused by plate motions are usually related to extensional zones (spreading centers) or to places where plates are sliding past one another and have a horizontal stress component. Spreading is generally symmetrical with respect to the spreading axis or rift, and if the orientation of the spreading axis changes, transform faults will separate the segments (figs. 12.11 and 12.12). Note that the relative motion of such transform faults (as shown by the half arrows in fig. 12.12) is opposite to that of the spreading motion.

Rock cracks after earthquake, California

Figure 12.12 Plan view of spreading axis displaced by transform faults

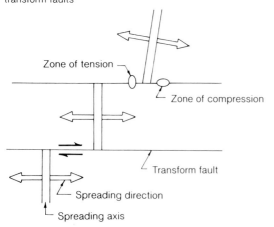

Horizontal stress components also appear when the leading edges of two plates are moving tangentially. This is the case in California where a right lateral stress (fig. 12.13) is being continually built up along the San Andreas and other faults and is sporadically released, making this area one of high seismic risk.[2]

2. In the case of right lateral motion (also called dextral motion), to an observer on one fault block, the other block appears to move to the right; in left lateral motion, or sinistral motion, the block appears to move to the left.

Figure 12.13 San Andreas fault

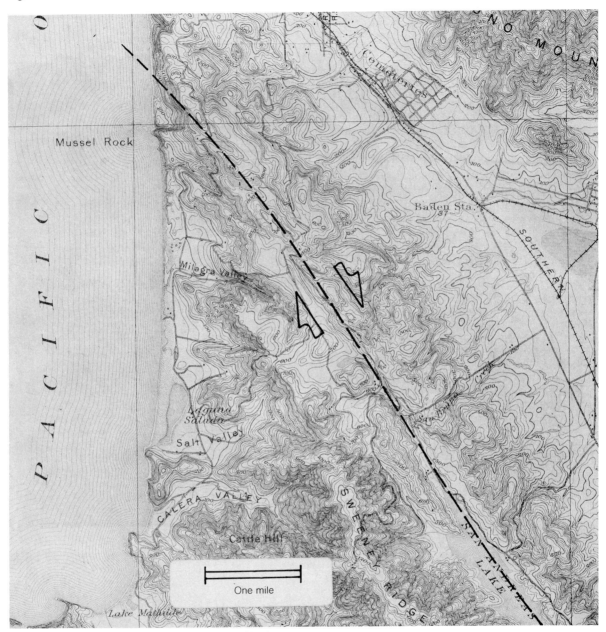

Parts of the San Francisco and San Mateo (Calif.) 1:62500
Quadrangles, U.S. Geological Survey.

Arrows mark the course of the San Andreas fault, looking south. Note placement of housing developments.

Deep focus earthquakes arise when leading edges of plates interact head-on and one is forced to depth. Resistance to the penetration of the subducting plate into the mantle is the cause of stress buildup.

Although most of the seismically active zones of the world are the result of interactions which take place at plate edges, a number of other active areas within the plates are known. Some—such as the Salt Lake City, Utah, area—are related rather directly to movements on known faults, but others cannot be pinpointed in this way either because the fault locations are not known in sufficient detail (as is the case for the bedrock faults covered by relatively unconsolidated materials in the New Madrid region of Missouri) or because the faults are so numerous that

Figure 12.14 Annual incidence and intensity of earthquakes

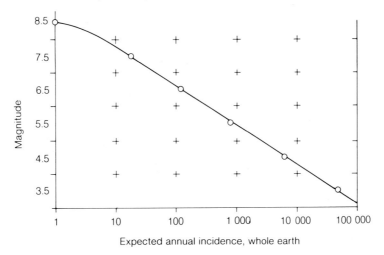

Expected annual incidence, whole earth

Drawn from data in B. Gutenberg and C. F. Richter, *Seismology of the Earth and Related Phenomena*, Princeton University Press, 1954.

the particular active ones have not been identified (as is the situation in eastern Massachusetts).

Historical reports of earlier earthquakes combined with continuous monitoring of seismic activity in more recent times have identified those areas of the world that are subject to earthquake activity and those areas that are relatively free of significant seismicity. Figure 12.14 is an estimate of the expected annual incidence of earthquakes of given magnitude over the whole earth, and figure 12.15 is a generalized summary of this kind of information for the conterminous United States presented as a map of earthquake hazard assessment. It should be noted that the quality of the information used to generate the map is quite uneven. The seismicity of California and Utah is well documented from the results of modern instrumentation, but the high risks assigned to some other areas are based on single, large, historical earthquakes (Cape Ann, Mass., 1755; New Madrid, Mo., 1811–12; Charleston, S.C., 1886) and only rather recently have serious attempts been made to document better the potential risks of these areas.

A particularly difficult aspect of seismic risk evaluation is the rate at which stress builds up and is released. If it is released more or less uniformly as it builds up, the result is a situation of high seismicity but low risk. Long-term stress accumulation unaccompanied by much if any seismic activity, however, leads to a condition of low seismicity but high risk.

It should be noted that the primary effect of an earthquake *per se* is generally restricted to more or less violent ground motion that is seldom, if ever, the direct cause of loss of life. Rather, high earthquake-related mortality results from *secondary effects* such as structural failure of buildings, disease in consequence of ruptured waterlines and sewers, fire from broken electric and gas mains, landslides, dam failures, or seismic sea waves. It is at least partly the responsibility of the engineer to site, design, and construct works that will eliminate, or at least reduce, the risk of such secondary effects.

Failure of structure foundations occurs when the passage of ground waves causes lateral and vertical accelerations or alternately expands and contracts unconsolidated fill (forcing it into and out of a suspenoid condition) or liquifies a clay mass causing it to flow. Obviously, bedrock foundations are to be preferred; landfills or channel fills along streams and swamps should be avoided if possible. Failure of structures occurs when they are unable to

Figure 12.15 Earthquake-shaking hazards in the United States

Expectable levels of earthquake-shaking hazards. Levels of ground shaking for different regions are shown by contour lines that express in percentages of the force of gravity the maximum amount of shaking likely to occur at least once in 50 years.

withstand the horizontal or the vertical accelerations imparted by passing earthquake waves. "Tied" structures represented by wood frame, steel frame, or reinforced concrete construction are superior to stone or brick construction on this count.

The location of structures with respect to geologic and topographic features is probably the most important single aspect of seismic risk amelioration. Certainly, known active faults should be avoided as should sites where flooding or landsliding could occur. The Mexico City earthquake in September 1985 had its focus about 500 kilometers to the west southwest. The extensive damage in the city was related to the amplification of surface waves in the soft lake sediment on which the city is built. Cities closer to the epicenter but founded on rock, for example Acapulco, suffered much less damage.

The actual risk from earthquake damage to be assigned to an engineering work will be related not only to regional seismicity but also to the details of its local distribution. Such engineering geology factors as the nature and structure of the bed rock,

whether foundations are on surficial materials and, if so, their nature, slope, and degree of consolidation and saturation should also be determined. Local information regarding earthquake risk at a particular site should always be sought and may usually be obtained from state and federal geologic surveys, the Nuclear Regulatory Commission, or engineering geologists and geophysicists in private practice or at universities.

ARTIFICIALLY INDUCED SEISMIC WAVES

Like fire, which is certainly a hazard but also a servant of mankind, seismic waves may be controlled and may yield useful insights of direct use in engineering. The more usual engineering applications are the determination of subsurface conditions by *seismic exploration* methods, the determination of the **rippability** of materials, and the prediction of *damage* as a result of blasting.

CASE HISTORY 12.1 *Anchorage, Alaska*

Figure 1 Anchorage school, Alaska

The buildings in the photograph were an Anchorage elementary school whose clayey foundation experienced liquefaction during the Alaskan earthquake of March 27, 1964. It should be noted that these framed wooden buildings, although destroyed, are for the most part still integral units. Loss of life is minimized by such construction.

Figure 12.16   Reflected and refracted rays

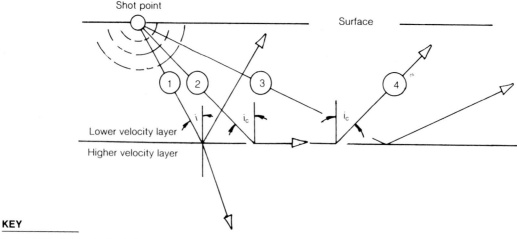

KEY

1	Reflected and refracted ray
2,4	Refracted ray
3	Reflected ray
i	Angle of incidence
i_c	Critical angle

Seismic Exploration

Artificial shock waves may be induced in the earth by explosive charges, dropping of weights, or hammer blows, and their components sensed by geophones. Energy from the shot point radiates outward and downward as seismic body waves until the waves encounter a boundary with material of different properties. Here they are *reflected* or *refracted* according to their angle of incidence and the velocity differences of the media (in direct analogy to the reflection and refraction of light at boundaries between materials of different index). For example (fig. 12.16), at small angles of incidence a ray may be divided into reflected and refracted components (ray *1*) and at large angles be totally reflected (ray *3*). At some intermediate *critical angle* the energy is propagated along the interface between the lower velocity material above and higher velocity material beneath (ray *2*) and reradiates energy upward (ray *4*) into the lower velocity medium at an angle equal to the critical angle of incidence, i_c. Since the rate of wave propagation depends upon the elastic constants of the material and is faster in more rigid bodies, energy propagation following a critical refraction path may thus return to a given point on

the surface before that of a wave following a reflection path.

Rays from a given shot point simultaneously follow these and other paths as shown in figure 12.17a. The seismic reflection relations, say for the arrival of a wave at point X'' in the figure, may be described as

$$t_1 = \frac{1}{V_1} \sqrt{x^2 + 4d^2}$$

where t_1 = reflected wave travel time; V_1 = velocity in upper layer ($< V_2$); x = horizontal distance between shot point and receiving station at X''; and d = thickness of upper layer. The travel time, t_1, and the distance, x, are measured and if either V_1 or d are independently known, the equation may be solved. For example, if the seismic velocity in an upper medium, say a soil, is known then the travel time of a reflected seismic wave may obviously be interpreted in terms of the depth to bed rock and a series of shot locations can be used to determine the bedrock contours. Caution, however, is necessary in dealing with irregular subsurface configurations since reflections may give misleading indications. In

Figure 12.17 Travel paths and times of seismic waves
$(V_2 = 2V_1)$

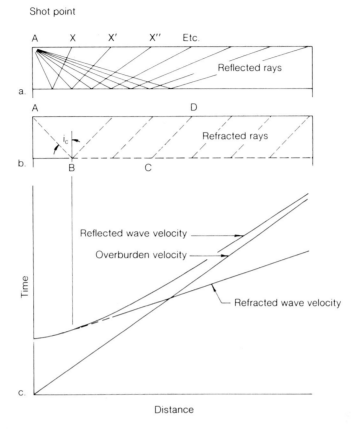

figure 12.18, the true reflection path from a dipping surface (solid lines) and the apparent reflection from a horizontal surface (dashed lines) are shown. Fortunately, they may be distinguished because of their different travel times.

The geometry of critical angle refraction is shown by the dashed lines of figure 12.17b. The travel path of a ray from the shot point to a geophone, say at D, is $ABCD$ and the travel time is

$$t_2 = 2\frac{AB}{V_1} + \frac{BC}{V_2}$$

$$= 2\frac{AB}{V_1} + \frac{x - 2d \sin i_c}{V_2}$$

$$= \frac{x}{V_2} + \frac{2d}{V_1} \cos i_c$$

where t_2 = refracted wave travel time, V_1 and V_2 = seismic velocity in the upper and lower layers respectively $(V_1 < V_2)$, d = upper layer thickness, and i_c = critical angle of incidence. Given the measured values of t_2 and x together with values for d and V_1, possibly from reflection work, the seismic velocity of the lower horizon, V_2, may be ascertained.

The *arrival times* of seismic waves following different paths to a series of observing stations have different but interrelated characteristics and travel-time curves such as shown in figure 12.17c (derived from figs. 12.17a and 12.17b) may be usefully employed. The arrival times for reflected waves lie on a hyperbola, nearly horizontal for points close to the shot point (vertical shooting) and becoming asymptotic to a straight line representing the overburden

Figure 12.18 Effect of dipping reflection horizon

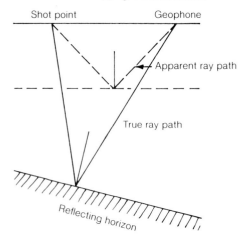

Figure 12.19 Rippability

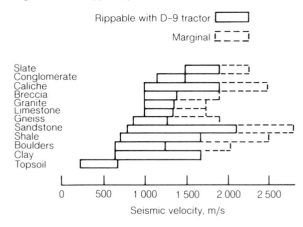

velocity at greater distances. Refracted wave arrival times, in contrast, lie along a straight line of lesser slope assuming horizontal layering and $V_1 < V_2$. This curve is tangent to the reflection hyperbola at that distance from the shot point for which the travel time in the lower layer is zero (point B in fig. 12.17b). Beyond this critical point the refracted wave will always be the first arrival at a surface geophone.

Rippability

Whether a material may be dug or ripped or must be blasted is a function of its cohesion and is related directly to the speed at which seismic waves are propagated through it. In a general way, longitudinal velocities below about 2 000 m/s for most materials indicate that they can be broken up by a ripper mounted on a D-9 tractor (fig. 12.19).

Blasting Damage

The effects of blasting on adjacent structures is a serious concern for engineers and adequate preventative measures must be taken in planning and performing work that involves the use of explosives. The accepted criterion for evaluating the potential for structural damage from blast vibration is the *maximum longitudinal velocity* of a particle in the direction of the blast-induced wave (fig. 12.20). The critical level depends upon the nature of the explosive, the material being blasted, the rock between

the blasting point and the structure, and the potentially affected structure itself, particularly with respect to the frequency of ground and structure motions.

A preblasting survey should always be conducted and should include a study of the geologic conditions at the site, the examination with photographs of adjacent structures, and the establishment of vertical and horizontal survey points. Detailed records should be kept of the size of charges, the blast point locations, the delays, and the results of vibration monitoring. It is often useful to perform some preliminary tests with undersized charges.

VOLCANIC ACTIVITY

Volcanoes are geologic safety valves that release excess pressure in subsurface magma chambers and mark the vent of a pipelike, or rarely tabular, conduit through which material is expelled from depth. Internal pressure in magma chambers builds up as solids form by crystallization and confine the residual liquid to a smaller and smaller volume. This internal pressure may exceed the confining pressure of the country rocks, and their failure allows magma to move upward to the surface. Alternatively the magma mass may work its way upward, say from a depth of 10 to 5 kilometers, thus migrating from a region of higher pressure to one of lower pressure

Figure 12.20 Effects of blasting vibration

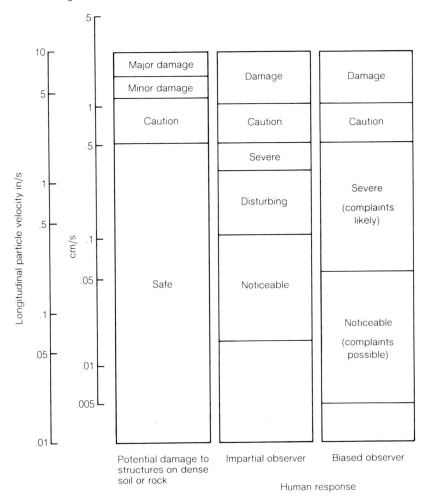

Drawn from data in U.S. Navy. *Foundations and Earth Structures*, NAVFAC DM–7.2, 1982.

that allows the separation of dissolved gases, especially water vapor, from the silicate liquid.

The nature of volcanic activity as seen at the surface depends upon the nature of the failure that allows material to escape and critically upon the kind of material that is expelled. *Higher silica content* of magma leads to *increased viscosity,* retention of volatiles, and consequently, to greater pressure buildup. On a few rare occasions the roof over a magma chamber has had sufficient strength to resist rupture until the internal pressures were very high and the consequent failure was explosive. This was the case for Krakatau, a volcano rising 820 m above the sea in the Sunda Strait between Sumatra and Java, which in two days in 1883 was explosively transformed into a caldera over 6 km in diameter and 300 m deep even when backfilled by debris. The eruption of Krakatau was not the largest in historic times, as may be seen from figure 12.21, and even larger explosive eruptions are recognized to have happened in prehistoric times.

The dormant volcanic cone of Mount St. Helens. Mount Hood
in the background

Figure 12.21   Large volcanic events in recent times

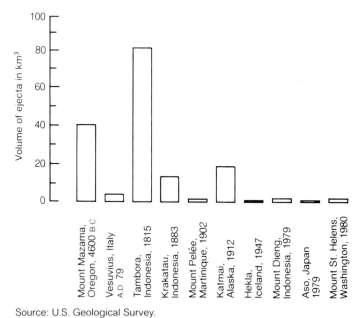

Source: U.S. Geological Survey.

CASE HISTORY 12.2   Mount St. Helens Eruption

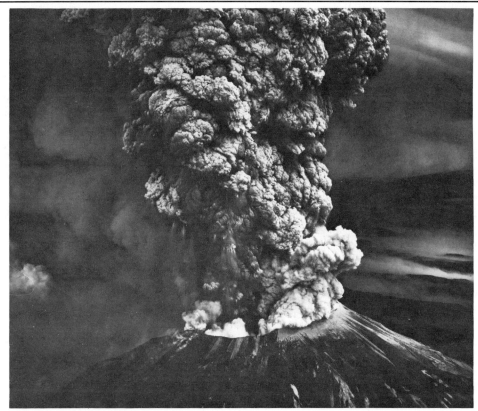

Figure 1   Mount St. Helens erupting

The recent series of volcanic eruptions at Mount St. Helens in Washington State began on March 27, 1980 with an eruptive plume that reached over 18 000 m into the atmosphere. The U.S. Geological Survey monitored the original and subsequent eruptions and reported an initial release of energy equivalent to 50 megatons of TNT, approximating the power of the largest Soviet Union nuclear bomb test of October 1961.

The photograph depicts the eruption of May 18, 1980, for which USGS scientists estimated that the minimum volume of ash and rock ejected amounted to 1.3 billion cubic yards or about 1 cubic kilometer. This would approximately equal the volume of material from the volcano's A.D. 1500 eruption and about one-third of the volume of the volcano's 1900 B.C. eruption. These volumes are comparable to the eruption of Mount Vesuvius in Italy that buried Pompeii in A.D. 79; the Mount St. Helens 1980 eruption produced much less than the 20 cubic kilometers of ash from the 1883 eruption of the Indonesian volcano of Krakatau. The volcanic eruption in 1815 of Tambora, also in Indonesia, produced the largest volume on record, 80 cubic km.

Considerable structural damage was caused by ash falls traveling as far as 200 km from Mount St. Helens. Mudflows due to sediment buildup interrupted navigation on the Columbia River. Escaping steam and gas triggered mudflows and blew down trees and vegetation.

Much of the driving force for the explosive expulsion of lava and other volcanic debris from a volcanic vent is due to the **exsolution** (separation) of dissolved gases from the liquid lava as it moves up the conduit and into a region of lowered pressure. The phenomenon, **vesticulation,** is the same as when the cap is removed from a bottle of warm carbonated beverage, which lowers the pressure with consequent effervescence. Essentially gas-free lavas will be extruded quietly on the surface whereas gas-rich lava may be expanded as a froth or be broken into tiny individual liquid droplets that cool and crack into minute glass shards suspended in the hot gas. This mixture is a suspensoid and although very hot may be heavier than air and can then flow down the volcano slope and across the surface as a **nuée ardente** (glowing cloud), burning and suffocating everything in its path. In 1902 such a cloud from the volcano Mount Pelée completely destroyed the town and ships in the harbor of St. Pierre, Martinique, some 8 km away, killing more than 25 000 people. Lower density suspensoids carry tiny volcanic fragments high into the atmosphere from whence they fall as volcanic ash covering with a layer up to several meters in thickness many square kilometers downwind from the source. Such an ash layer is readily weathered and is sometimes transformed into montmorillonitic clay or bentonite encountered as a stratum preserved by the deposition of later sediments.

Larger liquid blobs in the gas stream cool to cinders and fall near the vent where they often comprise most of the material in the volcanic cone. **Cinder cones** are a common accompaniment of a large shield-shaped volcano and they dot its flanks like small pimples; Diamond Head and the Punchbowl on Oahu Island, Hawaii, are cinder cones on the flanks of the island-forming Koolau volcanic range. Even larger liquid masses from softball to basketball size are sometimes spewed upward, cool during their air passage, and return to the surface. Collectively the solid ejecta of a volcano—measuring in size from micrometers to meters—is called **tephra.** Material below 2 mm in size is called *ash,*

Figure 12.22 Plateau basalts

Columbia River plateau basalts
NW United States

Deccan plateau basalts
India

debris between 2 and 64 mm is called *lapilli,* and fragments larger than 64 mm are called *bombs* or *blocks.*

Successive outpourings of lava and volcanic debris gradually build up extensive *lava plateaus* (fig. 12.22) or more commonly a conical pile having a crater in the top around the volcanic vent. The slopes of the pile are indicative of the kinds of material that was expelled since the lower viscosity,[1] basaltic lava tends to spread farther and have flatter slopes than higher viscosity, salic lava or cinders whose deposits have slopes of approximately 30°. As a pile increases in height the conduit is also lengthened and a later eruptive episode may bypass the longer lava- or debris-plugged primary conduit to break out on the volcano's flank.

1. Viscosity is the frictional resistance of a liquid to flow measured in poises (Poiseville, 1842): Air 1.8×10^{-4}, glycerine 83, mafic basaltic lava at 900°C $\times 10^4$, cane sugar at 109°C 2.8×10^5, silicic magma at 900°C 10^{12}, glacial ice 1.2×10^{12}.

Lava flow, Tepic, Mexico

Aztec pyramid ca. 2800 B.C. inundated by lava flow and later excavated

Because of their characteristic shape and deposits the locations of the volcanoes of the world are very well known. Like earthquakes, and for related reasons, they are mostly concentrated along edges of tectonic plates where rising magma at spreading centers and from partial melting of subducting plates provides their nourishment. Examples of volcanic areas related to spreading centers are the Azores, Iceland, and the Galapagos Islands. Volcanic chains related to collisional subduction zones are represented by the American cordillera from Alaska through the Andes, and the Japanese islands, and volcanoes from Italy through Asia Minor and the East Indies.

Although most of the world's volcanoes are related to plate edges a few are known in plate interiors. A classic case is the Hawaiian chain stretching 3 500 km in a northwesterly direction from the island of Hawaii along the Emperor seamounts (fig. 12.23). Present-day volcanic activity is restricted to the island of Hawaii and nearby submarine eruptions at the southern end of the chain. The relative age of the volcanic piles is shown by their degree of erosion and their absolute age measurement, which increases steadily northwestward. The explanation of these relations appears to be the presence of a *thermal plume* within the upper mantle of the earth at lat. 19°N, long. 155°W that periodically brings sufficient heat to the base of the Pacific plate to melt it. The plume position, marked today by the island of Hawaii, has remained fixed while the plate has moved over it, and earlier volcanoes riding atop the plate have simply been moved away from their source of lava. The direction of motion of the Pacific plate over the past 50 million years or more has thus been marked like a paper chase with dead volcanoes. The rate of motion of the plate is the age difference of rocks at any two points along the chain divided by their separation, and the periodicity of volcanic activity is the average separation of the volcanoes divided by the rate at which the plate has moved.

Many volcanoes, particularly those expelling basaltic rather than more salic magma, erupt in a relatively quiet fashion and often show patterns of behavior that become familiar to the inhabitants.

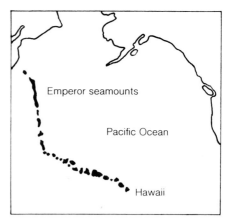

Figure 12.23 Hawaiian volcanic chain

Should, however, the need arise to monitor or predict the short-term behavior of a volcano a number of techniques have proven successful. For example, the upward movement of molten material in the conduit that precedes eruption generates seismic waves whose source may be tracked upward, and eruption is typically preceded by a gentle doming or swelling of the ground, which may be measured by sensitive levels called tiltmeters.

The eruption of a volcano, whether as a quiet extrusion of lava or as a catastrophic explosion, is a gas-driven phenomenon entailing gas solution and its ebullition (bubbling) in and from silicate melts or magmas as a function of their temperature, pressure, and composition.

The principal volatile or vaporizable component in a magma is water; it is not very soluble in molten silicates and significant quantities of H_2O are held in the magma solution only by relatively high external pressure. Up to about 10 weight % H_2O may be dissolved in melts of granitic composition, and up to 5 weight % in basaltic melts, at pressures of 5 kbar corresponding to a depth of about 20 km. Water will separate from the melt if the external pressure on a magma is lowered or if its internal pressure (vapor tension) rises. In the first instance, separation will result from the bodily movement of the magma upward to regions of lower pressure, and in the latter instance from the dissolved H_2O becoming confined to a smaller and smaller volume of

Figure 12.24 Solubility of water in granitic magma as a function of pressure

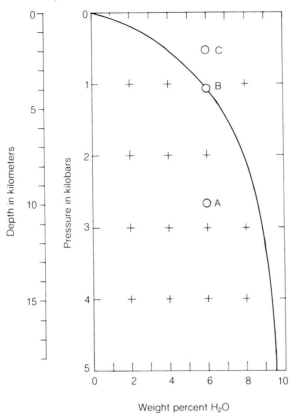

KEY

A   Undersaturated magma
B   Saturated magma
C   Oversaturated magma

Figure 12.25 Pressure-temperature relations for water retention in granitic magma

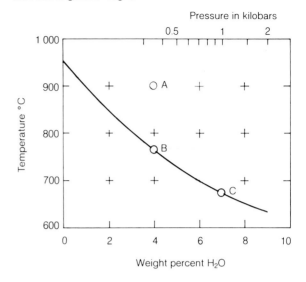

KEY

A   Undersaturated magma
B   Saturated magma
C   Second boiling when vapor pressure equals assumed confining pressure

magma as the melt is cooled and anhydrous (water-free) crystals are formed.

The dependence on pressure to confine the amount of H_2O dissolved in a magma is illustrated by figure 12.24. A magma containing 6 weight % H_2O is undersaturated with water at a depth of 10 km (point a); upward movement of the magma to 4 km (point b) brings it to the point of saturation with water; and a further rise (say to point c) would call for the ebullition of $6 - 4 = 2$ weight % water vapor. This does not appear to be a large change at first glance; but since 1 cubic centimeter of water yields 4 500 cm³ of vapor at 1 000°C the change in volume

is marked, and if the volume of the container is fixed the increase in pressure must be similarly large.

Should the upward movement be rapid (as it might be for magma rising in a volcanic conduit) the temperature will be little changed by transport though ebullition, being endothermic (heat-absorbing), will cause a temperature drop of a few degrees and minor heat losses may be expected through the conduit walls. A rapid upward movement also tends to maintain a high vapor pressure since the water vapor cannot diffuse into the wall rock at the rate that it is being produced.

The ebullition of water vapor from magma that is induced by an increase in the vapor tension of residual liquid as crystallization proceeds comes about because most of the crystals forming are anhydrous and the few that are water-bearing (amphibole, mica) abstract only a small portion of the water present. The change in water concentration and the pressure required to hold this water in solution in a cooling melt is illustrated by figure 12.25. Assume

Figure 12.26 Changing relationships in magma extrusion

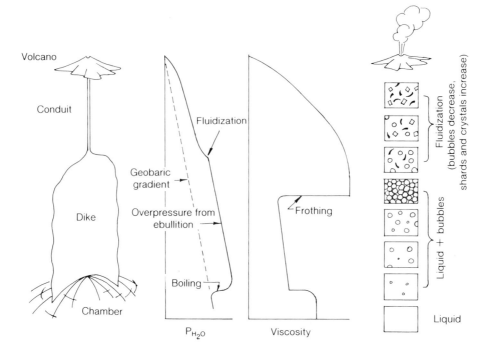

a magma to be at 900°C under a pressure of 1 kbar and to contain 4% H_2O (point *a*). At this point the magma is not saturated with respect to water so H_2O is retained in the melt as it cools to about 760°C (point *b*). Further cooling causes the crystallization of solid phases from the magma with a consequent reduction in the volume of liquid phase and an increase in the concentration of water in the residual liquid. H_2O concentration continues to increase with falling temperature until it reaches about 7% at 670°C when its vapor tension at 1 kbar equals the confining pressure. At this point boiling begins and continues as long as crystallization continues.

The expansion of the magma mass, assuming constant temperatures and pressure, that accompanies the ebullition of water vapor provides the driving force needed to cause the magma to move rapidly upward in a volcano conduit where it is subjected to less pressure, resulting in more ebullition, and causing more rise.

A representational diagram showing some of the features of a volcanic eruption is given in figure 12.26. Once boiling begins in a magma it may be anticipated that the various mixtures of liquid, solid particles, and gas will have varying physical properties, particularly in regard to their density and viscosity. A few gas bubbles distributed throughout a large volume of liquid have little effect on the bulk viscosity, but as the number of bubbles increases a condition of close-packed bubbles with interstitial liquid may be reached. The viscosity of this froth is greater than that of the precursor liquid plus bubbles. For example, should frothing occur during upward migration of lava in a volcano conduit, the conduit may become plugged resulting in explosive eruption.

Beyond the liquid–gas proportions required for a froth, bubbles coalesce and the system becomes one of fluidized particles and liquid droplets in a gas stream. Large and small, solid or liquid particles are entrained in the upward-rushing gas stream and more or less explosively ejected from the volcano.

The change from an upward-rising liquid to an upward-rising gas is marked by a distinct change in the intrusive mechanism. Liquid either rises permissively through cracks or generates them by liquid wedging while gas penetrates upward through a tubular conduit. The rock bodies that mark these two styles of intrusion are tabular dikes and sills for liquid intrusion and pipes for fluidized material.

SUMMARY

Earthquakes and volcanic eruptions are natural phenomena that cannot be controlled, but the risks associated with them and with other hazards can be more or less well calculated. In some instances prediction is possible and improved means of protection can be anticipated.

Loss of life and property damage resulting from a natural disaster have historically been largely due to siting in dangerous locations; for example, farming on the slopes of a dormant volcano, building on a river flood plain, or construction on or near an active fault line. Further, the use of inappropriate construction means such as brick or stone rather than timber or steel framing in earthquake-prone areas together with utility lines without safety devices has been a principal cause of loss of life. Much of the risk related to such disasters can be reduced by proper planning.

The destructive shock waves caused by earthquakes can be generated on a smaller scale and usefully employed as a means of subsurface investigation. The nature and geometry of rock bodies and their rippability, thickness of overburden, and other features of engineering interest can be assessed by this means.

ADDITIONAL READING

Ballard, F. M. 1976. *Volcanoes of the earth.* Austin: University of Texas Press.

Berlin, G. L. 1980. *Earthquakes and the urban environment.* vols. 1, 2, 3. Boca Raton, Fla.: CRC Press.

Bolt, B. A. 1978. *Earthquakes: A primer.* San Francisco: W. H. Freeman.

Bolt, B. A.; Horn, W. L.; McDonald, G. A.; and Scott, R. F. 1976. *Earthquakes, tsunamis, volcanoes, avalanches, landslides, floods.* New York: Springer-Verlag.

Cakmak, A. S.; Abdel-Ghafar, A. M.; and Brebbia, C. A., eds. 1982. Rotterdam: A. A. Balkema.

Coffman, J. L. 1979. *Earthquake history of the United States.* (1971–1976 supplement). U.S. Coast and Geodetic Survey Publication 41–1.

Decker, R. W., and Decker, B. 1981. *Volcanoes.* San Francisco: W. H. Freeman.

Green, J., and Short, N. M., eds. 1971. *Volcanic landforms and surface features: A photographic atlas and glossary.* New York: Springer-Verlag.

Iacopi, R. *Earthquake country.* Menlo Park, California: Lane.

Lomnitz, C. and Rosenbleuth, E., eds. 1976. *Seismic risk and engineering decisions.* New York: Elsevier Science.

Murphy, J. R., and O'Brien, L. J. 1977. The correlation of peak ground acceleration amplitude with seismic intensity and other physical parameters. *Bulletin Seismological Society of America,* v. 67.

Neumann, F. 1954. Earthquake intensity and related ground motion. Seattle: University of Washington Press.

Ollier, C. 1969. *Volcanoes.* Cambridge, Mass.: MIT Press.

Petak, W. J. 1982. *Natural hazard risk assessment and public policy.* New York: Springer-Verlag.

Richter, C. F. 1958. *Elementary seismology.* San Francisco: W. H. Freeman.

Sheets, P. D., and Grayson, D. K. 1979. *Volcanic activity and human ecology.* New York: Academic Press.

Simkin, T., et al. 1983. *Volcanoes of the world: A regional directory, gazeteer, and chronology of volcanism during the last 10,000 years.* Stroudsburg, Pa.: Hutchinson Ross.

Williams, H., and McBirney, R. 1979. *Volcanology.* San Francisco: Freeman Cooper.

Wood, H. O., and Neumann, F. 1931. Modified Mercalli intensity scale of 1931. *Seismological Society America,* v. 21.

13

Geologic Hazards: Slope Failure, Subsidence, and Flooding

INTRODUCTION

Earthquakes and volcanoes, discussed in the preceding chapter, represent striking geologic hazards but are by no means the only phenomena to be considered in the planning and construction of engineering works. Other hazards, discussed here, include the failure of slopes as seen in landslides and slumping, the change in ground level due to mining or removal of water or oil from subsurface formations, and the flooding that occurs when a stream overtops its banks. These are hazards of particular concern in engineering practice because, unlike an earthquake or volcanic eruption, they may well be caused by ill-conceived engineering activity.

SLOPE FAILURE

Natural topographic slopes usually represent *equilibration* between the inherent angle of repose of bed rock or surficial debris and the various geologic processes that tend to modify them. Slopes are not geologically permanent, but on the short time scale of human endeavor and in an undisturbed state they may be considered so except in special instances.

If a slope is in equilibrium the mechanical forces on any part of it are in balance. The principal force is that of gravity, which has components parallel and perpendicular to the slope surface (fig. 13.1). The normal component, σ_n, is a stress that dies out in an inward and downward direction (as previously discussed in chapter 11) and is opposed by the compressive strength of the material beneath. The downward shearing force parallel to the surface, τ, is opposed by frictional drag, the cohesion of the slope material, and the compressive strength of the downslope material, again dying out with distance. A number of circumstances may be considered to modify this equilibrium and may, if sufficiently great, cause the slope to fail. In figure 13.1, for example, changing the *angle* of the slope will change the balance of forces, *loading* of the surface at A will increase both σ_n and τ, or *removal* of material at B will reduce resistance to τ. In addition, the stability of the slope material may be substantially altered by *introducing water* or sewage effluent from upslope ponds, excavations, ditches, or disposal wells, or by *removing the vegetative cover* and thereby eliminating the water storage capacity afforded by vegetation.

Mechanisms of Failure

There are a number of distinct mechanisms by which slopes fail. In many instances initial failure by one mechanism triggers failure by another and the result may be complex. Usage varies as to the terminology and classification of slope failure but for this discussion three broad categories are used: falls, slumps and slides, and creeps and flows.

Falls are clastic failures in which the failed material moves downslope in a free fall. Slides and slumps are masses of rock or soil displaced along one or more surfaces of shear failure but without significant internal deformation of the failed segments. **Slumps,** or rotational slides, are spoon-shaped failures in cohesive materials that take place in a rather short time—from in a matter of seconds to within a few days. **Slides,** translational slides, or slab slides are failures, which are usually elongate elliptically in plan and are bounded by a planar slip surface beneath, that occur in all geologic materials at both rapid and slow rates. Slides in rock usually move on

Figure 13.1 Components of gravity in a slope

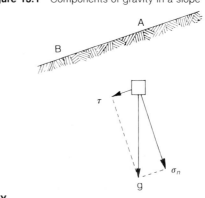

KEY

τ Shear stress parallel to slope
σ_n Stress perpendicular to slope
g Force of gravity

internal surfaces of weakness such as foliation, cleavage, bedding, or joint surfaces; whereas in cohesionless material failure is by shear along an internal surface at the angle of repose.

Creeps are slow plastic-to-fluid movements. **Flows,** often initiated by a slump or a slide in wet or saturated material, are relatively rapid mass movements involving internal deformation of the failed material. The slump or slide delivers the failed material downward onto a slope that is greater than that required for its stability, and the material then continues downward as a flow of viscous liquid or suspensoid.

The following are some particular terms applied to the various kinds of failure:

Falls Rockfall, soilfall, avalanche
Slides and slumps Rockslide, landslide, slab slide
Creeps and flows Rock creep, soil creep, solifluction, rock glacier; dry flow, mudflow, debris avalanche

Past failures of a slope may leave a record in the form of scarps, lobate bands, hummocky surfaces, chaotic debris, talus, linear hillslope scars, or displacement or misalignment of such things as roads or telephone poles. *Incipient failure* is most prominently marked by tensional cracking of soil, pavements, or structures and is often accompanied by small settlements.

Road cut failure, Garrard County, Kentucky

Large slump, Mam Tor, Derbyshire, England

Rock fall, Kentucky

Mudflow, Walnut Creek, California

268   Engineering Considerations

Landslide, Oregon

Slumps Failure by slumping is essentially restricted to cohesive earth masses, since consolidated rocks typically fail along planar surfaces, and there are no curved slip surfaces in clean, dry, cohesionless materials (sand and gravel). The driving force causing failure of a slope by slumping is the *weight* of a spoon-shaped segment. The resisting force is the *friction* and *shear strength* of the material at the slip surface. The shape and position of the *slip surface,* that surface along which rupture will occur in a slope failure, is influenced by the distribution of **pore pressure** of contained water and variations of shear strength within the mass. The important parameters affecting the stability of a particular slope are slope angle together with the nature of the underlying materials and their properties such as degree of saturation, unit weight, coefficient of friction, cohesion, and angle of shearing resistance.

Figure 13.2 Failure surface in slump

Earth mass

Firm material

The downward limit of the arcuate failure surface is
determined by the presence of firm material

Figure 13.3 Typical slump failure with associated flow

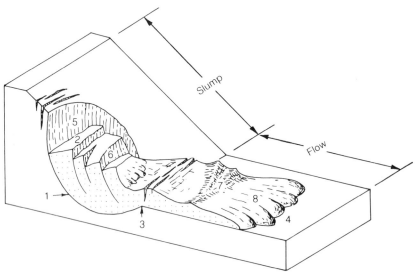

Key

1.	Surface of rupture	5.	Main scarp
2.	Head	6.	Zone of transverse faulting
3.	Foot	7.	Transverse ridges
4.	Toe	8.	Radial cracks

In order to examine the geometry of slump failure, an idealized slope will be considered to be long and identical at all cross sections. A spherical segment will be adopted as a simplified shape of the slip surface that is within the earth mass and at its limiting condition is tangent to an underlying firm material (fig. 13.2). The appearance and parts of a typical slump failure together with associated flowage are illustrated by figure 13.3.

Slides The slip or failure surface for *sliding* or *creeping* material is roughly planar, always equal to or less than the slope angle. The thickness of the failed slab or sheet is usually in the order of a few meters but may sometimes involve a much thicker section. In rock the slip surface usually follows a bedding or joint surface across which the rock is particularly weak in shear. Figure 13.4 shows cross

Figure 13.4 Slides

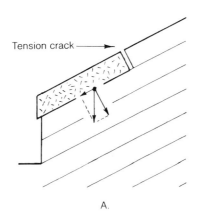

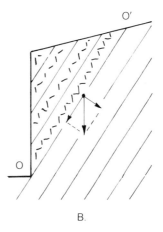

A.

B.

KEY

A.   Sliding block on inclined slip surface
B.   Limit of sliding failure O O′. Resolution of forces shown
      by arrows.

sections of incipient rock slides. The driving and resisting forces are equal (limit equilibrium) when
$\tau = f\sigma_n + B$ where f is static friction and B the shear
strength arising from any chemical bonds or surface
irregularities on the surface of potential failure. Note
that in figure 13.4a the slab is detached at a tension
crack and the uphill portion is left in an unstable
condition, perhaps to later slide. In figure 13.4b
sliding could occur on any parallel surface above but
not below 0–$0'$ along the potential slip planes illustrated. No tension crack, often a clue to incipient
failure, is formed. Stability for such slopes can be
improved by eliminating water from the potential
failure surfaces or by cutting back the slope to an
angle less than that of the detachment planes.

    The failure surface of a slide may emerge at
the surface as at $0'$ in figure 13.4b. The failed block
may also be bounded below and laterally by slip surfaces. Figure 13.5 illustrates a particular case in
which the failed block moves on an inclined bedding
or joint plane and the bounding surfaces of the slide
are failed vertical joints.

    The angle of repose of cohesionless materials,
about 30°, is their limit equilibrium and such materials as dry sand or gravel adjust rapidly to this

Figure 13.5 Slide of a jointed slab

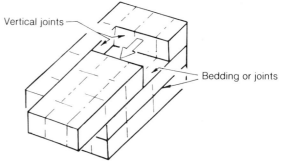

slope. For example, if the toe of a dry sand slope
represented by OAB in figure 13.6 is removed, the
portion $ABCD$ is unstable and will slide so that the
slope returns to limit equilibrium $O'C'$. Since the
presence of pellicular water in cohesionless material
modifies its mechanical behavior (see chapter 10)
the stability of slopes in such material is particularly
sensitive to changes in water content. Sand containing pellicular water will stand vertically but on
either drying or saturation flows to lie at about 30°.

Figure 13.6 Sliding of sand caused by removal of toe

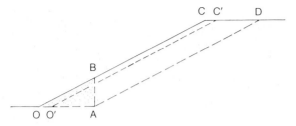

KEY

OC	Original surface
OAB	Removed causing sand ABCD to slide
O'C'	New surface

Excavation in sand cut by a water table may be particularly difficult or dangerous as indicated by figure 13.7. The original cut shown by the dashed line fails to the solid line because of adjustment of saturated sand to its angle of repose and removal of support from the vertically standing portion.

Flows The materials within slopes are necessarily under unbalanced stress and may be expected to respond according to the resistance to deformation of the material that forms them. Falls, slides, and slumps all involve the *detachment* of a rock or soil mass from its substrate and are thus clastic phenomena. Slope failure, however, may also take place by *plastic deformation* and *flowage,* either laminar or turbulent.

Cohesive substances such as clayey masses, many soils, organic matter, ice, and some shales are representative of common materials that deform plastically under their own weight and can, therefore, be expected to creep down a slope. The amount of contained water is usually critical to the degree of their plastic response. Plastic deformation becomes transitional to fluid flow with increasing admixture of water or air in the deforming mass. The presence of a creeping blanket of soil may be inferred from the presence of float blocks detached from an outcrop; downslope-bending rock layers; misaligned roads; turf rolls below boulders; sod lips at the top of cutbanks; and the displacement, tilting, or breaking of walls, posts, monuments, trees, or structures.

Figure 13.7 Failure in sand containing different amounts of water

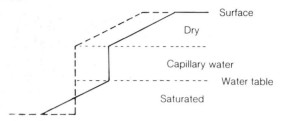

Earthflows, debris flows, and mudflows are a common style of failure in which tongue- or tear-drop-shaped bodies of unconsolidated material move downslope at rates ranging from a few millimeters per day to meters per second. Such flows are mobilized when water infiltrates and raises pore pressures until the plastic limit is exceeded. Most of the movement takes place by slippage along the boundaries, but internal deformation also takes place.

Flows of dry soil, mud, or debris often develop when the kinetic energy of a mass gained from a fall or slump is expended in downslope transport of non-cohesive debris or when water-saturated masses lose coherence. The fluidized mass moves downslope, often at high speeds, in a relatively straight course usually leaving a scarped head, a central channel scar, and a debris fan at the foot.

A special kind of flow is occasionally encountered in areas that have undergone continental glaciation in Pleistocene time (map 6, appendix II). Clays deposited by melt-water from the retreating ice have locally been brought to the state of thixotropy by the leaching of adsorbed cations. Such "quick" clays can be transformed instantly from a solid to liquid condition by sudden jarring such as accompanies earthquakes, blasting, pile driving, or even traffic vibration. The results of this transformation may be disastrous as shown by the intense ground disturbance at Anchorage accompanying the Alaskan earthquake of 1964 and by the death of about 70 people in Quebec, Canada, from this cause in recent years. Quick clays may be tested in the field by placing a molded lump in the palm of the hand and vibrating it by striking the edge of the hand sharply; if quick, the clay will liquify.

Figure 1 Pescadero Creek mudflow, California

The destruction depicted took place in the Pescadero Creek area of California and was caused by a mudflow related to a flood that moved down the valley from right to left. The soil in this particular area upstream from the building and under its foundation was readily liquified by the excess of water due to rainfall and flooding. Runoff began as floodwater, but the ground rapidly liquified and the stream bank slopes became unstable and failed. Large debris is noticeable within the mudflow.

The mudflow undermined the foundation of the portion of the building closest to the valley while the part farther away from the channel escaped damage. The collapsed part of the building remained fairly well intact but the removal of foundation material caused its complete subsidence. Retaining walls of concrete or rock would have minimized the damage.

Figure 13.8 Reduction of safety factor of a slope due to added water

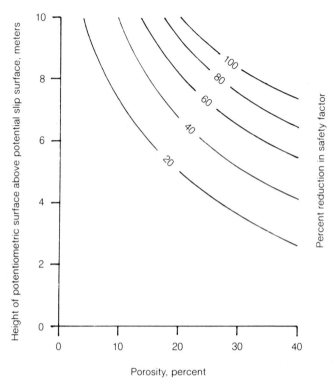

Water-Induced Slope Failure

It is often stated that slopes slump in consequence of their lubrication by water. This is a gross oversimplification of the matter with only the kernel of truth that slope failure may be somehow related to water added by rain, reservoir leakage, etc. Only a very thin film of water suffices to yield the full lubricating effect, and this much water is typically present in the sediments beneath a slope except perhaps in the uppermost few feet; the sediments are thus permanently lubricated and additional water has no further effect on this parameter. The principal causes of water-induced slope failure are to be found in the *increased weight* of newly saturated materials and the reduction of the angle of internal friction in consequence of *increased pore pressure.*

For a given slope and potential failure surface in a static condition subject only to differing amounts of water, all of the parameters that control the safety factor are constant except for the shearing resistance, $\overline{S}$, and W, the weight of earth and water in the potentially sliding mass. For noncohesive sediments (sand, silt), $\overline{S} = (P - hW) \tan \phi$ and for cohesive sediments $\overline{S} = c + (P - hW) \tan \phi$ where c = cohesion value, P = pressure per unit area of potential sliding surface due to overlying solids and liquids, h = potentiometric head, W = unit weight of water, and ϕ = angle of sliding friction for the slip surface. The very large reduction in $\overline{S}$ that occurs when porous sediment above a potential slip surface is saturated is partially offset by a concomitant increase in W. The net effect of saturation, however, is to reduce the safety factor as is illustrated by figure 13.8 which is calculated for 10 meters of silt (specific gravity of quartz = 2.65) that is saturated to different levels above the slip surface. Curves are drawn for porosities of 10, 20, and 30%.

Figure 13.9 Conditions for slope failure due to reservoir drawdown

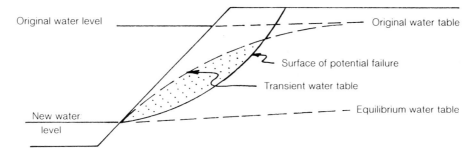

Original water level ——————— Original water table

Surface of potential failure

Transient water table

Equilibrium water table

New water level

KEY

▨ Saturated sediment above failure surface

Redrawn from K. Terzaghi in *Application of Geology to Engineering Practice*, Geological Society of America, 1950.

A slope that forms the edge of a body of water may fail if the level of the water is rapidly reduced by a drawdown of water faster than the sediment can be dewatered. The result of such a rapid drawdown is that the descending transient water lags behind its eventual equilibrium position, and the saturated sediment beneath this transient water table and above a surface of potential failure has an increased weight due to its contained water (stippled area in figure 13.9).

Earthquake-Induced Slope Failure

The stability of a slope is usually considered only in relation to the gravity driving force and resisting forces internal to the mass. However, *horizontal accelerations* may be caused by an earthquake and can increase the shearing stress along a potential slip surface to the point of failure. An earthquake (fig. 13.10) producing a horizontal acceleration n times that of gravity, n_g, produces a mass force, $n_g W$, through the center of gravity of the block on lever arm e. This increases the turning moment around the arc center by $n_g We$. If the safety factor of the slope under static conditions is

$$\frac{R\,\overline{S}\,\widehat{L}}{Wd},$$

the safety factor under conditions of horizontal earthquake acceleration will be

Figure 13.10 Effect of horizontal acceleration on slope stability

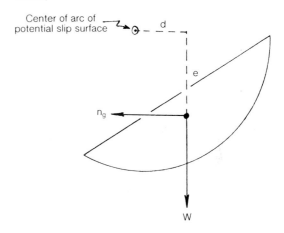

Center of arc of potential slip surface

d

e

n_g

W

KEY

W   Weight of potential slump segment
d   Moment arm
e   Lever arm
n_g Horizontal acceleration of n times gravity

$$SF = \frac{R\,\overline{S}\,\widehat{L}}{Wd + n_g\,We}.$$

The numerical value of n_g is, of course, related to the intensity of the earthquake (see chapter 12). Table 13.1 provides some approximate values.

Table 13.1 Approximate Values of Acceleration × Gravity, n_g

Intensity	n_g
VII	0.10–0.13
VIII	0.17–0.2
IX	0.3–0.4
X*	0.6–1.1
XI	(1.0)
XII	(1.9)

*Approximate intensity of the San Francisco earthquake of 1906

Figure 13.11 Pendulum

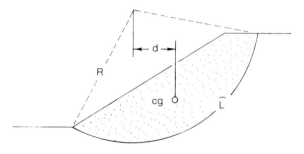

KEY

cg	Center of gravity
L	Curved rupture surface
R	Radius of rupture surface
d	Moment arm

Figure 13.12 Limit equilibrium

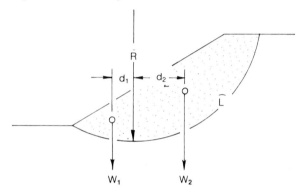

KEY

R	Radius of rupture surface
L	Curved rupture surface
W_1 W_2	Weights of pendulum components
d_1 d_2	Moment arms

The unconsolidated materials least prone to failure due to earthquake acceleration are low-sensitivity clays in a plastic state and dense sand, either dry or saturated. Those most susceptible to failure are slightly cemented granular aggregates such as loess and partly or wholly submerged loosely packed sand.

Investigation and Treatment of Slopes

Slopes whose failure could cause damage or disaster should be carefully examined in three dimensions. *Mapping* is done to delimit the potentially unstable area and *drilling* performed to provide information as to the nature and properties of subsurface ma-

terials. A series of potential rupture surfaces within the mass are next considered and their safety factors calculated. The surface with the lowest safety factor, the worst surface, is taken as the basis for preventative action.

Determining stability   Several methods may be used to ascertain the stability of an existing or planned slope. The goal is to determine the ratio of resisting to driving forces or *factor of safety*. Stable slopes will have a ratio greater than 1 (usually 1.5–2.0 but may be as small as 1.1); a condition of

Figure 13.13 Slice method

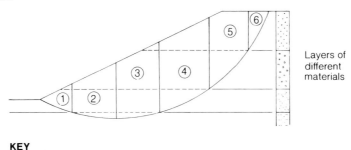

KEY

1–6   Individual vertical slices whose contributions to
       stability of the slope are calculated and summed.

limit stability exists when the safety factor is unity; and the slope is unstable if the value is less than 1.

The mass of earth that moves in a slump failure may be likened to a pendulum swinging around the center of the curved rupture surface, (fig. 13.11). The critical parameters may be seen to be the radius, R, the average shearing resistance per unit area of sliding, $\overline{S}$, the arc length of the circular segment representing the rupture surface, $\widehat{L}$, the weight of the earth mass, W, and the length of the moment arm, d. Combining these parameters

$$\text{Safety factor} = \frac{\text{resisting moment}}{\text{driving moment}} = \frac{R\,\overline{S}\,\widehat{L}}{W\,d}$$

Closer examination of figure 13.11 shows that a portion of the pendulum bob counteracts the general tendency for movement (fig. 13.12). The limit equilibrium for this failure surface in the slope may thus be written $W_1 d_1 + c\,\widehat{L}\,R = W_2\,d_2$ where c is the cohesion of the material on the potential failure surface.

A common means of determining the stability of a slope is the *slice method,* which is especially useful when the slope is composed of nonuniform materials arranged in layers. In this approach the earth mass is divided into vertical slices (whose widths are controlled by the intersection with the layers of the failure surface under examination, fig. 13.13), and the contribution of each slice is summed.

From figure 13.11 the driving and resisting forces are seen to be

Slice	Driving T_w	Resisting f	c
1	−	−	−
2	−	−	−
3	+	−	−
4	+	−	−
5	+	−	−
6	+	−	−

where T_w = tangential component of the total weight acting to cause rotation, f = solid friction, and c = cohesion of material at base of slice.

The actual failure surface within a mass is the surface that has the lowest safety factor. This surface cannot be determined uniquely by measurement but must be found by a series of approximations. To assist in its location the following general rules for determining this **worst surface** are useful:

Slope Less than 20° Appreciable curvature, dips steeply into base of slope, and may emerge or "daylight" outside of slope toe

Slope 20° to 35° Less curvature; usually emerges at toe of slope

Slope Greater than 35° Nearly flat curvature and may be approximated by a plane; emerges at toe of slope

Figure 13.14 Slope stabilization by removal of weight

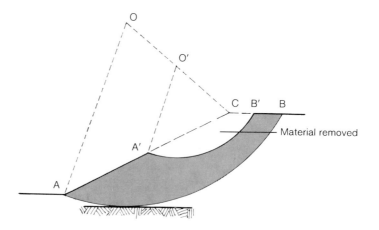

It should be noted as a practical matter that the stability of a slope near limit equilibrium that is underlain by a cohesive earth mass can be improved by several means. Internal friction may be increased by drainage or by limiting water access by ditching across the crest. Tangential forces may be reduced by flattening the slope angle or by removing weight as shown in figure 13.14. In the figure, O' is the center of arc $A'B'$ and O is the center of arc $A B$, which is at limit equilibrium for the wedge ABC. Removal of wedge $A'B'C$ will stabilize the slope.

As a first approximation, the geometry of figure 13.15 may be utilized to locate a test surface for safety factor calculations.

The various triggering influences (e.g., variability of water content, planned changes in slope configuration, and earthquake effects) are considered to complete the evaluation. At this stage it is possible to make a judgment as to future actions based on the results obtained while taking into account the history of similar situations in the local area. If the project is to continue but conditions are in delicate balance, a *monitoring* system should be

Figure 13.15 First approximation for worst surface

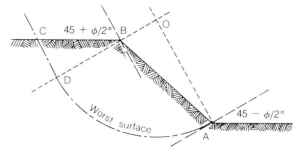

1. Construct perpendiculars at A and B to $45 - \phi/2°$ and $45 + \phi/2°$ at toe and shoulder of slope respectively to find O. (ϕ is angle of internal friction)
2. Swing arc AD
3. Construct CD parallel to OA

established. This might take the form of survey-controlled stakes, strain gauges, or observations of groundwater levels in wells.

Treatment for unstable slopes may be accomplished by avoidance—the best procedure when possible—prevention, or correction. Table 13.2 gives various methods that have been employed.

Table 13.2 Methods for Prevention and Correction of Landslides

Effect on Stability of Landslide	Method of Treatment	General Use		Frequency of Successful Use[1]			Position of Treatment on Landslide[2]	Best Applications and Limitations
		Prevention	Correction	Fall	Slide	Flow		
Not affected	I. Avoidance methods: A. Relocation	x	x	2	2	2	Outside slide limits	Most positive method if alternate location economical
	B. Bridging	x	x	3	3	3	Outside slide limits	Primary highway applications for steep, hillside locations affecting short sections
Reduces shearing stresses	II. Excavation:[3] A. Removal of head	x	x	N	1	N	Top and head	Deep masses of cohesive material
	B. Flattening of slopes	x	x	1	1	1	Above road or structure	Bed rock; also extensive masses of cohesive material where little material is removed at toe
	C. Benching of slopes	x	x	1	1	1	Above road or structure	
	D. Removal of all unstable material	x	x	2	2	2	Entire slide	Relatively small shallow masses of moving material
Reduces shearing stresses and increases shear resistance	III. Drainage: A. Surface: 1. Surface ditches	x	x	1	1	1	Above crown	Essential for all types
	2. Slope treatment	x	x	3	3	3	Surface of moving mass	Rock facing or pervious blanket to control seepage
	3. Regrading surface	x	x	1	1	1	Surface of moving mass	Beneficial for all types
	4. Sealing cracks	x	x	2	2	2	Entire, crown to toe	Beneficial for all types
	5. Sealing joint planes and fissures	x	x	3	3	N	Entire, crown to toe	Applicable to rock formations

[1] 1 = frequently; 2 = occasionally; 3 = rarely; N = not considered applicable.
[2] Relative to moving or potentially moving mass.
[3] Exclusive of drainage methods.

From A. W. Root in *Landslides and Engineering Practice*, Highway Research Board Special Report 29, E. B. Eckel ed. © Highway Research Board, National Academy of Science. Reprinted by permission of the National Academy of Science.

Table 13.2 Methods for Prevention and Correction of Landslides (continued)

Effect on Stability of Landslide	Method of Treatment	General Use		Frequency of Successful Use[1]			Position of Treatment on Landslide[2]	Best Applications and Limitations
		Prevention	Correction	Fall	Slide	Flow		
	B. Subdrainage:							
	1. Horizontal drains	x	x	N	2	2	Located to intercept and remove subsurface water	Deep extensive soil mass where groundwater exists
	2. Drainage trenches	x	x	N	1	3		Relatively shallow soil mass with groundwater present
	3. Tunnels	x	x	N	3	N		Deep extensive soil mass with some permeability
	4. Vertical drain wells	x	x	N	3	3		Deep slide mass, groundwater in various strata or lenses
	5. Continuous siphon	x	x	N	2	3		Used principally as outlet for trenches or drain wells
Increases shearing resistance	IV. Restraining structures:							
	A. Buttresses at foot:							
	1. Rock fill	x	x	N	1	1	Toe and foot	Bed rock or firm soil at reasonable depth
	2. Earth fill	x	x	N	1	1	Toe and foot	Counterweight at toe provides additional resistance
	B. Cribs or retaining walls	x	x	3	3	3	Foot	Relatively small moving mass or where removal of support is negligible
	C. Piling:							
	1. Fixed at slip surface		x	N	3	N	Foot	Shearing resistance at slip surface increased by force required to shear or bend piles
	2. Not fixed at slip surface		x	N	3	N	Foot	

Effect on Stability of Landslide	Method of Treatment	General Use		Frequency of Successful Use[1]			Position of Treatment on Landslide[2]	Best Applications and Limitations
		Prevention	Correction	Fall	Slide	Flow		
	D. Dowels in rock	x	x	3	3	N	Above road or structure	Rock layers fixed together with dowels
	E. Tie-rodding slopes	x	x	3	3	N	Above road or structure	Weak slope retained by barrier, which in turn is anchored to solid formation
Primarily increases shearing resistance	V. Miscellaneous methods: A. Hardening of slide mass: 1. Cementation or chemical treatment							
	(a) At foot		x	3	3	3	Toe and foot	Noncohesive soils
	(b) Entire slide mass		x	N	3	N	Entire slide mass	Noncohesive soils
	2. Freezing	x		N	3	3	Entire	To prevent movement temporarily in relatively large moving mass
	3. Electroosmosis	x		N	3	3	Entire	Effects hardening of soil by reducing moisture content
	B. Blasting		x	N	3	N	Lower half of landslide	Relatively shallow cohesive mass underlain by bed rock Slip surface disrupted; blasting may also permit water to drain out of slide mass
	C. Partial removal of slide at toe			N	N	N	Foot and toe	Temporary expedient only; usually decreases stability of slide

SUBSIDENCE

Over geologic time any given point on the earth's surface has probably experienced a number of vertical movements of hundreds or thousands of meters as evidenced, for example, by the presence of marine sedimentary rocks well above sea level. This movement continues today as the lithosphere adjusts to both internal and external forces and is particularly marked along gently inclined shorelines. Obvious causes are erosion and deposition of surficial materials, but large areas are also affected by internal motions of the lithosphere.

Currently there is regional subsidence along the coasts of the United States from Maine to North Carolina, in the area around Daytona Beach in Florida, and along the Gulf areas of Louisiana and Texas. Broad regional uplift is present in the mountain states of the West, in the Southeast from northern Louisiana to the Atlantic coast, and around the western Great Lakes. Some of these motions are tectonic, some are adjustments to loading (as in the Mississippian delta), some are rebounds from unloading glacial ice (as around the Great Lakes), but most have unknown causes. Vertical motion on a smaller scale may be the result of areally restricted tectonic movements in the crust associated with prefaulting strain, the vertical component of faulting itself, or swelling that precedes a volcanic eruption. These grand motions of the earth's surface are the province of the geologist and are rarely of concern to the practicing engineer. There are, however, a large number of both natural and human-induced phenomena in surficial materials or at shallow depths that take place on a scale and at a rate that make them of great engineering importance.

Causes of Uplift and Subsidence

Uplift and subsidence, with some exceptions, are not catastrophic but do place structures and buried utility lines under stresses that are not normally included in their design parameters. The causes of uplift and subsidence are categorized and the more important in relation to engineering practice are discussed in the following section:

Tectonic processes of faulting, folding, and volcanism resulting in elastic, plastic, and clastic strain

Erosion and deposition

Change of state with related volume changes accompanying chemical change, adsorption and dehydration, and freeze–thaw cycling

Addition or removal of substance such as change in amount of pore fluids, bulk fluid withdrawal, or removal of solid support

Miscellaneous processes including earth tides, isostacy (buoyant adjustments of subcontinental dimensions), and thermal expansion and contraction

Changes in fluid content   There are a number of ways in which changes in the fluid content of a granular, porous material can modify the overall rock volume and thus cause uplift or subsidence of the upper surface. The structure of some loose, dry, low density deposits with high porosity such as dry alluvium, dessicated mudflows, or loess whose specific gravity is below 1.3, whose field moisture is less than 10%, and whose clay content (montmorillonite) is about 12% may be likened to a house of cards with weak interparticle bonds of water molecules, clay minerals, or calcareous matter. Artificial addition of water, say for irrigation, may destroy the delicate intergrain connections, and the house of cards falls with a loss of volume more or less equivalent to the original porosity. This phenomenon is *hydrocompaction* and is mechanically related to the slaking of compaction shale, liquefaction (quick clay) phenomena, and mudflows. Obviously, the ideal conditions will be found in dry climates where a deep water table and high evapotranspiration rates leave a thick dessicated section between the root zone and the water table. The San Joaquin Valley of California is a classic area for irrigation-related hydrocompaction.

An uncemented granular mass having liquid-filled pores confined within an impermeable envelope (e.g., a saturated lens of sand within a clay stratum) is a model of a common geologic condition whose applicability is even broader when substitution of low permeability throughout the volume of

Table 13.3 Subsidence Related to the Withdrawal of Fluids

Location	Maximum Subsidence in cm	Area Affected in km²	Indicated Average Rate in cm/yr
Oil and gas withdrawal			
Goose Creek, Tex.	90+	10	15
Wilmington, Calif.	823	65	37
Bachaquero field, L. Maracaibo, Venezuela	140	480	7
Lagunillas field, L. Maracaibo, Venezuela	340	300	12
Niigata, Japan	80	50	50
Po delta, Italy		800	30
Groundwater withdrawal			
Osaka, Japan	175	280	12
London, England	21	518+	.07
Mexico City, Mexico	700	25+	50
Savannah, Ga.	20	130	4
Houston-Galveston, Tex.	120	2 600	1.6
Eloy-Picachaco, Ariz.	110		3
Las Vegas, Nev.	20		1.5
Los Banos-Kettleman City, Calif.	760	1 300	30
Tulare-Wasco area, Calif.	365	1 200	13
Santa Clara Valley, Calif.	330	650	6.6

From J. F. Poland and G. H. Davis in *Reviews in Engineering Geology*, vol. 2, D. J. Varnes and G. Kiersch *eds.*, Geological Society of America, 1969. Reprinted with permission of J. F. Poland.

interest is made for the impermeable envelope. The compactibility of the granular body under load in such circumstances is a function not only of the strength and state of packing of the component grains but also of the rate at which the contained fluid can escape; that is, the permeability of the stressed volume and its enclosing material. *Consolidation* of the mass under static load takes place by two separate but related processes; first the pore water is placed under stress and gradually expelled—primary consolidation—and later there are adjustments in the granular fabric in response to the applied load and lessened water content—secondary compression. Obviously, both the permeability and strength of the granular fabric are sensitive to the proportions of sand and clay.

Changes in volume are often associated with the **withdrawal of fluids** (e.g., water, brine and steam; petroleum and natural gas; helium, carbon dioxide, and solublized metallic ores) or with **fluid injection** often employed as a means of pressurizing oil fields or in the disposal of liquid wastes. The mechanics of the volume change appears to be the transfer of that portion of the overburden load originally carried by

hydrostatic pressure to the granular skeleton of the rock, which is then elastically compressed. The stresses on the liquid and solid portions of a liquid-filled granular mass at some depth below the surface were identified by Terzaghi (1925) as $p = p' + u_w$ where p = total stress = overburden = load = geostatic pressure, p' = effective stress = grain-to-grain load, and u_w = neutral stress = fluid pressure = pore pressure. This being the case, should u_w be decreased or reduced to zero by withdrawal of the intergranular fluid that portion of the overburden load previously carried by pore pressure will be transferred to the granular skeleton with consequent deformation, usually elastic, of the mass and subsidence of the surface. Of course the opposite is true and injection of fluids will cause reciprocal effects. The amount of subsidence will be related to the mineralogy; the sorting, packing and degree of cementation of the grains; the amount of prior consolidation; and the porosity and permeability of the mass. Instances of subsidence related to fluid withdrawal are widespread and some well-documented cases are identified in table 13.3.

Terzaghi (1926) developed a theory to explain the consolidation of the soil mass that accompanies subsidence by making a number of simplifying assumptions: (1) the soil mass is saturated and homogeneous, (2) coefficients of permeability and volume compressibility are constant throughout the period of deformation, (3) Darcy's law of fluid flow holds, (4) consolidation is one-dimensional with water leaving the stressed volume along vertical lines, (5) consolidation is entirely the result of changed water content (no secondary compression), and (6) both water and mineral grains are incompressible. Obviously one or more of these conditions may be violated in the field, however, the Terzaghi equation,

$$\frac{\partial U}{\partial t} = C_v \frac{\partial U}{\partial z^2},$$

gives a good approximation; and when about 90% of the excess pore water has been dissipated, the time required to complete consolidation is

$$t = \frac{H^2}{C_v}$$

where C_v = coefficient of consolidation, H = half-thickness of the compressible layer, z = depth below surface, U = pore pressure, and t = time. C_v, the coefficient of consolidation, is usually determined by a simple test in which the load on a sample compressed between two porous plates is increased incrementally and the change in thickness recorded. Such a test, run in a period of days, should be used to provide general rather than exact data for projections of real consolidation with an active lifetime of years.

Water withdrawal or drainage is often through a *point discharge,* and the concentration of flow may occasionally increase seepage forces to levels at which poorly coherent soil particles are eroded. If the axis of discharge is transected by a free face such as the head or wall of a gully, landslide, or excavation, the erosion forms a hole along the flow axis that may readily enlarge and propagate headward. The conditions for this kind of *piping* (Parker et al, 1964) are (1) sufficient water to saturate some part of the soil or rock above base level, (2) sufficient hydraulic

head to move the water through the subterranean route, (3) a susceptible surficial deposit or rock to convey the water, and (4) an outlet or potential sink for the flow.

The artificial recharge of fluids into reservoir rocks to replace groundwater losses, repressurize oil fields, or dispose of liquid wastes can, of course, stop or even reverse subsidence. The rate of recharge must be carefully controlled since recharge rates in excess of the ability of the rock to accept the fluid (permeability) will result in *hydrostatic pressures* in excess of the *lithostatic pressure* and cause fracturing or faulting. Small earthquakes at Denver, Colorado, have been documented as caused by overpressurization during deep-well waste disposal activities at the Rocky Mountain Arsenal nearby, and similar occurrences are known from oilfields in Colorado, Texas, Utah, and California (Piper, 1970).

Highly organic soils that form near the water level or in present or past saturated areas such as muskegs, bogs, marshes, or deltas present particularly difficult engineering problems because of their low bearing strength and the fact that their volume may be reduced by drainage or oxidation with consequent subsidence. Most of the mass of organic soils is in their water content so they have a specific gravity in the order of unity when saturated but less than 0.1 when dried. These soils also oxidize and shrink on drying and are a fire hazard. Drainage of organic soils is often done to provide for farming or urbanization. As a result wetlands subsidence has affected some 20 000 km² in the United States and is estimated to cost $58 million annually. The larger affected areas in the United States are parts of the Everglades in Florida and the delta areas east of San Francisco and near New Orleans.

Subsurface drainage through cavernous limestone or other soluble rocks may be accelerated by such activities as dewatering of mines and quarries or regional groundwater withdrawal. The head loss is made up by increased infiltration through sinkholes and consequent drawdown of the water table in the soil over the tributary area. Sinkholes are the means of entry of water and detritus to cavern systems and, when active, are seen as dry, closed depressions above relatively narrow, usually vertical

Figure 13.16 Subsidence related to mining

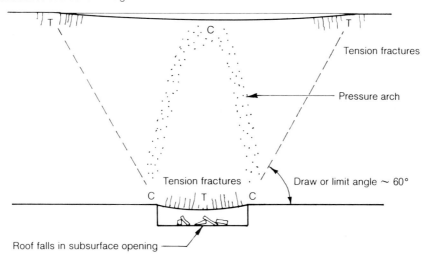

Tension fractures

Pressure arch

Draw or limit angle ~ 60°

Tension fractures

Roof falls in subsurface opening

KEY

T Area of tension
C Area of compression

conduits from the cavern below. Increased seepage through the soil is concentrated downward into the lower outlet and causes erosion with possible piping at the base of the soil which propagates upward, eventually resulting in a catastrophic collapse of the surface. It should be noted that the danger is the collapse of the overlying soil and not dissolution of the cavern roof. Mead (1936) placed sinkhole collapse hazards in perspective when describing the bearing strength of cavernous rocks in connection with dam construction in the TVA system. The rate of solution of limestone, though rapid as a geologic process, is too slow in terms of years to constitute a threat. The concrete of the dam is more vulnerable to solution and alteration than is the limestone.

Catastrophic subsidence involving another fluid, magma, or volcanic gas and debris is a common accompaniment of volcanic activity when large volumes of material are transferred from depth to the surface. Calderas and cauldron subsidences may represent volume losses of tens or even hundreds of cubic kilometers.

Loss of surface support   Loss of surface support and consequent subsidence is a natural result of underground mining and is today a major concern in the design and operation of mines, tunnels, or other subsurface excavations. Mine openings are usually supported while in use and may be back-filled as they are abandoned. However, some openings and unconsolidated backfill remain as causes of potential subsidence. Subsidence above openings in rock is typically slow, and decades may pass before the effects are seen at the surface. The mechanics of mine subsidence in the simplest view may be envisioned as the compound bending of a beam represented by the overlying rock as it sags into the underground opening (fig. 13.16). As in any bent beam the convex side is placed under tension, T, and the concave side compressed, C, as indicated in the figure. The result is a *pressure arch* over the opening that gradually works upward to the surface. The upward progression of the arch involves an increasingly broader area of deformation, and surface effect will be felt over an area many times the size of the opening depending upon the depth to the opening and the rock properties. Structures and buried utility lines will be tilted and placed under stress, possibly even tension followed by compression. In the mine, roof rock under tension spalls into the opening, and having an increased volume because of increased voids, the amount of potential subsidence is thus the mined volume minus the void space generated.

RIVER FLOODING

Most river valleys have been excavated by the streams that occupy them and thus the dimensions and configuration of a stream channel and a valley floor represent the *time-average size and shape* required by the stream to accommodate its discharge.

Flood Plains

The trunk and tributary streams serving to remove surface water from a watershed ordinarily accomplish this function by flow in well-established channels; occasionally, however, these channels are inadequate and the excess water spreads from the channels across the valley floor as a **flood.** The valley floor will be used only during occasional high-discharge conditions and for a relatively short period, but this extra capacity is an integral and essential part of the earth's surface water transport system and reduction in its area or modification of its shape may lead to serious consequences. The **flood plain** is thus an active component of the stream and is used as a safety valve when the need arises.

Unfortunately, the flood plains of rivers are among the more desirable places for agricultural and urban development. Farms are developed on the fertile flood-plain soil, cities rise on the level ground near river transportation and water supply, and highways and rail lines utilize the easy grades. As a result, valley floors have often become choked with the works of civilization with a consequent deterioration in their capacity to handle flood waters. In the United States each year about 200 people are killed in floods and property damage is in the order of 3 billion dollars. The frequency and devastation of flooding will continue to rise unless avoidance or engineering control measures keep pace with urban and agricultural development.

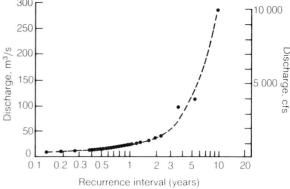

Figure 13.17 Flood frequency curve

From L. B. Leopold, U.S. Geological Survey Circular 559, 1968.

Drainage Basins

Amelioration and control of flooding requires that the flooding process be understood in order that integrated flood-control programs are intelligently planned and conducted. Such programs will typically involve large portions of drainage basins and require cooperation among a number of political entities.

For any specific drainage basin the primary input data is derived from rainfall records, gauging stations on the rivers, historical flood records, and information regarding the type and extent of vegetation cover. Most areas have accurate rainfall records and the U.S. Geological Survey, NOAA Weather Service, and other government agencies maintain gauging stations for flow records. From this data the statistical probability of flooding (see chapter 12) and flood frequency curves such as shown in figure 13.17 can be prepared.

The general factors and relationships that are operating in a drainage basin may be visualized by considering various model conditions. Assume first a modest rainfall or snow melt affecting a restricted upstream area. Before the storm the ground surface is unsaturated, the vegetation is in leaf, and the stream channel half-filled; thus significant quantities of water can be absorbed into the ground and transpired from the vegetation, and the runoff fills but does not overtop the stream banks. Now change any of the initial parameters in a way to increase the

James River in flood, Virginia

Flood damage in Mississippi

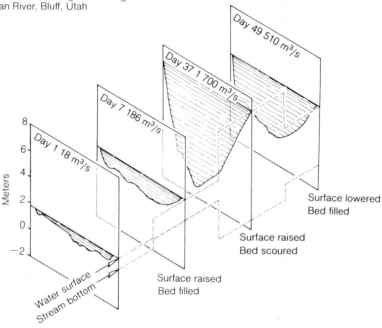

Figure 13.18 Changes in stream regimen during the 1941 flooding of the San Juan River, Bluff, Utah

Based on data in L. B. Leopold and S. T. Maddock, U.S. Geological Survey Professional Paper 252.

runoff—by a construction activity, urbanization, a sudden rain that quickly saturates and thus seals the upper portion of the soil, a previous rainy spell that sealed the soil pores, a change in season or temperature so the soil is frozen or leaves absent and transpiration reduced, or a higher initial level of the stream. The result may be the overtopping of the stream banks and a localized flash, or upstream flood.

Greater than average discharge results in an increase in both stream cross section and velocity. Increased velocity engenders increased turbulence that, in turn, results in the suspension of stream bed sediments thereby enlarging the channel section and causing the significant erosion associated with streams in spate. At the same time the increased turbulence decreases the effective discharge of the stream. Figure 13.18 gives a pictorial view of the changes in stream regime needed to accommodate flood conditions.

Downstream these floodwaters enter the trunk stream and increase its discharge. Before flooding, however, the tributary may have contributed only a few percent of the trunk discharge, and the increased water volume may not be enough to take the main stream overbank, thus restricting the flooding to an upstream area. Finally, imagine that a condition may exist in which a number of tributaries are at bank-full stage and that the sum of their discharge is greater than can be handled by the main channel; therefore, the main channel banks are overtopped resulting in downstream flooding although none exists upstream.

Downstream valleys are often broad with more or less well-developed flood plains and the channel is often bordered by natural levees formed from sediments deposited by previous floods. Downstream flood waters will usually rather quietly fill the lower ground between these levees and the valley walls and be temporarily stored. Sheet floods that fill the entire valley with rapidly flowing water are rare and usually the consequence of a sudden release of ponded water such as a dam failure.

Flood Plain Management

Management of the flood plain is the modern approach to the control of flooding and its effects and involves a complex interaction of engineering, geology, agriculture, forestry, and planning. The main

Figure 13.19 Hydrologic factors operating in a drainage basin

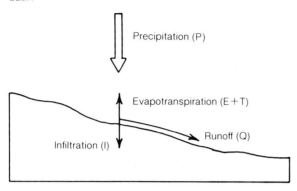

Figure 13.20 Flood hydrographs

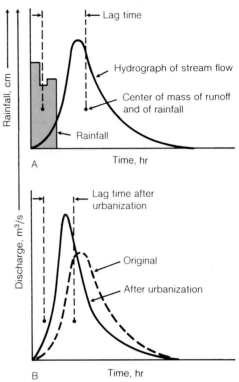

From L. B. Leopold, U.S. Geological Survey Circular 559, 1968.

aspects of flood plain management involve the following:

Drainage basin analysis, flood prediction, flood statistics, frequency prediction, and the collection of stream-gauging data

Flood control, both structural and nonstructural

Flood proofing to minimize damage and loss of life

Economic considerations, regional and urban planning

All of these areas are engineering-related and it is appropriate to consider them from that aspect.

The general factors and relationships operating in a drainage basin are shown in figure 13.19 and may be expressed as $P = Q + (E + T) - I \pm S$ where P = precipitation, Q = runoff, E = evaporation, T = transpiration by plants, I = infiltration into soil and rock, S = water held or discharged from dam storage; and $Q = crA$ where c = coefficient of runoff, r = intensity of rainfall, A = area of the drainage basin.

Examination of the parameters of these two equations will quickly show that although many are fixed by climate and geography (P,E,r,A) a number may be modified by human activity.

The value of c, the coefficient of runoff, is greatly increased and infiltration, I, concomitantly reduced as surfaces are sealed by housing, parking lots, and roads in consequence of urbanization. Farming and construction work likewise increase c because of the removal of vegetation, particularly grasses and forest duff, that would ordinarily slow surface flow and enhance infiltration. Transpiration,

T, is a function of the leaf area of plants and is reduced by the removal of forests and grasslands.

It is usually difficult to significantly rectify these human-induced parameters even though their deleterious consequences are well understood. Rather, engineering solutions in the way of flood control dams upstream from flood-prone areas ($+S$) and flood control structures such as flood walls and levees at the flood-prone point are built to accommodate an increase in Q.

In a typical flood sequence the rainfall or snow melt begins at time zero and the hydrograph of stream flow rises toward the peak or flood (fig. 13.20a). There is a lag time due to the absorption of water into the soil and rock that, in turn, become saturated and the runoff increases. The lag time is the time distance on the graph between the center of mass of the rainfall and runoff. If the region is

Figure 13.21 Comparison of the rainfall–runoff relationships for preurban and urban conditions, Nassau County, New York

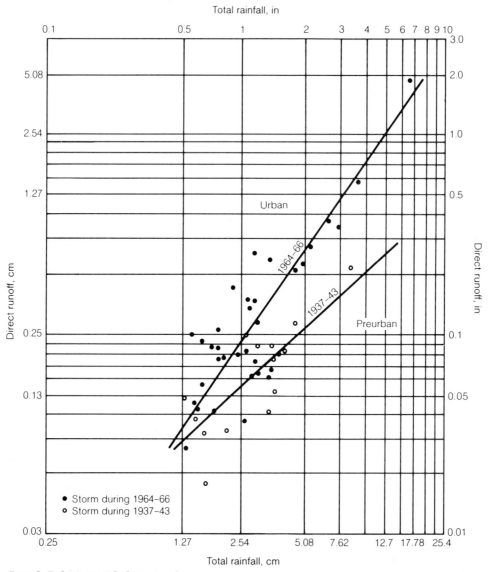

From G. E. Seaburn, U.S. Geological Survey Professional Paper 627 B, 1969.

developed or urbanized the lag time is decreased due to instant runoff from paved streets, parking lots, houses, and ploughed fields and the initial peak discharge is greater and occurs earlier in the flood sequence (fig. 13.20b) increasing the severity of the flood. A comparison of rainfall–runoff relationships for preurban and urban conditions for one subarea of a stream are shown in figure 13.21 and the runoff–frequency relationships for a similar area are shown in figure 13.22. The effect of human development and activity is evident in both cases.

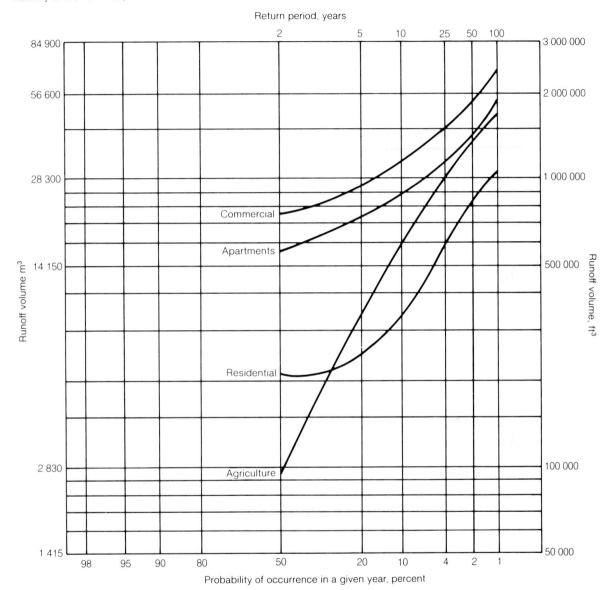

Figure 13.22 Runoff volume-frequency curves of a small tributary of the Fox River, Illinois

Figure 13.23 Bank storage of flood waters

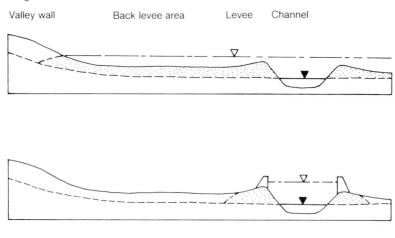

Valley wall   Back levee area   Levee   Channel

KEY

 Bank storage

▼ Normal stream level

▽ Flood level

--- Water table

Flood Control

Flood control measures fall into two broad categories, either structural controls or nonstructural methods. Structural flood control involves the creation of fixed engineering structures or earthworks to control the discharge, sediment load, and the shape of the channels and rivers carrying the runoff. Channels may be straightened and modified by unlined earthworks or lined with concrete or rock retaining walls (channelization). The lined channels produce almost instant runoff, which quickly conveys water from an urban area but can add to the magnitude of the peak flood surge at downstream points. Unlined channels are subject to erosion but do have the advantage of allowing water to soak into the surrounding flood plain thus partially maintaining the lag time between precipitation and the peak flood surge as well as reducing the height of the flood crest.

Rising flood waters cause the water table in the vicinity of the river to rise and the consequent transfer of surface water of the river to groundwater in its banks may provide a means for the storage of a significant quantity of the flood waters (fig. 13.23). Because equilibrium is seldom attained, the water table will usually have a relatively steep outward-facing surface that migrates away from the stream as flood waters rise. With falling flood levels the stored water is gradually returned to the channel, thus prolonging the flooding episode. It should be noted that channelization by artificial levees greatly restricts the amount of water that may be stored in the banks and hence causes flooding to be more intense although of shorter duration.

Flow can also be controlled by earth levee banks along the course of the stream to confine the flow to the main channel and prevent overbank flooding. Many cities along major rivers are protected by concrete floodwalls that have road access gates, which allow normal traffic through the wall but can be quickly closed by floodgates in an approaching crisis (fig. 13.24).

One of the more effective flood control measures is the use of storage dams. These usually take

Figure 13.24 Types of channel improvement

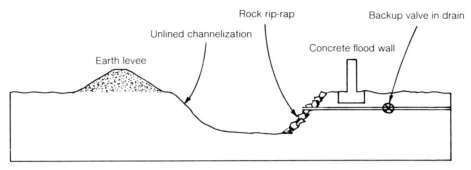

Figure 13.25 Multiple-use dam

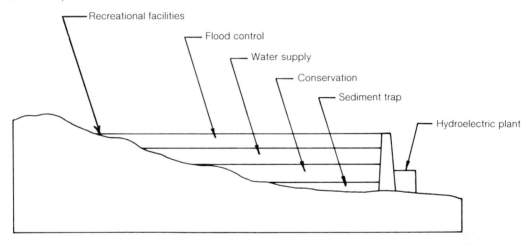

the form of multiple-use reservoirs utilizing a portion of their capacity for flood control, the major storage for water supply, and a reserve for conservation and drought backup. Many such dams are equipped with a hydroelectric power generation capability and engineered recreational facilities thereby helping cost-effective economics (fig. 13.25). Compromises in operation are necessary in multiple-use dams since the maintenance of high water for recreational purposes, water supply, and a head for efficient power generation is not compatible with an adequate storage capacity for flood control. In practise the reservoir level can be lowered prior to the flood season and when the initial flood stage arrives water can be stored until the downstream areas are clear and then released.

Flood proofing is being increasingly introduced into urban areas where flooding historically occurs. The methods include the construction of levees around individual buildings or small communities, the use of elevated floor levels, and the construction of watertight buildings. Construction materials that are less susceptible to flood damage are employed and building design places the damageable contents above flood level.

Zoning and the control of land use coupled with building regulations are widely used to minimize flood damage. A series of flood zones can be delineated for any portion of a river course and its flood plain from the data derived from drainage basin analysis and historical records (fig. 13.26). The most frequently flooded area should be kept clear or used

Figure 13.26 Flood zoning

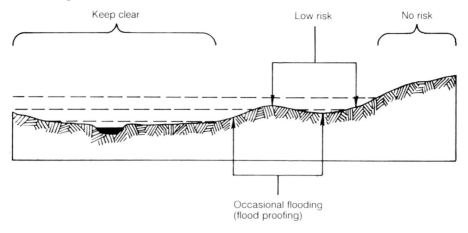

Keep clear • Low risk • No risk • Occasional flooding (flood proofing)

Figure 13.27 Idealized summary of floodplain hydrology management

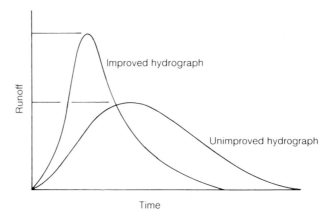

Runoff • Improved hydrograph • Unimproved hydrograph • Time

for such flood-insensitive activities as park areas, playing fields, or parking lots. The area of occasional flooding can be protected by flood proofing and the area of low risk is generally acceptable for standard construction and urbanization.

The general aims in floodplain management (fig. 13.27) are directed toward increasing the flood protection of an urbanized area to accommodate increased peak flood surge and the decrease in lag time for runoff. The key point is that areas under the improved and unimproved runoff-time curves are the same, but the distribution of flood peaks is different.

SUMMARY

Natural disasters may have natural causes, but in many instances their onset can be traced to the unthinking acts of engineers who, in an effort to improve a situation, actually worsen it. Natural slopes, for example, represent an equilibrium established over a very long time and can be expected to respond to loading or cutting. Similarly, stream channels and valleys are the expression of a natural equilibrium that is adjusted to accommodate the often large differences in a stream's discharge. Modification of this discharge by dams or flood control structures or by changes in channel configuration must be expected to produce responses in the amount and distribution of water flow and accompanying erosion and sedimentation.

The cause–effect relationships of hazards such as slope failure, subsidence, and flooding are reasonably well established, but each new situation must be carefully evaluated.

ADDITIONAL READING

Attewell, P. B., and Taylor, R. K., eds. 1984. *Ground movements and their effects on structures.* New York: Chapman & Hall.

Brunsden, D., and Prior, D. B., eds. 1984. *Slope instability.* New York: John Wiley & Sons.

Chowdbury, R. N. 1978. *Slope analysis developments in geotechnical engineering,* v. 22. New York: Elsevier Science.

Costa, J. E., and Baker, V. R. 1981. *Surficial geology: Building with the earth.* New York: John Wiley & Sons.

Eckel, E. B., ed. 1958. *Landslides and engineering practice.* Highway Research Board special paper 29, NAS-NRB Pub. 544. Washington, D.C.: National Academy of Science.

Holzer, T. L., ed. 1984. *Man-induced land subsidence.* Boulder, Colo.: Geological Society of America.

Huang, Yang H. 1983. *Stability analysis of earth slopes.* New York: Van Nostrand, and Reinhold.

Lofgren, B. E. 1969. Landslides due to the application of water. In *Reviews in Engineering Geology.* V. 2. ed. D. J. Varnes and G. Kiersch. Boulder, Colo.: Geological Society of America.

Pestrong, R. 1974. *Slope stability.* New York: American Geological Institute, McGraw-Hill.

Poland, J. F., and Davis, G. H. 1969. Land subsidence due to the withdrawal of fluids. In *Reviews in engineering geology.* V. 2. ed. D. J. Varnes and G. Kiersch. Boulder, Colo.: Geological Society of America.

Schuster, R. L., and Krizek, R. J., eds. 1978. *Landslides, analysis and control.* Washington, D.C.: Transportation Research Board, National Academy of Science.

Scullin, C. M. 1983. *Excavation and grading code administration, inspection and enforcement.* Englewood Cliffs, N.J.: Prentice-Hall.

Sharpe, C. F. S. 1968. *Landslides and related phenomena: A study of mass movements of soil and rock.* New York: Cooper Square.

———. 1950. Mechanisms of landslides. In *Application of geology to engineering practice* (Berkey Volume). Boulder, Colo.: Geological Society of America.

Trifunac, M. D., and Brady, A. G. 1975. On the correlation of seismic intensity scales with the peaks of recorded strong ground motion. *Bulletin Seismological Society of America,* v. 65, no. 1.

Voight, B., ed. 1978. *Rockslides and avalanches.* Parts 1 and 2. New York: Elsevier Science.

Zaruba, Q., and Mencl, V. 1969. *Landslides and their Control.* New York: Elsevier Science.

14

Pollution and Waste Disposal

INTRODUCTION

The nature and control of pollution comprise a topic far beyond the scope of this text. This chapter addresses only some general concepts and a few situations that are strongly related to the geologic environment: water quality, siltation, hazardous waste disposal, and deep-well injection.

The disposal of hazardous liquid wastes by deep-well injection is fundamentally the reverse of petroleum production, so the description here serves also to present important aspects of this industry. Underground disposal of solid waste involves the mining of cavities and follows the precepts presented in chapter 18. Information regarding siltation pollution can be extracted from chapters 5 and 7.

WATER QUALITY

Undoubtedly, the most serious large-scale and long-term pollution problems having a strong geologic context are those related to surface and underground water. The drainage of all water falling on the earth, unless returned to the atmosphere by evaporation, is either by surface streams or by underground movement. Water supplies are obtained from both of these sources and any dissolved or suspended impurities will cause a deterioration in quality. With the quantity of water on earth remaining essentially constant and the scale of man's activities and population steadily increasing the problem of an adequate supply of clean water becomes more acute with time. In many parts of the world deterioration of water quality has affected whole countries and water is unhealthy if not undrinkable.

The problem is not local but regional. For example, where water supplies are obtained from a river for a succession of cities down the fall of the stream the return of water waste in each case makes the need for filtration and purification for those users farther downstream increasingly important. Sewage and waste water generated in an urban center is typically treated by using settling and chemical treatment ponds where solids and dissolved substances can be removed with the help of mechanical and earth filtration before return to the river. The river itself should be kept flowing so that sunlight and bacteria can enhance the purification process, and in rivers controlled by dams periodic discharges of water should be made to ensure that the water does not become stagnant. Cities represent *point sources* for potential pollution along a river; but intermediate stretches may receive pollutants from *diffuse sources* so the water entering the municipal system must also undergo treatment.

Surface and underground water can easily become polluted with various chemicals or effluents from overloaded septic sewer systems; infiltration from poorly designed sanitary landfills; runoff from fertilized fields, feed lots, and surface dumps; acid rain; tailings ponds; effluents from isolated industrial plants, and many other point and diffuse sources.

Many large coastal cities of the world have for many decades been disposing their sewerage into the oceans through outfalls or dumping untreated garbage by barge or scow into an offshore bay or the open ocean with little reference to such factors as tides, currents, or bottom conditions. New York City and Boston are examples as well as such cities as Sydney, Australia, where the refuse from meat plants was attracting sharks and creating a hazard at swimming beaches. Sewer outfalls around Boston Harbor have provided nutrients for algal growth that periodically fouls the beaches and shore. Studies of bottom sediment composition off the mouth of New York Harbor after several decades of dumping revealed a dramatic increase in base metals such as copper, lead, and zinc in the sediment due to leaching from the dumped refuse. In other cases metal cans and other items have been returned to shore by winds, tides, and currents.

SILTATION

Among the more serious kinds of pollution of the earth's surface and one of direct interest to geologists and engineers is *siltation*. All natural drainage systems and coastlines involve silt and sand transport and tend to approach an equilibrium, but when any of the factors involved in this natural equilibrium are altered the sediment–water system will either be subject to *deposition* or excessive *scour*. Two important factors in siltation are the supply of sediment to the system and the diminution of the volume or velocity of the water needed to carry the sediment. If the supply of sediment increases or the flow of water decreases, then deposition will take place. If the converse is the case, scour will occur.

The more serious problems occur with deposition or siltation. The multiplication of earth-moving activities in the fields of mining, agriculture, road building, and general construction has dramatically increased the supply of sediment to drainage systems, and at the same time the flow of water has been decreased by dams and by greater demands for water from drainage systems with only portions being returned. These human-induced changes in the natural equilibrium of sediment transport by streams and longshore drift cause the silting of rivers and harbors and the interruption to navigation on waterways. The problem is usually tackled by *maintenance dredging,* in itself a costly procedure and only a temporary solution, or by the construction of structures to modify water movement along river

Figure 1   Siltation, gold mill tailings, western Australia

Figure 2   Siltation, gold mill tailings, Venezuela

Siltation problems may be aggravated by the disposal of waste from mining and milling activities, which remove the valuable substances from the ore but often generate large volumes of fine-grained rock waste. The silts and slimes should be held and dewatered in specially designed ponds or returned underground to fill mined cavities. To the detriment of the environment this is not always done.

Figure 14.1 Waste production

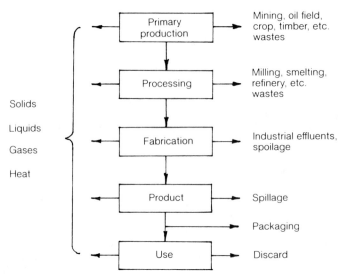

banks and coasts. In general, any structure on a river or coast that is built without regard to the hydrologic and engineering conditions can be expected to add to siltation or erosion problems.

WASTE PRODUCTION

Waste in the form of heat or matter is an inherent part of any process, and although the amount of waste may be reduced by good management and recycling, it can never be eliminated. Waste production is thus an integral aspect of any process and an understanding of materials, their rates of transformation, and the products generated is essential to rational waste management and disposal. Sooner or later wastes must be returned to the geologic environment. Consequently, the geologic features of the disposal site are critical determinants for nonpolluting disposal.

The accumulation of waste materials will, of course, occur in any circumstance in which their production rate exceeds the rate of their transformation to innocuous or reusable substances and it may be noted that such accumulation is not limited to human activities; coal, for example, formed because natural oxidative processes fell behind the rate

of accumulation of decaying vegetation. Humans, however, are such efficient converters of substances for their own benefit that their rates of waste generation typically far exceed the natural transformation rates available to convert the multitudinous waste materials to nonhazardous, nontoxic, or nonobnoxious forms.

The disposal of most waste materials does not ordinarily require extensive geologic insight and is readily handled by procedures following well-established principles. These methods are typically least-cost approaches designed to modify the waste in such a way as to allow it to reenter the cycles of inorganic and organic processes of the earth, to sequester hazardous or toxic substances in localized sites, or to reuse all or part of the waste material. Briefly, these methods might be termed *disperse* and *dilute, concentrate* and *contain,* and *recycle.*

The physical and chemical natures of waste embrace a gamut of solid, liquid, and gaseous substances including rock, sewage, vegetation, industrial effluents, household trash, garbage, and heat. To appreciate the complexity of the waste management problem it is valuable to consider the variety of waste-producing steps that enter into the making of some particular product (fig. 14.1). To appreciate

Waste pond, barite workings, Tennessee

Waste pile, oil shale workings, Scotland

the magnitude of the problems that must be resolved if we are not to irreversibly foul our nest, consider the wastes generated in the making of a number of products—for example, household electricity, an ordinary pencil, gasoline, breakfast food, paper, copper wire, or hydrochloric acid.

It is not the purpose of this text to explore the ways and means of waste management or to discuss the engineering aspects of landfills, sewage plants, and similar well-understood means of pollution control or waste disposal since the geologic principles involved are straightforward. Rather, it seems important to address the special problems that must be resolved in the disposal of toxic or hazardous substances by their emplacement in secure surface situations or at depth within the earth in deep wells or in mined cavities.

DISPOSAL OF HAZARDOUS WASTES

Deep-well injection is commonly employed as a means of disposal of liquid wastes that cannot otherwise be safely or economically treated. The method is inherently costly and complex and will therefore be employed only under special circumstances. However, it is so closely related to the production methods for oil, gas, or water; the recovery of solid materials by solution mining; and the use of subsurface storage for petroleum products that the technology is well understood and warrants a general description.

The use of mined cavities for long-term storage of high-level radioactive wastes has been much discussed and is the method most likely to be adopted. Obviously, a special mine geometry, relatively impermeable rock in a geologically stable locale, and a carefully designed monitoring system are essential.

Disposal of Hazardous Solid Wastes

Subtitle C of the Federal Resource Conservation Act of 1976 sets forth a number of general requirements for solid waste disposal including regulations for the construction and operation of storage and disposal facilities, the registration of waste generators, and the establishment of a manifest system to track wastes from their point of generation to final disposal. Under this act *hazardous solid waste* is broadly defined as material

> which because of its quantity, concentration or physical, chemical, or infectious characteristics may: (a) cause, or significantly contribute to, an increase in mortality or an increase in serious irreversible, or incapacitating reversible, illness; or (b) pose a substantial present or potential hazard to human health on the environment when improperly treated, stored, transported or disposed of, or otherwise managed.

Recent estimates put the annual production of hazardous wastes in the United States at 150 million tonnes; using a bulk density of 1.5, this amounts to 100×10^6 m^3 or enough to cover a standard township to a depth of one meter.

The Environmental Protection Agency (EPA), which is charged with regulatory supervision of this Act, regards groundwater contamination from land dumping or landfills as the most serious disposal problem. Its regulations, therefore, focus on a philosophy of *total containment* (i.e., prevention of leachate from reaching the groundwater as a result of a lack of knowledge regarding transportation, degradation, adsorption, or synergistic effects of complex pollutants in the ground). Only rarely will subsurface conditions be such that total containment is possible, and most landfills for solid hazardous wastes will probably have to be lined. Two designs for lined landfills suggested by the EPA are shown diagrammatically in figure 14.2.

Radioactive Waste

High-level radioactive waste, having a toxicity more than 1 000 times the acceptable toxic level, presents a particularly difficult problem of disposal since it remains a potent hazard to health for long periods

Figure 14.2 Landfill designs for containment of hazardous waste

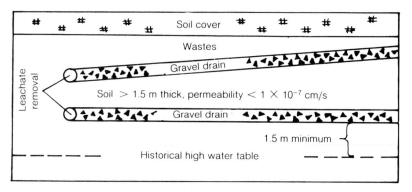

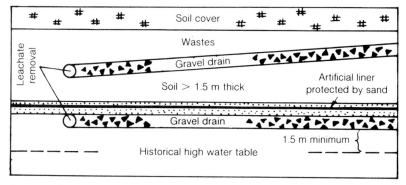

Source: U.S. Environmental Protection Agency.

of time. Ionizing radiation from decaying radioelements may cause burns or anemia from contact, ingestion, or inhalation and, more seriously, damage genetic material.

The basis of nuclear power generation is the spontaneous decay of ^{238}U into a number of daughter products accompanied by the liberation of heat. The decay products are generated in widely different amounts and are themselves radioactive with half-lives ranging from fractions of seconds to thousands of years. The total energy flux from the decay of a particular daughter isotope is given by the area under its decay curve (fig. 14.3a), and some level of activity must represent the environmentally acceptable decay rate (as shown by the dashed line). Obviously, the area under such a curve depends upon both the initial concentration and decay rate (half-life, $t_{1/2}$) of a particular isotope. Figures 14.3b and 14.3c contrast the kinds of decay curves that may be anticipated. Because of these variables, the disposal of radioactive waste will be a function of the material (as solids, liquids, or gases), the initial concentrations of the particular daughter isotopes, and time.

Figure 14.3 Radioactivity decay curves

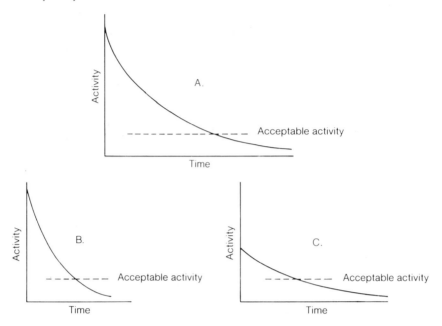

KEY

A. Initial activity eventually decays to an acceptable level,
    perhaps earth background
B. High initial activity, rapid decay
C. Low initial activity, slow decay

Figure 14.4 is the flow sheet for nuclear fuel usage and identifies the sources of high- and low-level wastes that must be disposed of. The recycling of spent reactor fuel by a reprocessing plant reduces, but does not eliminate, wastes with high levels of radioactivity.

The waste materials resulting from nuclear power generation are of essentially two kinds: (1) the *depleted fuel* materials that are very long-lived emitters of nonpenetrating alpha particles and (2) *fission products* that typically emit penetrating radiation and have relatively short half-lives. Actual materials include not only the fuel rods but failed or obsolete equipment, metals, glass, concrete, combustible trash, resins, and liquids. Very **low-level liquid wastes** and gaseous wastes may be safely dispersed in the environment but higher level liquids and gases present serious disposal problems and will

therefore usually be *transformed to solids* by calcination, vitrification, adsorption, or distribution through an asphalt, metal, or polymer matrix. Predisposal waste-handling procedures should allow high-level wastes to cure for 5–10 years on-site to reduce their short-term activity before transforming them to solids for disposal. Decay in the disposal site for a few hundred years will reduce the activity to that of a uranium stockpile, but a quarter million years is needed for the decay to background of long-lived radioactive species.

The best strategy for the disposal of such highly dangerous material, aside from its complete removal from earth, is *burial* in an immobile form at depth within rocks that are impervious and dry, chemically stable, and that will not be subjected to significant mechanical stress. Such rocks would include massive salt deposits, thick shale or volcanic ash beds, or crystalline igneous rocks located away

Figure 14.4 Flow sheet for nuclear fuel usage

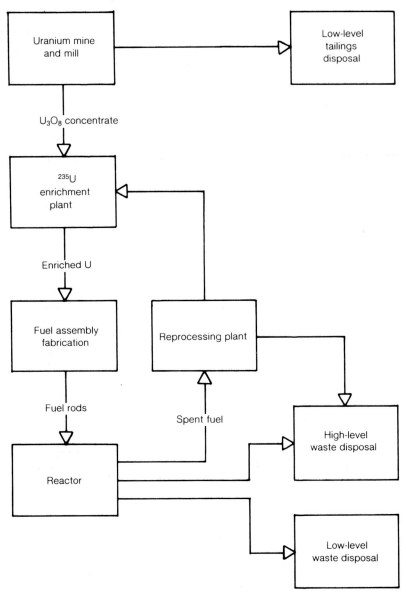

Source: Based on USAEC Report WASH–1174–1973.

from seismically active (earthquake-prone) areas. Rock salt deposits, either as massive bedded formations or salt domes, are considered to be particularly favorable materials because they are abundant, usually occur in regions of low seismic activity, are relatively dry and impermeable, have a self-healing capacity should fractures form, are a good shielding material, and have a high compressive strength and higher thermal conductivity than most rocks.

A number of potential underground disposal sites, which take the kind and location of appropriate rocks into account, have been identified and are shown on map 19 in appendix II. Naturally, detailed site studies including drilling for geologic and hydrologic data, dimensional information, and rock-mechanical testing must be carried out before mining and disposal take place.

The nature of radio-decay is such that self-heating of the material always takes place and in extreme cases may cause the material to melt. It should be noted that the self-heating process is also self-damping because of the physical expansion of the material and consequent separation of atoms. With the possible exception of plutonium, no radioactive wastes from nuclear power generation are sufficiently enriched so that the critical mass required for an explosion can occur. (Natural uranium contains about 0.5% ^{238}U and is enriched to about 3% for use in thermal reactors; bomb-grade uranium must be better than 97% ^{238}U.) Naturally, radioactive wastes will not be concentrated for disposal to the point at which significant heating occurs. The disposal problem is thus one of placing containers of radioactive wastes, which have been solidified or dispersed through an inert carrier (glass, concrete, etc.), in safe permanent storage equipped with means to monitor the facility.

The disposal arrangements may be such as to allow accessibility to the stored containers in order that they may be maintained in sound condition or recovered at some future time, or may be for permanent inaccessible burial. The tunnel system in use at the Nevada Weapons Testing Site is an example of the latter system. The excavation of either type of storage would follow the precepts of more usual

underground mining, (chapter 18). For inaccessible disposal the mining plan should provide for the formation of numerous isolated disposal pockets to preclude accidental interchange of the waste. Pockets disposed in a hexagonal array would allow least mining for most separation and a disposal mine plan might be as shown in figure 14.5.

Disposal of Hazardous Liquid Wastes

The most complete understanding of the physics and technology relating to fluids deep underground is provided by the petroleum industry and is generally described in terms of *reservoir engineering*. Since deep-well injection as a means of disposing of hazardous liquid waste must, in most ways, be reciprocal to processes of oil and gas recovery and directly related to the means employed for repressurization and secondary recovery from subsurface reservoirs, the literature of reservoir engineering provides the obvious information for study.

Reservoir engineering Subsurface reservoirs suitable for disposal of liquid wastes will have the same general characteristics as those that held petroleum for millions of years and, in practise, may well be an exhausted oil field for which many of the reservoir characteristics are known. Additionally, the effects of injection may be documented if the reservoir has been produced or undergone secondary recovery by means of a water drive. Typically, such reservoirs are bodies of permeable rocks whose contained fluids are constrained from migration by impermeable barriers in the way of a concave-upward cap, a gouge-filled fault zone, or an up-dip decrease in permeability. Some particular geometries for fluid entrapment are shown in figure 14.6.

The major kinds of reservoir *energy* available for oil production are the compression of fluids within the reservoir, the gravitational energy difference between upper and lower portions of the fluid, and the energy of compression and solution of dissolved gases. The energy of fluid compression is generally of minor importance; gravitational forces tend to cause segregation of fluid phases of different density. The gas content of oil reservoirs is the principal source of energy for the expulsion of oil. Petroleum

Figure 14.5 Plan for mine disposal of radioactive waste

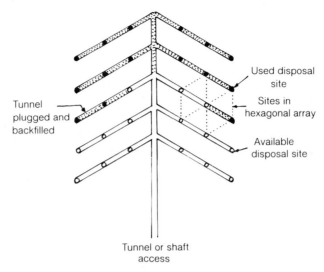

Tunnel plugged and backfilled

Tunnel or shaft access

Used disposal site

Sites in hexagonal array

Available disposal site

Figure 14.6 Distribution of fluids underground in permeable horizons

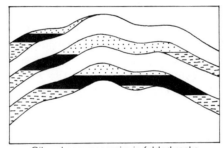

Oil and gas reservoirs in folded rocks

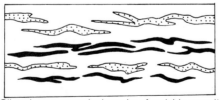

Oil and gas reservoirs in rocks of variable porosity

KEY

■ Oil saturated
▨ Gas saturated
▤ Saltwater saturated

Figure 14.7 Reservoir pressure vs depth for Louisiana Gulf Coast wells

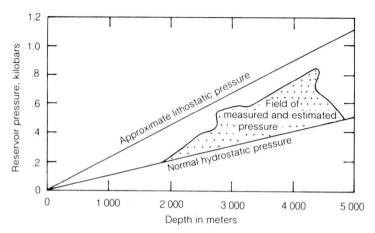

fluids will be depleted in a reservoir used for waste disposal and, particularly because of the absence of free or dissolved gas, may present a somewhat simpler hydraulic picture.

Fluids underground completely fill the pore spaces provided between grains or within joints in rocks, and if these pores are connected the fluid will move under the influence of a pressure gradient. This flow through porous media follows the relation of Darcy (chapter 6) if the reasonable assumptions are made that the fluid is a single-phase substance, that there is no physical–chemical interaction of the fluids and the grains, that pressure gradients and hydraulic heads are equivalent and interchangeable, and that the system is one of laminar flow.

Studies of multiphase systems (oil-dissolved gas–water) in the petroleum industry have shown the permeability coefficient to be dependent only on the average fluid viscosity, and the system can thus be considered in terms of a single-phase fluid. Chemical interaction of petroliferous fluids and their containing rock is limited to nil but could be important for caustic fluids. Friction factors in fluid movement reduce the transition range for laminar to turbulent flow to a Reynolds number of about 1 to 10, which is one to two orders of magnitude above measured values, and hence laminar flow will dominate.

Under normal conditions the fluid pressures in an underground reservoir will be hydrostatic (i.e., in equilibrium with a saltwater column extending from the reservoir level to the surface). Reservoir pressures in excess of hydrostatic, which arise from rearrangement of sediment material (principally clays) in consequence of burial and lithification, are not uncommon. However, in no instance will the pressure exceed the lithostatic head, which results from the weight of the overlying rock column and which will be 2.2 to 2.4 times the hydrostatic head. Reservoir pressure-depth curves for some typical Gulf Coast wells that illustrate this over-pressuring phenomenon are given in figure 14.7.

The pressure-production history of an oil field typically follows the relations shown in figure 14.8. Should such a documented field be used for waste disposal it is apparent that waste equivalent in volume to the cumulative oil produced (4.4×10^7 barrels, in this instance) could be safely injected over a period of about 18 years at injection pressures no greater than the initial pressure. That these parameters are only approximate, however, can be recognized when it is remembered that oil or oil–gas solutions do not have the same viscosity or surface tension as aqueous solutions and that production may have modified the initial pressure distributions.

Figure 14.8 Oil field pressure-production history

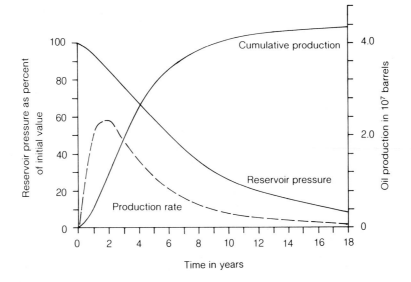

Oil fields that have employed pressure maintenance operations for primary recovery typically have *injection wells* peripheral to the field—gas injection wells to the gas cap to drive the oil down-structure or water injection wells near the water–oil boundary to drive the oil up-structure. In secondary recovery operations, however, the injection wells are commonly interspersed with *recovery wells* throughout the field in such a way as to have injection wells surrounded by recovery wells. A typical layout of wells is shown in figure 14.9a and the steady-state equipressure contours between two wells in the network in figure 14.9b. Sequential positions of a liquid injection front are shown in figure 14.9c wherein the sudden necking to the production well should be noted.

Such a field is a good model for a waste injection system if *monitor wells* are substituted for recovery wells and if a number of incidental but important conditions are taken into account to ensure safe operation. Among these conditions are the requirements that leakage to aquifers above the reservoir or to the surface cannot occur through either "lost" holes or those in the system, that injection rates do not exceed the pressure-containment capacity of the reservoir, that chemical interactions

Figure 14.9 Secondary recovery

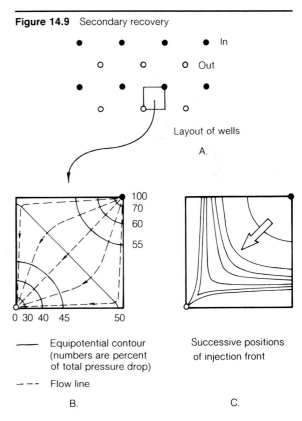

Layout of wells

A.

Equipotential contour (numbers are percent of total pressure drop)

--- Flow line

B.

Successive positions of injection front

C.

with the well casings and reservoir rocks are taken into account, and that the potential effects of plugging of pores by suspended or precipitated solids are considered.

Deep-well disposal system   A general model of a deep-well waste disposal system may thus be described as a closed reservoir of permeable rock, once probably containing petroleum fluids but now saturated with salt water of density 1.01 to 1.10 g/cm^3, into which the waste fluid is injected under sufficient pressure to displace this connate water. Pressurization of the reservoir by injection to levels below that of the lithostatic head will accommodate some waste before venting of the reservoir is required. When this predetermined pressure is reached recovery wells are opened and salt water is removed from the reservoir at points distant from the waste injection location. This water is highly saline and suitable arrangements for its disposal must be made. As injection proceeds, the slug of waste gradually displaces the water without significant mixing because of its laminar flow at low velocities and gradually increases in volume to fill the reservoir. Recovery wells are used to monitor the reservoir mechanics, first to measure pressurization and later to provide for chemical sampling.

SUMMARY

The subject matter of pollution and waste disposal far exceeds the scope of this text. In a broad sense, waste may be considered as the unusable, or perhaps only unused, product of a given process whose safe and environmentally acceptable disposal must be accomplished. Ancient mankind could live with, or even on, the waste products generated by primitive activities, but the modern world generates wastes of such kind and scale that their disposal is essential.

The geologic environment currently and foreseeably represents the only practical disposal site for wastes. Here they may be diluted into large volumes of water (a bad practice) or contained in concentrated form in stable, closed underground volumes.

Recycling of waste materials is an important aspect of waste management and every possible reuse should be investigated. Waste, after all, is a material for which the production costs have been paid, and like ore in the ground, has a zero cost base. Recycling of some metals such as gold, silver, copper, lead, and aluminum is regularly practised. Mine wastes may be used to backfill the mine excavation or as aggregate, saline oil-field brines could be a source of salt and heavy metals, and exhausted uranium reactor fuels can be transformed into plutonium fuel.

ADDITIONAL READING

Blasewitz, A. G., ed. 1983. *The treatment and handling of radioactive wastes.* New York: Springer-Verlag.

Brunner, D. R., and Keller, D. J. 1971. *Sanitary landfill design and operation.* U.S. EPA Report SW-65 TS.

Cheremisinoff, P. E., and Young, R. 1975. *Pollution engineering practice handbook.* Ann Arbor, Mich.: Ann Arbor Science.

Donaldson, E. C.; Thomas, R. D.; and Johnston, K. H. 1974. *Subsurface waste injection in the United States: Fifteen case histories.* U.S. Bureau of Mines Information Circular 8636.

Duckert, J. M. 1975. *High-level radioactive waste: Safe storage and ultimate disposal.* Washington, D.C.: U.S. ERDA.

Fox, C. H. 1969. *Radioactive wastes.* Washington, D.C.: U.S. Atomic Energy Commission.

Galley, J. E., ed. 1968. *Subsurface disposal in geologic basins: A study of reservoir strata.* American Association of Petroleum Geologists. Memoir 10.

Link, P. K. 1982. *Basic petroleum geology.* Tulsa, Okla.: Oil & Gas Consultants International.

McClain, W. C., and Bradshaw, R. L. 1970. Status of investigations of salt formations for disposal of highly radioactive power reactor wastes. *Nuclear Safety.* v. 11.

Muskat, M. 1949. *Physical principles of oil production.* New York: McGraw-Hill.

Pierce, W. G., and Rich, E. I. 1962. *Summary of rock salt deposits in the United States as possible storage sites for radioactive waste materials.* U.S. Geological Survey Bulletin 1148.

Platt, A. M., and Bartlett, J. W. 1976. Alternatives for managing post-fission nuclear wastes. In *Perspectives on Energy.* 2d ed, ed. L. C. Ruedisili and M. W. Firebaugh. New York: Oxford University Press.

Pojasek, R. B. 1979. *Toxic and hazardous waste disposal,* v. 1–6. Ann Arbor, Mich.: Ann Arbor Science.

Purdue Industrial Waste Conference Proceedings, 34th, 1979; 33rd, 1978; 32nd, 1977; 31st, 1976; 30th, 1975. Ann Arbor, Michigan: Ann Arbor Science.

U.S. Energy Research and Development Administration. 1976. Alternatives for managing wastes from reactors and post-fission operations in the LWR cycle. Washington, D.C.: ERDA-76-43.

Williams, R. E. 1975. *Waste disposal in the mining, milling and metallurgical industries.* San Francisco: Miller, Freeman.

15

Engineering and Environmental Planning

INTRODUCTION

The *selection of a site* for an engineering facility and the planning of the nature and timing of the steps to be followed in order to reach a particular goal is a complex process. Much of the process is, of course, nongeologic, but for many projects the contributions of geology are so interwoven with other actions that it is necessary to briefly consider a generalized picture. Figure 15.1 suggests the sequence of events involved in the decision-making process and helps to identify points of geologic input. Note that at each stage some approaches are eliminated and others are modified until an optimum choice can be recommended.

The objective should be stated in broad terms in order that a sufficiently wide selection of means to surely include the optimum solution will be considered. For example, if the objective is flood protection of a town it might be accomplished by building dams or levees, diversion or dredging of the stream, relocation or local zoning of the endangered town, or some combination of these options.

Technologic feasibility includes a knowledge of the engineering art in the way of the possible and

Figure 15.1 Flow sheet for decision making

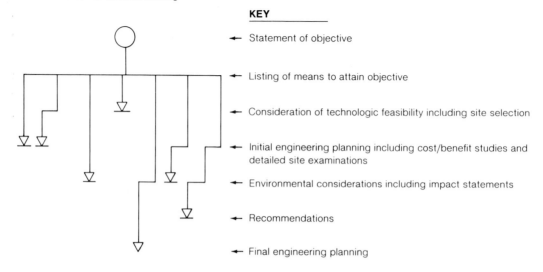

KEY

← Statement of objective

← Listing of means to attain objective

← Consideration of technologic feasibility including site selection

← Initial engineering planning including cost/benefit studies and detailed site examinations

← Environmental considerations including impact statements

← Recommendations

← Final engineering planning

*Steps represent modification of earlier positions

a weighing of the engineering, geologic, and environmental parameters at each site. A good engineering design for one location may well be inadequate or inappropriate at another.

Initial engineering planning is conducted to determine the best among several possible choices. Such planning requires detailed information about the bed rock, the surface and groundwater, the soil, and the other geologic features of the site as well as considerations of engineering design. It is common at this stage to draw up a balance sheet of the monetary costs and benefits of the project. Intangible effects should also be considered but seldom can be reduced to a monetary basis.

Projects that use federal funds and may have *environmental consequences* are subject to the federal Environmental Policy Act of 1969, which directs all agencies of the federal government to

identify and develop methods and procedures which will insure that presently unquantified environmental amenities and values are given appropriate consideration in decision-making along with economic and technical considerations.

The document in present use for this purpose is an *environmental impact statement* that must be filed as part of any licensing procedure. The importance of the environmental impact statement lies in its ability to discriminate between otherwise equally workable engineering approaches by forcing the selection of the plan least damaging (most beneficial) to overall environmental quality.

Considerations of purpose, location, cost, geology, and environmental aspects provide a rational means for the selection among multiple choices and establish the bounds within which final engineering planning may be conducted.

ENVIRONMENTAL SITE SELECTION

The public awareness and concern about the quality of the environment makes it imperative that the engineer be sensitive to these concerns in the planning, construction, and operation of the project. Not only must the engineer conform to various regulations and tenets of good practise on the job but should also give serious thought to those important and often unforseen consequences that arise when construction activities do not anticipate a natural hazard or upset some natural balance. Much of engineering and environmental geology is concerned with the assessment of the risks at a particular site associated with particular natural hazards such as swelling clay, siltation potential, earthquakes, floods, or landslides. This material was previously discussed in chapters 12 and 13 and should be carefully considered in the earliest stages of any engineering site planning and design activity.

Following site selection and preliminary design, consideration must then be given to the environmental impact of the work both during its construction and after its completion. There is an extensive literature dealing with the impact of human activities upon the environment and it is not the purpose of this text to deal extensively with the topic. Some of the deleterious consequences of poor or unthinking practice must, however, be discussed and some of the accepted planning approaches used by engineers and environmentalists pointed out.

Among the less obvious and more important environmental aspects of many engineering works are their effect upon the **hydrologic cycle** in the way of changing the quality and quantity of ground and surface waters, excessive siltation, or disturbance of the biotic habitat. Such hydrologic changes are often subtle secondary consequences related only through a tenuous chain of relationships to the engineering work and hence difficult to anticipate. What, for example, is the connection between the construction of the Aswan High Dam on the Nile in upper Egypt with the loss of fisheries in the eastern Mediterranean?

The selection of an engineering site may be dictated by the exigencies of a particular engineering situation, but the choice is commonly partially or wholly controlled by the *topography* and *geology* on both a regional and a local basis. *Landscape evaluation* and identification of *usage priorities* are part of the recognized discipline of regional planning and engineers should, and often must, work closely with various planning groups.

Land-Use Planning

Basic to land-use planning is the recognition that there is a limit to the land supply, that all land is not the same, that particular characteristics of the land may be of more importance to society than its geographic location, and that the needs of future generations must be considered. In any land-use planning there should be an initial examination of issues, goals, and objectives; a summary and analysis of pertinent data; a land-classification map; and a report that indicates the best possible use of the land after considering and weighing the pros and cons of each possible action. The initial examination, which should be done in collaboration with concerned citizens and agencies, typically considers such things as population trends; regional infrastructure; natural resource conservation; and protection of scenic, cultural, and historic sites.

The pertinent data required for intelligent planning encompasses a wide range of material; the following outline, which follows Keller (1982), suggests the nature and scope of the data needed:

Present conditions
  Population and economy
  Current land-use
  Existing plans, policies, and regulations at local, state, and federal levels
Constraints
  Land potential
    Physical limitations. Hazard areas, soil limitations, steep slopes, . . .
    Fragile areas. Wetlands, dune areas, beaches, scenic areas, unique wildlife habitats, . . .
    Resource potential. Woodlands, farmlands, mineral deposits, . . .
    Capacity of community or regional facilities. Water, sewer, roads, schools, . . .
Estimated demand
  Growth of population and economy
  Future land needs. Minimum 10-year projection
  Facilities needed

Land-classification map A land-classification map is the heart of a land-use plan and provides in an integrated and condensed form a view of various political levels of land-use policy, an investment guide, a framework for planning of construction and avoidance of geologic hazards, a means of coordinating regulatory policies and decisions, and a means of estimating equable land taxation.

Land-classification mapping typically begins with a *soil map*, which provides site information on the distribution of soil types and some data on their physical characteristics.[1] Other properties of interest may be determined by appropriate testing and added to the map. The result is a *base map*, which is then overlaid by other constraints such as the locations of slide-prone slopes, swampy areas, or mineral resource areas. Next the current land use information is superimposed and finally the projected plan of utilization.

The land-classification map must include information on hazards and usage presented in an understandable fashion since the goal is to assist, and not confuse, the planner and the public. Care must be taken to present the information in a useful and clear-cut form that directly or indirectly emphasizes present usage, hydrologic conditions, and engineering properties of the soil and bed rock. A useful approach is the drafting of *colored transparencies* that may be superimposed on the map to show negative factors (in red) and positive factors (in green). Negative factors would include the geologic hazards of unstable slopes, flood and earthquake risk, and poor drainage; and positive factors might indicate the capability of an area to support a particular activity such as a solid waste disposal system, high-yield agriculture, or to provide stable foundations.

A process called *physiographic determinism* has emerged in recent years as a working philosophy for site selection and evaluation. Basically, the concept is to discover a plan already provided by nature rather than lay down an arbitrary design. A graphic approach for this "design with nature" may be used whereby the site selects itself with respect to the parameters employed. For example, consider the routing of a highway through an area in which optimum construction conditions (foundations, grades, fill, and cuttings) and minimum social disruption are desired. When the pertinent factors have been identified, three or more grades of the particular value are plotted as shades of grey on transparencies. One transparency is used for each factor, and when the

transparencies are superimposed, there will always be a relatively lighter area on the map that will define the optimum location.

A particularly difficult aspect of planning and site selection has to do with *scenic values*. Determination of such values is necessarily subjective and the results often anomalous; most people, for example, consider natural cliffs to be scenic, roadcuts a neutral feature, and the high-wall of a contour strip-mine to be ugly although the composition and geometry of each may be identical. An approach to this problem is to use a form such as figure 15.2 and insert the numerical values in the formula $SQI = LC + NC - MC$ where SQI = scenic quality index, LC = landform characterization, NC = natural characterization, and MC = human-made characterization.

Environmental Impact Statements

Current federal and state statutes stemming from the National Environmental Policy Act of 1969 require that any engineering work utilizing federal funds be authorized only after an acceptable case has been made that the project will not adversely affect the environment. The case is made by means of an environmental impact statement and guidelines have been promulgated by the Council on Environmental Quality in order to assure that both the letter and spirit of the Act are followed.

A statement of environmental impact that provides a basis for intelligent judgment as to the overall effects of a given project has become an integral step in the planning and licensing process. Typically such statements are both generated and reviewed by interdisciplinary teams of engineers, physical scientists, and social scientists.

Development of the statement Various approaches to the *development* of adequate impact statements have been used and only one is described here. Leopold and others (1971) described an approach that leads rather directly to a relative weighting of critical items to be considered and an outline of the topics to be covered in a particular statement. First, a matrix is set up incorporating a

1. Published by the Soil Conservation Service, U.S. Department of Agriculture.

Figure 15.2 Evaluation of scenic values of a landscape

1. Landform characterization
   A. Convex landforms

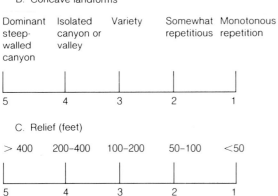

Dominant mountain or hill	Isolated mountain or hill	Variety	Somewhat repetitious	Monotonous repetition
5	4	3	2	1

   B. Concave landforms

Dominant steep-walled canyon	Isolated canyon or valley	Variety	Somewhat repetitious	Monotonous repetition
5	4	3	2	1

   C. Relief (feet)

> 400	200–400	100–200	50–100	<50
5	4	3	2	1

2. Natural characterization
   D. Percent of land covered with indigenous vegetation and natural rock or soil

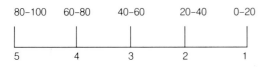

80–100	60–80	40–60	20–40	0–20
5	4	3	2	1

   E. Miles of natural stream per square mile.

>8	4–8	2–4	1–2	<1
5	4	3	2	1

3. Man-made characterization
   F. Number of buildings per square mile

<5	5–10	10–20	20–40	>40
5	4	3	2	1

   G. Miles of road per square mile

<1	1–2	2–4	4–8	>8
5	4	3	2	1

listing of all of the *existing characteristics* and conditions of the environment on one axis and a listing of *proposed actions* that may have environmental impact on the other axis. The former may be organized as follows:

A. Physical and chemical characteristics
   1. Earth
   2. Water
   3. Atmosphere
   4. Processes
B. Biological conditions
   1. Flora
   2. Fauna
C. Cultural factors
   1. Land use
   2. Recreation
   3. Aesthetic and human interest
   4. Cultural status
   5. Human-made facilities and activities
D. Ecological relationships
E. Others

The proposed actions are subdivided into the following:

A. Modification of regime
B. Land transformation and construction
C. Resource extraction
D. Processing
E. Land alteration
F. Resource renewal
G. Changes in traffic
H. Waste emplacement and treatment
I. Chemical treatment
J. Accidents
K. Others

A small portion of the extended matrix is shown in figure 15.3, which may be used to show its manner of use.

The next step is to select the actions that are likely to be significant in the particular project. As an example, assume contour strip mining in eastern Kentucky that, in the action portion of the matrix, will involve blasting and surface excavation. These actions are then considered in relation to their possible effect on existing conditions and the relations

Figure 15.3 Portion of a matrix for assessing environmental impact

	Resource extraction						
Processes	Blasting and drilling	Surface excavation	Subsurface excavation	Well drilling	Dredging	Clear cutting (lumbering)	Commercial fishing and hunting
Floods							
Erosion							
Deposition							
Solution							
Sorption							
Compaction							
Stability							
Earthquakes							
Air movements							

of concern identified by a diagonal slash in the appropriate box. In the case of contour strip mining, blasting may affect slope stability and serve as a human-induced earthquake; surface excavation may cause changes in runoff (floods), the rate of erosion, the amount of deposition (siltation), and in slope stability.

Once the interactions are identified they are evaluated for their magnitude and importance on a numerical scale of 1 to 10. *Magnitude* is used in the sense of degree, extensiveness, or scale. For example, the spoil from surface excavation may be in a form and so placed that erosion and extensive siltation could occur and thus be considered an effect of a large magnitude. The *importance* of each environmental impact must include the consideration of the consequences of changing the particular condition or other aspects of the environment and usually involves value judgment. For example, the effect of a proposed change in stream regime would be a prediction of flood-risk and elsewhere in the matrix, of its effects on aquatic life, recreation, health, and safety.

The numerical values determined for the magnitude and importance of each interaction are placed respectively above and below the diagonal in each appropriate box and in the completed matrix will both identify the critical interactions and provide a numerical basis for evaluation. Recognition of particularly deleterious interactions at this stage (for example, the dependence of flooding on siltation) may warrant modification of the engineering plan to remove or reduce the magnitude of the effect.

The text The *text* of an impact statement should address itself to a detailed discussion of those columns (proposed actions) or rows (existing characteristics) in which a number of actions were marked and discuss in detail those interactions marked by larger numerical values in particular boxes. The guidelines of the Council on Environmental Quality (Federal Register, 1971) give the following points to be covered:

1. A description of the proposed action including information and technical data adequate to permit careful assessment of impact
2. The probable impact of the proposed action on the environment
3. Any probable adverse environmental effects that cannot be avoided
4. Alternatives to the proposed action
5. Relationships between local short-term uses of human environment and the maintenance and enhancement of long-term productivity
6. Any irreversible and irretrievable commitments of resources that would be involved in the proposed action should it be implemented

7. Where appropriate, a discussion of problems and objections raised by other federal, state, and local agencies and by private organizations and individuals in the review process and the disposition of the issues involved

All of these points can be covered as a discussion of the completed matrix, and the text is primarily a discussion of the reasoning used to assign the numerical values of magnitude and importance employed.

GEOLOGIC SITE EXAMINATION

The geologic conditions of engineering importance were discussed in earlier chapters and the principles developed should be applied in each evaluation. Of particular importance are the nature, attitude, degree of cohesion and fracturing of the bed rock, the nature and thickness of the soil, and the surface and subsurface hydrologic conditions. The following checklist is provided to suggest the nature and range of observations that may be important in developing design criteria:

Bed Rock

Rock type(s)
Physical properties

Density	Permeability
Strength	Solubility
Porosity	Alteration

Structure
  Spacing and attitude of bedding
  Attitude, movement direction and fracturing related to faults, and evaluation of capability of further movement
  Geometrical description of folded rocks
  Spacing and attitude of jointing and other physical discontinuities
  Attitude, nature, and contact phenomena of igneous rocks or unconformities

Soil

Classification	Geometric properties
Physical properties	Thickness
Cohesive or noncohesive	Zoning
	Density
Angle of internal friction	Compressibility
Permeability	Shearing strength
Plastic index	Sorting or gradation

Surface and Groundwater

Quantity	Recharge zone
Seasonal variation	Relation to aquifers and aquicludes
Quality	
Geometry	Spring lines, sink holes, etc.
Flow lines	
Water table	

Geologic properties of importance to structural design may be of little concern to a contractor who is bidding for excavation work. To an engineer, the ease and means of excavation of material are of greatest importance. The qualitative system (fig. 15.4) following Royster (1978) is a useful approach, which may be modified as appropriate, based on field investigation coupled with past experience. The principal assumption is that geologic formations or other lithologic units are by definition characterized by some degree of homogeneity in their makeup and properties and may therefore be traced and predicted from point to point. In particular regions the examples given may thus be replaced with lithologic units.

The relative ease of excavation may also be determined by seismic testing (see chapter 12). Scraping, blading, ripping, and excavating by bucket-wheel or shovel are all accomplished with ease in materials having seismic velocities below 0.5 km/s and blasting will usually be required above 1.5 km/s (table 15.1).

Figure 15.4 Excavation index

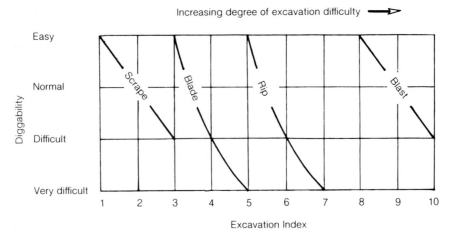

KEY

1 Relatively dry sand or silt, some clays.
2 Moist gravel, sand, silt and most clays.
3 Moist clay with less than 25% small disseminated rock fragments. Some highly weathered shales. Some organic materials.
4 Clay with 25–50% small, disseminated rock particles or minor pinnacle or boulder content. Some moderately weathered shales. Colluvium with minor boulder content. Sanitary landfill material. Alluvial boulders.
5 Clay with more than 50% disseminated rock fragments. Slightly weathered shale. Scree or colluvium with heavy boulder content. Saturated clay, silt, sand or gravel.

6 Very slightly weathered shale. Thin, slabby and well-jointed limestones and siltstones. Saprolite.
7 Thin-bedded chert with clay partings. Thin-bedded limestone or siltstone with interbedded shale.
8 Weathered igneous and metamorphic rocks. Friable sandstone. Medium- to thick-bedded limestone with solution cavities or clay or shale partings. Soils with rock pinnacles or large boulders.
9 Hard shale. Thin- to medium-bedded sandstone, siltstone or limestone with interbedded shale. Soil with numerous rock pinnacles or large boulders.
10 Thin- to thick-bedded sandstone, siltstone and limestone. Fresh igneous and metamorphic rocks.

Table 15.1 Limiting Seismic Velocities for Excavation, km/s

Means	Easy	Marginal
Dragline crawler	0.5	0.75
Bucket-chain excavator	0.55	0.8
Walking dragline	0.65	0.9
Pick and shovel	0.7	0.95
Tractor-scraper	0.7	1.3
Stripping shovel	0.75	1.15
Bucket-wheel excavator	0.8	1.05
Loading shovel	0.9	1.5

SUMMARY

Today geologists and engineers must broaden their concern with projects that modify the earth's surface to include the larger concerns of the general populace. Our environment is an inheritance that must be preserved for future generations and should be modified only when the benefits of a change are clear and outweigh all forseeable deleterious consequences. Every change of existing conditions generates a ripple of interrelated changes and these must each be evaluated before shovel is put to ground.

The importance accorded to environmental planning can be seen in the inclusion of environmental specialists in engineering firms, the many environmentally-oriented citizens groups, and the plethora of recently enacted environmental laws; for example:

National Environmental Policy Act, 1969
Water Quality Improvement Act, 1970
Noise Control Act, 1972
Resource Conservation and Recovery Act, 1976
Surface Mining Control and Reclamation Act, 1977
Clean Water Act Amendments, 1977
Clean Air Act, 1977
Uranium Mill Tailings Radiation Control Act, 1978
Refuse Act, 1980.

ADDITIONAL READING

Cheremisinoff, P. E., and Morresi, A. 1979. *Environmental assessment and impact statement handbook*. Ann Arbor, Mich.: Ann Arbor Science.

Golden, J., et al. 1979. *Environmental assessment and impact statement handbook*. Ann Arbor, Mich.: Ann Arbor Science.

Hackett, J. E., and McComas, M. R. 1969. *Geology for planning in McHenry County*. Illinois State Geological Survey Circular 438.

Hendricks, D. W.; Vlachos, E.; Tucker, L. S.; and Kellogg, J. C. 1975. *Environmental design for public projects*. Fort Collins, Colo.: Water Resources Publication.

Hough, B. K. 1969. *Basic soils engineering*. 2d ed. New York: The Ronald Press.

McHarg, I. L. 1969. *Design with nature*. New York: Natural History Press.

Organization for Economic Cooperation and Development. 1979. *Environmental impact assessment*. Washington, D.C.: OECD Publications.

Rau, J. G., and Wooten, D. C. 1980. *Environmental analysis handbook*. New York: McGraw-Hill.

UNESCO. 1976. *Engineering geological maps*. Paris: The UNESCO Press.

16

Exploration Methods

INTRODUCTION

As discussed in the previous chapter, every engineering work has a definable purpose and must be carried out at some selected site. The selection is sometimes dictated by nontechnical circumstances, but it is more likely that a location will be chosen following a careful investigation whose goal is to simplify the engineering problems and optimize the effectiveness of the work. Once the objective of an activity has been clearly defined, the criteria that must be met are established and exploration can be intelligently planned and conducted.

The purpose of exploration is to locate the site that will best serve the intended objective, including the consideration of all pertinent political, socioeconomic, and technologic factors. The technologic need can be for sound foundation conditions, drainage, ease of excavation, best routing, avoidance of hazards, or a host of others.

Political, social, and economic factors that bear on site selection are beyond the scope of this text. Engineering criteria depend upon the nature of the particular project and may be anticipated to be diverse. For each case, however, these criteria help to identify the geologic information that is needed. The geologic contributions involve both the location of sites at which optimum geologic conditions exist and the detailed analysis of conditions at that spot.

EXPLORATION CONSIDERATIONS

Observations made on the surface at an engineering site or the extrapolation of subsurface conditions from nearby areas may provide the essential facts for project design. More probably, however, the subsurface must be tested indirectly by such means as seismic or electrical probing or directly by test pits, trenches, or drill holes. The information required will usually be the nature and physical properties of the soil and rock, location of the water table, and presence and spacing of any discontinuities. Representation should be on maps and true-scale cross sections carried to such depth as to exceed the dimensions and effects of the proposed work. Only when the details of subsurface conditions are known can the planning of such things as footings, drainage or water barriers, excavation procedures, or retaining walls be intelligently and safely planned. This kind of site study may be included within an exploration program designed to identify and compare *potential construction sites* for a facility within some given region.

Dams, surface and subsurface disposal areas, port facilities and shoreline protection, tunnels and airports are representative of works that involve existing regional or local needs that will pertain to site selection. The particular items to be compared in choosing one site above another will, therefore, depend on nonengineering as well as engineering factors, and the considerations outlined in chapter 15 should be taken into account.

Further complications of exploration effort arise when the desired target is a mineral deposit, which must first be discovered. The remainder of this chapter discusses this general problem and implicitly includes features of interest to the study of other engineering mining sites.

EXPLORATION FOR MINERAL DEPOSITS

A mineral deposit is a natural occurrence of some mineral, sediment, rock, or fluid whose composition, physical properties, and location make it useful to man. Broadly speaking, mineral deposits are any geologic materials processed and used by man including water, fill and aggregate, dimension stone, gems, industrial minerals, fossil and nuclear fuels, and ferrous and nonferrous metallic minerals. Such materials are simply ordinary or unusual geologic materials formed by everyday geologic processes, and their discovery is generally accomplished by the interpretation of geologic events and associations. The term **ore** implies that the material may be recovered at a profit, and although usually restricted to metallic deposits, the concept of *profitability* is central to the working of any mineral deposit.

The object of exploration is to reduce a large area without defined characteristics to one or more small areas in which the characteristics are sufficiently defined to warrant further detailed study. Prospecting for mineral deposits is broadly controlled by the kind of material sought, constraints of geography, national politics, land rights, and geologic considerations. Once selection of material(s) and area is made a program employing appropriate prospecting methods, usually in conjunction or sequence, is mounted.

Geologic, geophysical, and geochemical exploration tools are not usually employed alone but in some preplanned combination; for example, airborne geophysical reconnaissance followed by geological and geochemical ground checking, geochemical soil sampling at geophysical traverse stations or geologic mapping and geochemical stream sampling.

Prospecting methods fall naturally into those employing remote sensing, geologic studies, and geochemical and geophysical methods. These approaches are applicable to the discovery of any mineral deposit, but for simplicity, the following discussions will generally assume the target to be a metallic **ore deposit.** Exploration plans are normally made on a large scale (i.e., worldwide, continent-wide, state- or country-wide), and steps are then taken to reduce this area to one or more promising targets upon which a decision is finally made to drill. This is partly due to the high cost of drilling but also due to the fact that drilling is relatively slow in rock and a single drill hole tests only a small area.

Geologic Field Work and Mapping

On-site examination is still the cheapest but probably the slowest method of geological exploration. Field work to discover suggestive rock types or structures or to find indications of mineralization may be used as the first exploration step but more often follows more sophisticated techniques. In most areas of the world, maps and air photographs are available to guide the work of the field geologist and with them the regional geology may be assessed on a general basis and a reasoned approach to the field work planned.

Geologic mapping usually takes the form of reconnaissance traverses and sampling to locate favorable rocks or geologic structures, and on location of these, detailed mapping is commenced. The main tools are the Brunton compass, a geologic hammer, a notebook, a hand lens, and sample bags; the means by which rocks are sampled, labeled, and observations positioned on a map or an air photograph. An easy method of noting a sample location on an air photograph is to push a pin through the location on the photograph, locate the pin hole on the back, and make notes on the unglazed paper backing. In some areas, mapping will produce relatively quick results, but in others of complex geology or hostile climate it can take years.

Geophysical Methods

Mineral deposits, by their special nature, often have physical properties that are sufficiently distinctive to be useful in their discovery and delineation. The properties that have been found to be particularly useful are density; magnetic susceptibility; seismic velocity; and electrical resistance, conductance, and potential difference. The principal *geophysical prospecting methods* are thus divisible into gravimetric, magnetic, seismic, and electrical methods. Geophysical observations can be made from the ground, air, or at sea, and though equipment varies in the three different situations, the geophysical principles remain the same.

In prospecting, a physical property or field is sampled in some regular way in order to discover any irregularities, or anomalies, that may be ascribed to the presence of a deposit of interest. Some generalities of importance are that some fields such as the earth's magnetic field may not be constant in time; gravimetric data requires corrections to be made for topography; and electrical properties are greatly affected by the water content of rocks. Typically, geophysical methods sample an electrical, magnetic, or gravity field whose intensity decreases as an inverse square. Consequently, the signal decays rapidly with depth to an anomalous body and with the height of airborne equipment above ground. Penetration, even with sufficiently sensitive apparatus, is also limited by sender–sensor geometry to about one-third to one-half of their separation.

Magnetic methods   Magnetic methods depend on the fact that different rocks, and in particular, some buried **ore bodies** locally distort the normal earth's magnetic field and produce an anomaly of either a positive or negative kind thus indicating a possible exploration target. The most magnetically distinct anomalies are due to the presence of iron ores that contain magnetite or other minerals with a degree of permanent (remanent) magnetism. Many metallic ore bodies also either contain slightly magnetic minerals or they weather to oxides of iron that are detectably magnetic in a magnetometer traverse (fig. 16.1). The simplest magnetometer traverse is conducted by a tripod-mounted magnetometer that can be easily hand carried.

The earth's magnetic field shows both local and regional variation in orientation and strength as a natural consequence of its interaction with materials of different magnetic permeability and with

Figure 16.1 Magnetic anomaly over the Pea Ridge iron deposit, Missouri

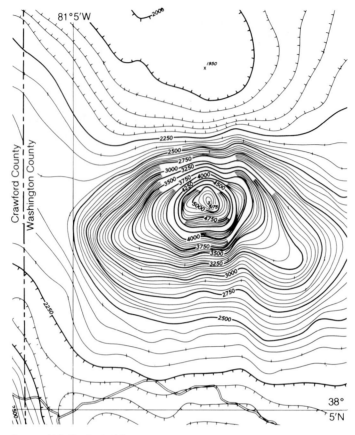

Source: U.S. Geological Survey.

the earth's internal and external electrical fields. Magnetic anomalies of prospecting interest are superimposed on this irregular and slowly varying background. Since the magnetic field lines are *inclined* to the surface except at the equator, magnetic measurements should distinguish between horizontal, vertical, and total intensity as well as polarity and direction. Care is taken not to take readings near surface magnetic features such as metallic structures, fences, power lines, or vehicles and not to carry magnetic objects.

An airborne magnetometer can be combined with other types of airborne sensors, some of which are trailed behind the aircraft as drogues, and with recording equipment mounted in the aircraft. This kind of geophysical work is much more expensive

than ground traversing but can cover vast areas of ground rapidly. Besides geophysical instruments these aircraft exploration units require complex onboard navigation systems. Further, the data that they gather are complex and require careful interpretation because of the airplane's varying altitude above the terrain. Similar recording equipment can be mounted in ships with the detection equipment trailed astern by cable for offshore work.

Seismic methods Seismic prospecting is used extensively in the oil industry for the detection of buried geologic structures that may contain oil or gas; it also provides information for structural interpretation in mineral exploration and about the subsurface conditions at engineering sites. The principle

Seismic source truck

is that artificial shock waves are induced in the earth and detected by strategically placed receivers called geophones. Analysis of travel times of reflected or refracted shock waves allows an interpretation of the underlying rock structure (see fig. 12.16).

The same principle is employed in offshore exploration for potentially oil-bearing structures with one ship dropping the explosive charges and another ship, on a parallel course, towing a recording cable connected to equipment on board. Almost all major offshore oil fields have been found in this way.

Probing by means of seismic waves, often generated by a sledge hammer blow or by dropping a weight, is commonly employed for small-scale engineering site studies. The nature, depth, and configuration of hidden bed rock, the presence of subsurface cavities, and some physical properties of the rock may be determined by this means.

Gravimetric methods   The **gravitational pull** on an object at or near the surface of the earth varies as a function of the density of the rocks in the crust at a particular point. Very sensitive instruments called gravimeters are used to measure variations in the gravity field from point to point, and the resultant map may be interpreted in terms of underlying rock type and structure. Careful corrections must be made to the measured data because of the differences in field strength as a function of distance from the earth's center and deficiencies or increments of mass due to irregular terrain.

Electrical methods   There are a number of electrical properties of earth materials that are readily measurable and are utilized in exploration methods. These methods may be divided into *passive methods* in which the properties of the ground are directly measured (self-potential) and *active methods* in

Self-potential apparatus in use

which the effects of a stimulating electrical field applied by direct contact or inductance are measured
(induced potential, resistivity, electromagnetic
methods). The resistivity method, employing a
Wenner-Gish-Rooney or Megger instrument, is
sometimes used for engineering measurements and
may be taken as an example of procedure. The apparatus consists of batteries or a hand-cranked generator, a double commutator, four electrodes, a
milliameter, and a potentiometer or a direct reading
ohmmeter. The current through the milliameter,
current electrodes, and ground is periodically reversed to obviate polarization at the electrodes, and
the current through the potential (sampling) electrodes is simultaneously reversed so that d.c. poten-
tials are read on the potentiometer. The electrode
configuration used for mapping (fig. 16.2) is moved
by steps along a *traverse line* and provides constant
depth penetration. Increasing separation of the elec-

Figure 16.2 Arrangement of components for resistivity
measurements

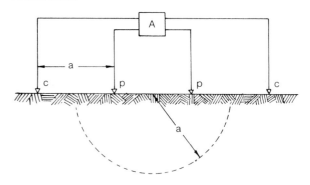

KEY

A   Generator, commutator, meters
c   Current electrodes
p   Potential electrodes
a   Approximate depth of effective measurement

Figure 16.3 Resistivity contours and geologic section over a gypsum deposit, Cape Breton Island, Nova Scotia

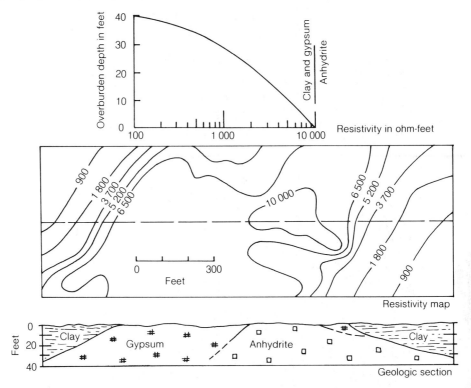

Resistivity map

Geologic section

trodes provides increasing penetration and if symmetrically done allows "resistivity drilling" to be performed. The electrical relations are given by

$$\rho = 2\,\pi\,a\,\frac{V}{I} = 2\,\pi\,a\,R$$

where ρ = resistivity, a = resistivity of hemisphere of radius a = length of current path = electrode spacing, V = voltage, I = current, and R = resistance. An example of resistivity mapping is given in figure 16.3 where the thickness of clay-overburden on gypsum and anhydrite has been determined. Measured and contoured resistivity values have been independently correlated with rock-type overburden thickness.

A summary of the more commonly used geophysical exploration methods is given in table 16.1.

Geochemical Prospecting Methods

Metallic ore deposits may be relatively small and easily hidden under surface cover. One million dollars of gold will fit into a small wastebasket and about $17,000 of high grade tin concentrate will fit in an oil drum. Fortunately, the deposit will typically be subjected to weathering and erosion and the resulting dispersion of its constituents greatly enlarges the target. Prospecting by following glacial erratics or stream-transported boulders upstream to their source has been employed for hundreds of years and the use of panning as a guide to locate deposits of gold, tin, or diamonds is a much-used technique. In recent years, however, the identification of dispersed constituents by the use of sensitive and selective *chemical testing* of soil, stream silt, water, and vegetation has been employed with considerable success.

A metallic deposit in the zone of weathering releases ions that are dissolved and transported in the groundwater and surface waters. Exchange reactions, principally adsorption, build up low-level "contamination" in the soil and stream particles in the vicinity of the ore deposit source, and this anomalous metallic content may be found by appropriate tests.

Table 16.1 Summary of the Four Major Geophysical Methods

Method				Field	Geologic Application	Action and Control
I. Gravitational	A. Torsion balance			Oil	Anticlinal structures; buried ridges; salt domes; faults; intrusions	Spontaneous Action No Depth Control
	B. Pendulum C. Gravimeter				Salt domes; buried ridges; major structural trends	
II. Magnetic				Oil, mining	Anticlinal structures; buried ridges; intrusions; faults; iron ore, pyrrhotite, and assoc. sulfide ores; gold placers	
III. Electrical	A. Self-potential			Mining	Sulfide ore bodies	Reaction to Energizing Fields Control of Depth of Penetration
	B. *Galvanic* application of primary energy	1. Potential distribution of second'y field measured	(*a*) Equipot., potential profile (*b*) Resistivity (*c*) Pot.-drop ratio	Mining, civil eng., oil	General stratigraphic and structural conditions; bedrock depth on dam sites; groundwater; oil structures; sulfide ore bodies; highway problems; elec. logging	
		2. Electromagnetic field meas.		Mining	Sulfide ore bodies	
	C. *Inductive* application of primary energy			Oil, mining	Faults; anticlinal, etc., structure; sulfide ore bodies	
IV. Seismic	A. Refraction			Oil, civil eng.	Salt domes; anticlinal etc., structures; faults; foundation & highway problems	
	B. Reflection			Oil	Low-dip structures; buried ridges; faults	

From C. A. Heiland, *Geophysical Exploration.* © 1946 Prentice-Hall, N.J.
Reprinted by permission of Prentice-Hall, Inc.

Many analytic techniques are employed in geochemical prospecting; the more usual being emission spectrography and various colorimetric tests that employ sensitive metallo-organic reactions capable of detecting the presence of low levels of adsorbed metallic ions. It should be noted that the testing for one metal may lead to the discovery of the deposit of another since many deposits are polymetallic.

Geochemical test results are plotted on maps and the data examined to discover chemical anomalies related to ore. The recognition of significant anomalies will usually require a statistical examination of a large amount of data in order to distinguish the anomalous data from the chemical background. This is usually done by using a *distribution plot* constructed from all of the data taken in the sampling area. The plot has a *logarithmic abscissa* in recognition of the lognormal distribution of the elements in nature. Figure 16.4 shows such a plot and a graphical differentiation of the populations present.

Portable geochemistry laboratory

Figure 16.4 Distribution curves for mixed background and anomalous populations

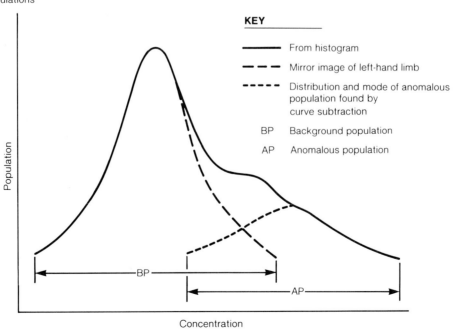

KEY

——————	From histogram
– – – –	Mirror image of left-hand limb
- - - - -	Distribution and mode of anomalous population found by curve subtraction
BP	Background population
AP	Anomalous population

Population

Concentration

Figure 16.5 Geochemical reconnaissance map, SE Maine

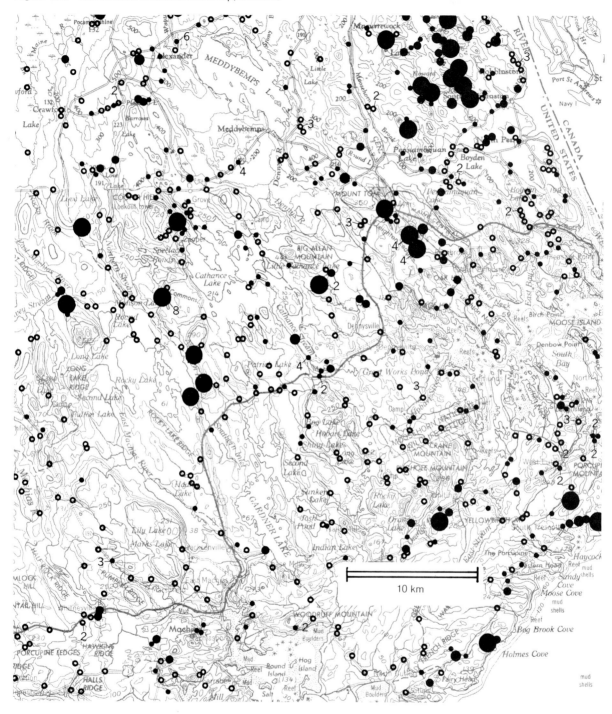

From E. V. Post, et al., *Mineral Investigations Field Studies Map MF–301*, U.S. Geological Survey, 1967.

KEY

Amounts of heavy metals (Cu + Pb + Zn + Cu)
extracted from stream sediments.
Large circles > 20 parts per million
Numbers are ppm copper

Geochemical prospecting is employed both in reconnaissance studies of large areas, often in conjunction with the collection of geologic or geophysical data, and in detailed site examinations. A reconnaissance study typically employs stream silts as sample materials and usually makes detailed examinations by collecting soils on a regular grid. Figure 16.5 is a portion of *Mineral Investigations Field Studies Map MF–301, U.S. Geological Survey, 1967,* which shows the results of reconnaissance for heavy metals in stream sediments in southeastern Maine. The size of the circles indicates the amount of metal found by colorimetric testing at each sampling site, and the clustering of anomalous values suggests locales for more detailed exploration. Figure 16.6 suggests an idealized geochemical approach to finding ore in which an area of interest is located by stream sample reconnaissance and the ore body is located by detailed soil sampling.

Remote Sensing Methods

Remote sensing, as the term is currently employed, may be defined as the use of electromagnetic radiation for the collection of information about an object without having physical contact with it. Aerial photography is an early form of remote sensing. Recent developments in remote sensing include capacities to use radiation outside the visible portion of the spectrum, employ electro-optical scanners in the place of film, and apply computer processing for the formation and interpretation of images. These are new and powerful techniques in the arsenal of the ore finder. Particularly notable is the routine coverage of very large areas through the use of earth satellites as photographic platforms and active airborne systems at radar frequencies that are unaffected by weather conditions.

Electromagnetic radiation moves at the speed of light with a harmonic wave motion, and electromagnetic waves of all wavelengths travel at the same velocity (nearly 300 000 km/s in a vacuum). The relationship is $C = \lambda \nu$ where C = velocity of electromagnetic radiation, a universal constant; λ = wavelength, commonly expressed in nanometers ($1nm = 10^{-9}m$), micrometers ($1\mu m = 10^{-6}m$), centimeters, or meters; ν = frequency in cycles per

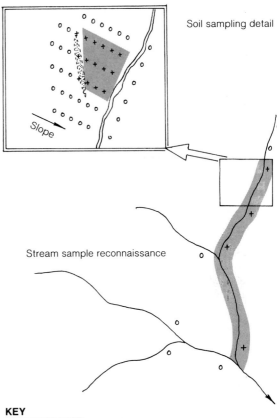

Figure 16.6 Idealized geochemical approach to ore finding

Soil sampling detail

Slope

Stream sample reconnaissance

KEY

+ Anomalous concentration
o Background concentration
ᵌ Ore zone

second termed hertz, Hz. Electromagnetic waves encompass a very extensive spectrum of wavelengths ranging from about 10^{-12} m (gamma rays) to 300 m (radio waves). That portion utilized by one or another remote sensing technique is shown in figure 16.7.

The interaction of electromagnetic radiation with matter as observed in its *selective absorption* and *reflectance* is quite varied and allows different information to be obtained through the use of different wavelengths. The principal limitation is in the relatively narrow bands of incident wavelengths that are available in a practical way; namely, solar radiation and radar waves.

Figure 16.7 Wavelengths utilized for remote sensing

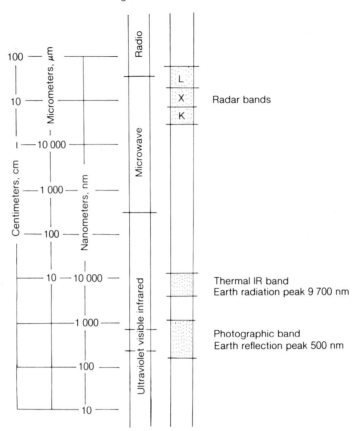

Visible radiation and portions of the ultraviolet and near infrared spectrum, about 0.25 to 1.2 μm, may be detected by photographic means by the use of different film emulsions. The most commonly used films are ordinary black and white; *infrared* black and white, which shows reflected solar radiation; infrared color, once called camouflage detection film; and **ultraviolet.** Ultraviolet film has poor resolution because of scattering of atmospheric dust and gas molecules but sometimes shows carbonate rocks as especially bright because of ultraviolet-stimulated fluorescence.

Aerial photography from kites, balloons, and aircraft has been widely used for many purposes that range from map making to crop inventories and provides the geologist or engineer with detailed views

at different scales of areas of interest. Among its more useful aspects is the use of photo pairs or sequences to provide a three-dimensional image, which greatly assists in identifying the boundaries of geologic formations, locating faults and other linear features, and showing topographic relief. Human eyes are coupled in such a way that near objects appear three-dimensional because each eye sees a slightly different view and the brain combines them to perceive an object at an appropriate distance (fig. 16.8a). This *stereoscopic vision* may be imitated if two sequential photographs of the same object are taken from an aircraft (fig. 16.8b) or space platform. When these paired photographs are viewed through a stereoscopic viewer, they are blended into

Figure 16.8 The geometry of stereoscopic vision

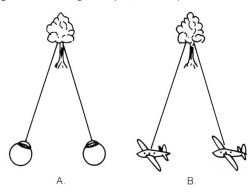

A. Different views of an object as received by two eyes are blended to a three-dimensional image by the brain.
B. Different views of an object acquired photographically, perhaps from an airplane, can also be blended into a three-dimensional image, usually through the use of a stereoscope.

A.   B.

Table 16.2 The Infrared (IR) Spectral Region

Wavelength in μm	Name	Remarks
0.7–300	IR	
0.7– 0.9	Photographic IR	
0.7– 3	Reflected IR	(Primarily reflected solar radiation)
3 – 14	Thermal IR	(Primarily inherent radiant temperature)

Table 16.3 Landsat Wavelengths

Band	Wavelength Coverage, μm	
4	0.5–0.6	Visible
5	0.6–0.7	
6	0.7–0.8	Photographic
7	0.8–1.1	Infrared

a single three-dimensional scene at the apparent distance of the original. The system may be quantified if the separation between the eyes or the photographic locations and the convergence angle at the viewed object are known.

Photography and scanner imagery in the infrared portion of the spectrum have proven to be particularly useful in geologic applications since both solar reflectance and inherent thermal radiation of materials may be observed. The terms used to designate the various infrared subregions are given in table 16.2.

Photographic and scanner imagery from satellites in earth polar orbit have been routinely available since the launching of the first **Landsat** satellite in 1972. The program provides coverage in four different wavelength regions (table 16.3) at a scale of 1:3 369 000.

Side-looking airborne radar (SLAR) image, lower Amazon basin

The satellite track (fig. 16.9) is the centerline of a 185 km-wide image swath that is displaced westerly on each orbit. The displacement is 160 km at the equator and 120 km at latitude 40° north and south, and the entire pattern repeats every 18 days. Stereoscopic viewing is possible because of the partial overlap of images on successive days.

Landsat is operated by the United States in the international public domain—images may be taken by Landsat anywhere in the world and may be purchased by anyone through the U.S. Geological Survey, EROS Data Center, Sioux Falls, South Dakota.

Imagery in the *microwave* region can be obtained by radar systems that illuminate the terrain and record radar echoes using an antenna mounted on an aircraft. The antenna axis in the usual configuration is side-looking; that is, angled downward

from the side of the plane so that it sweeps the terrain parallel to the flight path. The strength of the return signal is a function of surface roughness and absorption of radiant energy by various earth materials. The product looks remarkably like a black and white photograph but differs significantly in that the longer wavelengths provide less resolution and shadows always fall away from the "camera." The system is particularly capable of enhancing linear features and can provide images at night or through rain or cloud cover.

Modern remote sensing systems all employ *digital image processing* for enhancement, comparison, quantification, and storage. Input data may be directly from the remote sensing scanner or scanned from a photograph or other image. Basically, each picture element, or **pixel,** is assigned an *x-y* coordinate location on the image and a numerical value for its brightness. Pixel dimensions vary with image scale and program; they are 79 m × 79 m, about

Figure 16.9 Landsat orbit paths for a single day

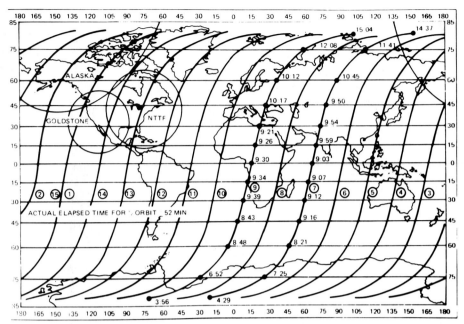

From *Landsat Users Handbook,* Goddard Space Flight Center
Document No. 76SDS–4528, NASA, 1976.

one and a half acres, for Landsat. Scales of brightness (gray scales) usually have 64 or 128 divisions. Once stored, computer programs can call up wanted aspects of the digitized image. The counting of pixels in various levels on the gray scale provides percent area information, or gray-scale values may be assigned colors. In this *false color mode* an area of known conditions brought to a distinctive color will provide a color reference to other areas of identical conditions (fig. 16.10).

Aerial photography, side-looking radar, and Landsat imagery are generally produced at rather small scales because they are public or quasi-public programs with many purposes to be served. Such imagery is excellent for geologic reconnaissance, but the scales are inadequate for detailed work. However, the availability of light aircraft, the wide range of photographic emulsions and handheld scanning systems, coupled with advanced television and microcomputer technology bring the availability of self-generated, sophisticated remote sensing within the budgetary limits of even small exploration groups.

Drilling Methods

The drilling of holes by one or another means is a very common aspect of engineering and mineral resource work. Drilled holes are used to recover natural underground fluids (water, petroleum, natural gas, steam) or artificially fluidized materials (sulfur, dissolved metals); to inject fluids for underground storage or disposal; to emplace blasting charges in excavation, mining, or quarrying; to provide drainage; and to probe the nature, shape, and size of underground bodies. The latter purpose is usually served in the final stages of a mineral exploration program when the decision must be made as to whether or not a potential ore body is economically viable. In engineering practise drilling is also often used to assess foundation conditions or gather other subsurface data.

There are many kinds and sizes of drilling equipment, but except for flame-piercing methods, all either make holes by rotating an auger or abrasive bit or by hammering a tool. **Augering** is restricted to drilling such soft or unconsolidated

Figure 16.10 False color matching

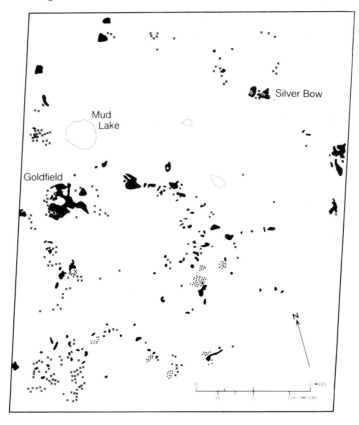

KEY

⌬ Hydrothermally altered limonitic areas
▓ Silica-rich, light-colored volcanic rocks
◢ Light-colored volcanic rocks with some hydrothermal
    alteration
x Approximate location of mines and prospects

From Rowan, et al., U.S. Geological Survey Professional
Paper 883, 1974.

materials as soil or coal; *rotary* and *percussion* drilling are typically employed for rock penetration. Rotary drilling and some percussion methods rely on a rigid drill stem to couple the driving mechanism and bit, which requires time-consuming pulling, uncoupling, recoupling, and lowering to recover the bit, hence the tall derrick of an oil rig. Cable-tool percussion drilling is done by lifting and dropping a heavy bit hung on a cable.

In most drilling a fluid such as air, water, or a special mud is circulated down the drill stem and up the hole to cool the bit and float out rock cuttings. In cable-tool drilling the hole must be bailed.

The samples recovered from most drilling operations are *chips* or *cuttings* brought to the surface (and possibly fractionated) by the returning drilling fluid where they may be caught on a screen for study. The more common exploration drilling, however, recovers a solid rock cylinder, or *core,* by employing

Exploration diamond drilling

an annular diamond-studded bit backed by a core barrel; the core is recovered after each few meters of advance. **Diamond drilling** has certain advantages also in the mobility of the equipment and its ability to drill holes at any angle. The hole and core diameters of standard diamond drills are given in figure 16.11.

Drills should be operated by well-trained and technically competent personnel in order that drilling rates are maintained at high levels with minimum downtime and that complete core recovery is assured. The location of drill holes, however, is the function of the engineer-geologist. He or she must also determine the location of collars; the bearing and plunge of the hole and its total length; examine the recovered core; and do such mapping and interpretation as may be required.

Drilling to delineate subsurface conditions will be done by using some regular pattern of holes spaced to provide the desired information. Drilling

Figure 16.11 Standard diamond drill dimensions

O.D. set bit, mm

- 122.1 PQ wireline
- 92.7 NC
- 75.7 NX
- 52.9 BX
- 48.0   AX
- 37.7   EX
- 29.8   XRA

18.7
21.5
30.1
42.0
54.7
69.5
83.1

Nominal core diameter, mm

for exploration purposes, however, is not done on a preplanned grid; rather the information obtained from the completed holes is used to determine the next location. For example, consider an idealized drilling sequence of a mineralized vein seen in outcrop at a single point; observation at the outcrop provides strike, dip, and width information, and drilling is needed to determine the lateral and vertical extent of the vein. The first hole should be located to test the vein at depth by drilling against the dip (fig. 16.12a). This hole should be located to intersect the vein at an angle in excess of 45° and at a depth such that the square root of vein thickness × depth = 100 or more. If depth extension is shown, the next holes should test lateral extent by drilling at points along strike to again provide square root of thickness times length in excess of 100 (fig. 16.12b). If all three holes make good intersections, the indicated volume is then at least 30 000 cubic units (fig. 16.12c). The next sequence of holes should be planned to bracket the target by stepping out along strike and down dip at distances successively doubling the previous intersections. Finally, the body should be delineated by filling in the pattern at regular spacing.

GEOLOGIC EVALUATION

Of primary geologic concern in the *economic assessment* of a mineral deposit (or engineering excavation) is the total amount of material present. This involves measurement of the *volume* of the deposit, determination of *tonnage* (volume × bulk density), and establishment of the concentration of the substance of value per tonne by **assay** (grade). For bulk deposits of essentially "as is" material such as limestone, glass sand, or gypsum the assay may be for impurities that reduce value.

The volume of a mineral deposit may be easy to calculate in the case of geometrically regular bodies such as tabular masses, cylinders, or spheres for which simple measurement mensuration formulae are available.[1] The volume of an irregular body is more difficult to obtain but may be found as the sum of those small regular units which comprise it. Easiest means is to divide the volume into a series

Figure 16.12 Drilling sequence

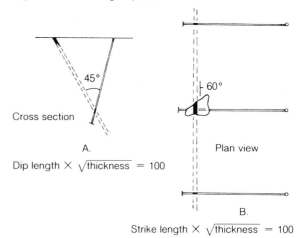

Cross section

A.

Dip length × $\sqrt{\text{thickness}}$ = 100

Plan view

B.

Strike length × $\sqrt{\text{thickness}}$ = 100

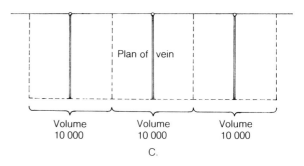

Plan of vein

Volume 10 000   Volume 10 000   Volume 10 000

C.

KEY

A. Cross section showing drill hole (solid lines) intersecting a vein (dashed) to test depth extension
B. Plan view of drill holes intersecting a vein to test its strike extension
C. Spacing of drill holes to block out ore

1. The areas, A, of some plane figures are
   triangle = ½ of base × height
   rectangle = length × width
   circle   = π × radius squared (π = 3.1416)
   The volumes, V, of right prisms and cylinders are the area of their base × their height.
   The volumes of uniformly tapering solids such as cones and pyramids and their frustrums are the average of the terminal areas × their height.
   The volume of a sphere = 4/3 π radius cubed.

Figure 16.13 Measurement of irregular volumes

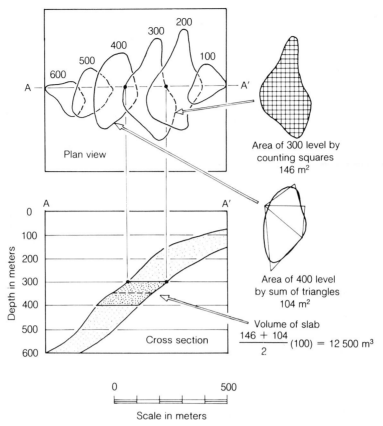

Area of 300 level by counting squares 146 m²

Area of 400 level by sum of triangles 104 m²

Volume of slab
$$\frac{146 + 104}{2}(100) = 12\,500\,\text{m}^3$$

Plan view

Cross section

Depth in meters

0
500
Scale in meters

of plates of equal thickness but having irregularly shaped parallel surfaces of a different area. The area of the surfaces may be found by any of several means,[2] and the average area of the bounding surfaces of a plate × its thickness is the plate volume; the sum of all plate volumes is the volume of the body. Figure 16.13 is a graphical example.

The actual measurement of a body requires information in three dimensions. To obtain this, two-dimensional surface data is typically extrapolated to

2. The area of an irregularly-shaped plane surface may be determined
   by measurement by planimeter, a mechanical device that integrates the area within a perimeter
   by the sum of areas of triangles
   by superimposing a regular grid and counting squares
   by weighing (an analytic balance may be used to weigh paper cutouts of areas)

depth by drilling coupled with geologically reasonable assumptions. In this context it is particularly useful to know a great deal about the nature of geologic boundaries and the shapes of rock masses. *Geologic boundaries* range in character from knife-edge sharp to completely gradational over extended areas. The former are represented by readily mapped bedding surfaces, faults, intrusive contacts, or sharp-walled veins; and the latter by gradational sedimentary bedding, igneous aureoles, or wall-less veins. For such fuzzy contacts, an arbitrary boundary or "assay wall" must be established by sampling or other means before a volume can be measured.

Calculation of volume is facilitated by the careful plotting of maps, the construction of sections, and the making of models. Unfortunately, three-dimensional information is not only difficult

Figure 16.14 Block diagram of a faulted vein showing orientation of standard views

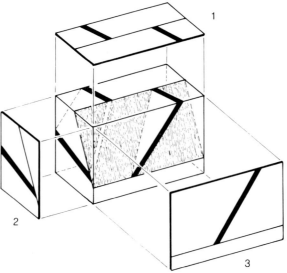

KEY

1 Top view, plan view, map view
2 End view, side view, elevation, cross section
3 Front view, front elevation, longitudinal section

to obtain but often defies the simple representation needed to visualize and calculate the volume of an irregular body. The usual means of representation is to generate a series of vertical or horizontal sections through the body that are parallel and evenly spaced. This series of plans or cross sections may then be used as is, transferred to glass to make a *sectional model,* or transformed into an *isometric block diagram.* Figure 16.14 shows the terminology used for slices in different orientations, and figure 16.15 shows serial plans and sections for a complex body.

Isometric block diagrams may be generated following the steps of figure 16.16. First the plan information is transformed from a rectilinear to a rhomboidal (60° and 120°) mesh, and then the levels are individually traced or plotted with appropriate vertical displacement.

Volume, when measured, will be transformed to tonnage by multiplying it by the bulk density of the ore (weight per unit volume). Because ore is a rock perhaps containing a significant percentage of

Figure 16.15 Serial plans and sections

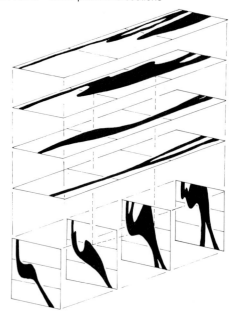

Figure 16.16 Steps in making an isometric block diagram

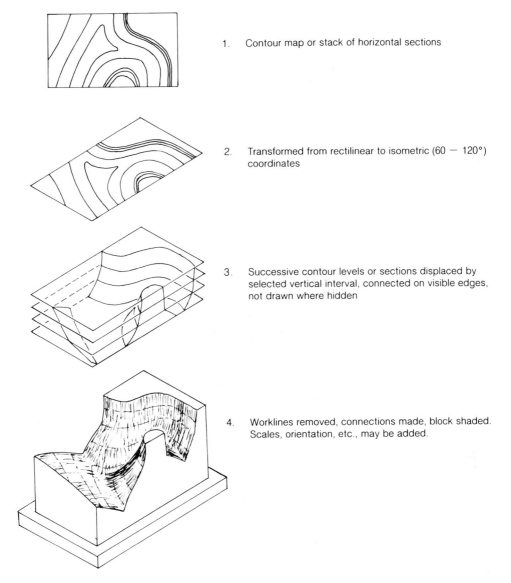

1. Contour map or stack of horizontal sections

2. Transformed from rectilinear to isometric (60 — 120°) coordinates

3. Successive contour levels or sections displaced by selected vertical interval, connected on visible edges, not drawn where hidden

4. Worklines removed, connections made, block shaded. Scales, orientation, etc., may be added.

voids, bulk density is properly measured on undisturbed samples or by weighing material removed from a regular volume.

Determination of the value per unit weight or grade of both the various parts of a deposit and its overall average requires careful sampling and assay coupled with an appreciation of the variability of both natural content and analytic data. Sampling is the process whereby a small portion of a body is taken as representative of some larger volume and should be done in as regular and mechanical manner as possible. Biased results because of contamination, salting (deliberate upgrading), or variable rock hardness must be guarded against. The responsibility for meaningful sampling ordinarily lies with the geologist.

Figure 16.17 Precision and accuracy

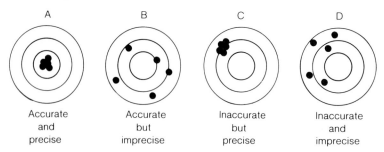

Typical samples are rock chips, cuttings removed from shallow channels, excavated materials from pits or trenches, or portions of diamond drill core. The pattern by which the samples are collected may be random, but calculation is easier and chance of preselection is reduced if a regular pattern is adopted. The spacing must be such that any significant variation in grade will be detected so it cannot be greater than the width of the anomalous zone. The accepted dimensions may have to be found experimentally by comparing the assay results from nested samples of successively closer spacing.

Large samples may be reduced in volume without change in quality by splitting or successive coning and quartering. *Splitting* is done with a device that divides the sample into two equal parts, one of which is discarded. In *coning and quartering* a sample is poured in a conical pile onto a canvas or rubber sheet, the pile is divided into quarters with a spatula, and alternate quarters discarded. The sample is then rolled by lifting alternate quarters of the sheet to rehomogenize it, poured to form a cone, and quartered again. This sequence is continued until the wanted amount is reached.

Samples, once obtained, are subjected to assay by an appropriate method. Choices as to assay cost, speed, precision, and accuracy should be made based upon competent advice and may be done either by an in-house laboratory or by reputable assayers. In either instance, however, the explorationist should submit disguised samples of known content and resubmit portions or earlier samples at irregular intervals to guide interpretation of the laboratory data.

When the volume, bulk density, and grade of a deposit have been found its *in-place value* may be readily determined as follows:

total tonnage = volume × bulk density
commodity tonnage = total tonnage × grade
in-place value = commodity tonnage × value per unit

EVALUATION OF ANALYTIC DATA

It is important in the evaluation of analytic data to grasp the importance of its statistical variability. *Accuracy,* or correctness; *precision,* or repeatability; and *bias,* or average difference, are statistical terms used to describe analytic data. A simple way of visualizing them is to consider the bullet holes in the targets of figure 16.17. High precision is shown by a small scatter and high accuracy by the (near) coincidence of the group average with the bullseye. Bias would describe the separation of the group average from the bullseye or from another group.

Assume the correct value (bullseye) to be 5.00 and repeated measurements to be as follows:

	a	b	c	d
	5.10	5.80	5.90	7.20
	5.20	4.70	6.20	5.20
	5.00	4.00	6.10	6.30
	4.90	6.40	6.00	5.80
	5.10	4.90	6.10	6.50
Σ	25.30	25.80	30.30	31.00
n	5	5	5	5
$\overline{x}$	5.06	5.16	6.06	6.20

where Σ = sum, n = number of determinations or population, and $\overline{x}$ = average for the set.

The precision (repeatability) of the measurements is represented by $\overline{x} \pm \sigma$ where

$$\sigma = \pm \sqrt{\frac{\Sigma d^2}{n-1}}.$$

σ defines the limits within which two-thirds of the measurements will fall (95% will fall between $\overline{x} \pm 2\sigma$); d^2 is the squared difference of each measurement from the average of the set, $\overline{x}$.

The precision of data sets may be compared by using the coefficient of variation, C, which states the error as a percentage of the average of the set

$$C = \frac{\sigma}{\overline{x}} \times 100.$$

For the measurements given, the important statistical parameters are

set	$\overline{x}$	σ	C	bias
a	5.06	0.114	2.3%	0.06
b	5.16	0.95	18.4	0.16
c	6.06	0.114	1.9	1.06
d	6.20	0.75	12.1	1.20

Bias = difference from true value of 5.00.

SUMMARY

Many of the evaluative operations applied to mineralized ground are overlapping; for clarification the following sequence of events would probably mark the identification of an ore body. At each stage samples are taken, maps and sections updated, and refinements of tonnage and grade calculations are made.

1. Location of mineralization followed by large-scale geologic mapping and sampling of visually distinct materials
2. Trenching or pitting to obtain a clearer view of the geometry and to obtain large samples
3. Trial drilling of a few holes located to provide information as to the extent of the body; sampling of core; construction of cross sections
4. Pattern drilling to clearly define boundaries of the deposit and establish its internal variability
5. Calculation of its in-place value

Much of the philosophy and methodology used in the search and evaluation of mineral deposits parallel those employed in the preconstruction phases of engineering works. A project's location may be dictated by particular rock or soil types or conditions and detailed engineering evaluation of materials by sampling and measurement is absolutely essential.

Exploration methods involving the acquisition of physical, chemical, or remote-sensed information are usefully employed in site searches; volume determinations are often required; sampling is a perfectly general procedure; and the statistical examination of any sampled data is necessary.

ADDITIONAL READING

Dobrin, M. B. 1960. *Introduction to geophysical prospecting.* 2d ed. New York: McGraw-Hill.

Hawkes, H. E.; Webb, J. S.; and Rose, A. W. 1979. *Geochemistry in mineral exploration.* New York: Academic Press.

Heiland, C. A. 1946. *Geophysical exploration.* Englewood Cliffs, N.J.: Prentice-Hall.

Lacy, W. C., ed. 1982. *Mineral exploration.* Florence, Ky.: Van Nostrand Reinhold.

Lillies, T. M., and Kiefer, R. W. 1979. *Remote sensing and image interpretation.* New York: John Wiley & Sons.

Peters, W. C. 1978. *Exploration and mining geology.* New York: John Wiley & Sons.

Richason, B. F., Jr. 1978. *Introduction to remote sensing of the environment.* 2d ed. Dubuque, Iowa: Kendall-Hunt.

Siegel, B. S., and Gillespie, A. R. 1980. *Remote sensing in geology.* New York: John Wiley & Sons.

Siegel, F. R. 1974. *Applied geochemistry.* New York: John Wiley & Sons.

Simpson, R. B. 1966. *Radar: Geographic tool.* Annals of American Association of Geographers.

Smith, J. J. 1968. *Manual of color photography.* Washington, D.C.: American Association of Photogrammetry.

Townshend, J. R. G., ed. 1981. *Terrain analysis and remote sensing.* Winchester, Mass.: Allen & Unwin.

Ward, F. N.; Lakin, H. W.; Canney, F. C.; et al. 1963. *Analytical methods used in geochemical exploration by the U.S.G.S.* U.S. Geological Survey, Bulletin 1152.

17

Industrial Rocks and Minerals

INTRODUCTION

Humans have always contrived uses for raw materials; indeed, levels of their discovery and application are used to define human cultural development in the Stone Age, the Bronze Age, and the Iron Age. Today nearly every kind of mineral and rock is somehow incorporated in our complex culture and economy, and the mere listing of geologic materials and their uses is a formidable task. Some sense of this complexity may perhaps be grasped by noting some of the uses of the relatively minor mineral fluorite (fluorspar), CaF_2, which is used in various grades (metallurgical, acid, and ceramic) in the manufacture of iron, steel, ferroalloys, aluminum and magnesium metal, glass and enamel, welding rods, cement, and hydrofluoric acid. Considering then only the hydrofluoric acid produced and referring to figure 17.1, it may be seen how the combination of hydrofluoric acid with other materials of geologic origin yields a new group of products and intermediates.

It is obviously impossible to present more than a few examples of the nature and uses of industrial minerals in an introductory text. Since industrial minerals are both won from the earth and used in

Figure 17.1 Uses of hydrofluoric acid

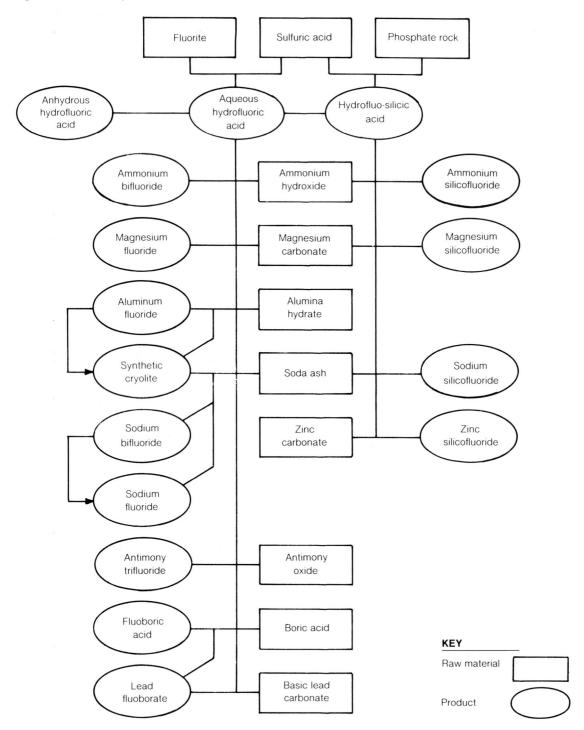

Figure 17.2 Limestone and dolostone usage in the United States (1963)

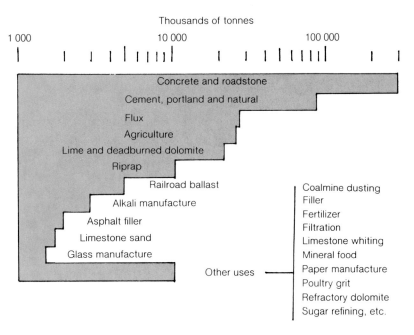

Thousands of tonnes

1 000   10 000   100 000

Concrete and roadstone
Cement, portland and natural
Flux
Agriculture
Lime and deadburned dolomite
Riprap
Railroad ballast
Alkali manufacture
Asphalt filler
Limestone sand
Glass manufacture

Other uses ——

Coalmine dusting
Filler
Fertilizer
Filtration
Limestone whiting
Mineral food
Paper manufacture
Poultry grit
Refractory dolomite
Sugar refining, etc.

the works of practising engineers, some mention must be made, however, of these important resources. The following sections deal briefly with examples of uses of calcareous rocks (limestone and dolostone), gypsum and anhydrite, clay products and construction aggregate. Later sections touch on some of the aspects of the fossil fuels, which provide the principal energy base of our economy.

CALCAREOUS ROCKS

The abundant carbonate rocks limestone and dolostone with their many useful physical and chemical properties in both raw and manufactured forms are among the more widely used raw materials. The list of their myriad uses is nearly interminable and includes agricultural applications (liming, fertilization, composting); construction uses (aggregate, portland cement, mortar and plaster, soil stabilization, dimension stone); and chemical and industrial uses in metallurgy, chemicals manufacture, sanitation, pulp and paper processing, ceramics, foods and petroleum technology. An indication of the dimensions of the industry is suggested by figure 17.2.

Calcareous rocks are found, usually as sedimentary deposits, virtually in every state of the United States and in every nation of the world (see map 4, appendix II). There is no significant industrial distinction for many purposes between high-calcium rocks such as limestone, chalk, or marble composed dominantly of calcite, $CaCO_3$, and calcium-magnesium rocks such as dolostone or dolomitic marble composed of dolomite, $CaMg(CO_3)_2$; however, either may be preferred or required in a particular application.

Many of the uses of calcareous rocks, a few of which will be briefly described, may be sensitive to the level of purity of the raw material and in practice the production of particular mines or quarries tends to be dedicated to providing material for one or a few finished products. Table 17.1 gives an idea of the range of impurity content of representative United States limestones.

The impurities in calcareous rocks are present as a natural consequence of the geologic processes forming the rock and may be uniformly distributed throughout the rock, vary in amount from bed to bed, or be randomly scattered. Of the principal contaminants silica and alumina are present as clay, silt, or

Table 17.1 Compositional Range for Representative United States Limestones

Limestone	Range	Limestone	Range
CaO	30–55%	SO_3	.01–.3
MgO	.5–21	P_2O_5	.04–.2
CO_2	33–46	Na_2O	.01–.3
SO_2	.1–20	K_2O	.01–.7
Al_2O_3	.04–6	H_2O	.2–1.6
Fe_2O_3	.05–2		

After R. S. Boynton, *Chemistry and Technology of Lime and
Limestone.* © 1967 John Wiley & Sons, N.Y. Reprinted by permission of
John Wiley & Sons, Inc.

Figure 17.3 P-T relations for calcite and dolomite

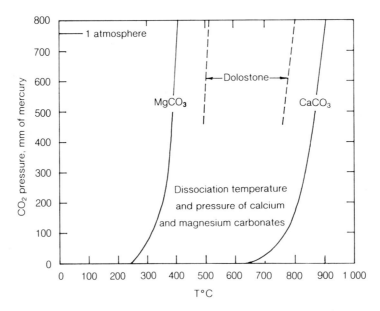

sand particles or as nodules; iron may substitute for
magnesium in dolomite or be found as iron oxides
or sulfides (pyrite, FeS_2); and phosphorus and sulfur
occur in phosphate, sulfide, or sulfate minerals.

A considerable amount of limestone and dolostone is converted to *quicklime* by heating (fig.
17.3).

These quicklimes are converted to hydrated limes in
mortars, plasters, and chemical products by adding
water.

$$CaO + H_2O \rightarrow Ca(OH)_2 + 15{,}300 \text{ cal/g.mol}$$

$$MgO + H_2O \rightarrow Mg(OH)_2 + 10{,}000 \text{ cal/g.mol}$$

$$CaCO_3 \xrightarrow{898\,°C} CaO + CO_2$$

Calcite   Quicklime   Carbon
  dioxide gas

$$CaMg(CO_3)_2 \xrightarrow{500–800\,°C} CaO \cdot MgO + CO_2$$

Dolomite   Dolomitic   Carbon dioxide
quicklime   gas

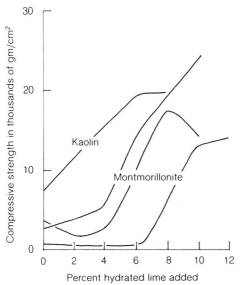

Figure 17.4 Compressive strengths of some lime-stabilized soils

Some critical variables in lime burning are the quality, size, and gradation of the limestone, and the rate, time, and temperature of calcination.

Among the many applications of hydrated lime, its use in the stabilization of clay-bearing soils is of particular importance in civil engineering. Calcium ions have the ability to displace H^+, Na^+, and K^+ from clay particle surfaces, and in the presence of CO_2,[1] to react with the clay to form a noncrystalline calcium silicate cement. Figure 17.4 shows how the addition of hydrated lime improved the bearing strength of some of the test soils. In general, lime will react readily with all clay-bearing soils regardless of their grain size; however, soils with a low plasticity may require another additive such as fly ash.

The largest single use of limestone is in the production of portland cement.[2] This material was developed in 1824 by Joseph Aspdin in England, but not until the latter years of the nineteenth century

was a standardized high-quality product developed and widely adopted. Cementitious or hydraulic limes, which have the property of hardening and setting when water is added, have been used since earliest times; the first documented application was the plaster covering of the pyramids of Egypt. The Romans developed structural uses for lime cements (pozzolan or puzzolan) by combining quicklime with a volcanic ash from Pozzuoli, Italy, and other materials of similar composition. Natural cements containing appropriate amounts of lime, silica, alumina, and iron oxides were later found to have appropriate hydraulic properties and were widely used.

Portland cement is manufactured from these same constituents but as a chemically controlled mixture of raw materials that is fed into a rotary or moving bed kiln and heated to temperatures up to 1 500°C. This firing removes carbon dioxide and water and results in a glassy clinker, which is then crushed and stored away from water. This powder is a reactive mixture; when water is added, it yields the following approximate proportions of calcium compounds:

Ca_3SiO_5	45%
Ca_2SiO_4	27
$Ca_3Al_2O_6$	11
$Ca_4Al_2Fe_2O_{10}$	8
$CaSO_4$	2.5

The sources of the various constituents are not critical and include such varied natural and industrial by-product materials as

Calcium   Limestone, marble, shells, marl, slag
Aluminum   Clay, shale, slag, fly ash
Silicon   Quartz sand, basalt, quartzite, fullers earth
Iron   Iron oxides and sulfides, flue dust
Sulfur   Pyrite, gypsum

As used in concrete, portland cement is the binder for fine aggregate that in turn fills the interstices of the coarse aggregate used to give bulk to the mixture. The proportions of these three components and especially the cement-aggregate ratio, are important to the strength of the final product (fig. 17.5).

1. The CO_2 content of soils may reach 10–20% of the soil gas (compared with about .05% in air) due to bacterial fermentation.
2. The name comes from the similarity in appearance of the set material to portland stone, a popular limestone used widely for building in the British Isles.

Figure 17.5 Compressive strength of portland cement—aggregate mixtures

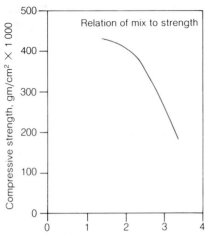

Relation of mix to strength

Total volume of aggregate per one volume of portland cement. Aggregate in proportion of 1 fine to 2 coarse.

GYPSUM AND ANHYDRITE

Calcium sulfates are found in extensive bedded deposits formed by evaporation of saline water in company with limestone, dolostone, rock salt, and clays. Either gypsum, $CaSO_4.2H_2O$, or anhydrite, $CaSO_4$, may be the primary sulfate deposited, and natural hydration–dehydration reactions have often transformed one into the other. Commercial gypsum deposits are found worldwide in rocks ranging in age from Cambrian to Recent.

Plaster products, principally dry-wall sheets for use in building, are the main use for gypsum. Processing involves grinding and heating of gypsum to produce the hemihydrate plaster of Paris, named after deposits at Montmarte near Paris, France.

When rehydrated, the hemihydrate returns to gypsum and sets hard with an increase in volume and evolution of heat. The newly formed gypsum is in the form of needlelike crystals that interlace within the gypsum sheet and penetrate the surface paper to form a strongly bonded, fireproof laminate that may be easily sawn or nailed.

Gypsum is also widely used as a land plaster, fertilizer, and soil conditioner. Its agricultural value lies mainly in its ability to modify the pH of alkaline soils; correct deficiencies in sulfur; and to increase the amount of exchangeable calcium in the soil. This latter property is particularly important in improving land reclaimed from the sea, or land impaired by flooding with salt water, by alteration of the deleterious sodium-bearing clay. Calcium ions displace sodium ions from clay

$$Na\text{-clay} + gypsum \rightleftharpoons Ca\text{-clay} + Na\ salts$$

and the reaction will go to the right if soluble salts are removed by efficient drainage.

Other important uses of gypsum are as a retarder in portland cement, as the raw material for sulfuric acid production, as an ingredient in the manufacture of chalks and crayons, and as a filler or filtering medium. Anhydrite has limited use compared with gypsum, being principally used as a raw material in the manufacture of ammonium sulfate and sulfuric acid.

Gypsum and anhydrite are low unit-value materials that are widely distributed. In consequence, their exploitation depends upon cheap mining, limited need for benefication of the ore, and low transportation costs.

$$CaSO_4 \cdot 2H_2O \xrightarrow{128\,°C} CaSO_4 \cdot 1/2H_2O \xrightarrow{163\,°C} CaSO_4 \xrightarrow{red\ heat} CaSO_4$$

Gypsum — Hemihydrate — Soluble anhydrite — Anhydrite

Table 17.2 Commercial Clays and Their Uses

Clay Designation	Use
High-grade clays	
Whiteware clays	Porcelain, china, whiteware, pottery, high-grade tile, sanitary ware
Kaolin or china clay	
Ball clay	
Paper clay	
Refractory clays	
Fireclays	Brick, liners, and bonding material
Glass-pot clay	
Flint clay	
Graphite fire clay	
Pottery clays	Stoneware, pottery, glazed ware
Low-grade clays	
Vitrifying clays	Roofing tile, common brick, enameled brick, hollow tile, conduits,
Terra cotta clays and shales	paving brick, pressed brick, glazing
Pipe clays	
Brick clays	
Loess and glacial clays	
Adobe	
Gumbo clays	
Slip clays	

After R. B. Ladoo and W. M. Myers, *Nonmetallic Minerals*, 2d. ed.
© 1951 McGraw-Hill Book Co., N.Y. Used with permission of the
McGraw-Hill Book Co.

CLAY PRODUCTS

Clays are widely distributed over the world with members of the kaolinite group tending to dominate in tropical and subtropical regions and those of the montmorillonite group (smectites) being more common in higher latitudes. Clays may be concentrated either as residual deposits formed by weathering on various bed rocks or as sedimentary deposits in beds. In the United States, the principal deposits of high-grade clays are found in the belt of Cretaceous and younger rocks that extends from New Jersey to Georgia and Alabama, thence northwesterly through western Tennessee and Kentucky and again southwesterly across Arkansas and Texas. Low-grade clays suitable for products such as tile and brick are exceedingly widespread, abundant, and available for local exploitation.

The principal properties of clays of commercial concern are their fineness of grain, color, purity, plasticity, and volume change both as a function of water content or firing. Specifications for these and other properties will naturally be controlled by the purpose for which the clay is intended and may thus be expected to vary widely.

The mineralogy of clay was previously discussed in chapter 1 and the properties of clay mineral aggregates were described in chapter 11. In addition to these aspects of this important substance is its wide use as a raw material in a variety of industrial applications. One large area of clay usage is in the manufacture of brick, ceramics, and refractories during which the clay is heated until the particles lose their chemically combined water and coalesce to form a stony mass. Table 17.2 provides a listing of some of the names used commercially for these clays and a sampling of their uses.

Table 17.3 Melting Points of Seger Cones

Cone No.	T°C	Cone No.	T°C	Cone No.	T°C
022*	590†	02	1 110	19	1 510
021	620	01	1 130	20	1 530
020	650	1	1 150	26	1 650
019	680	2	1 170	27	1 670
018	710	3	1 190	28	1 690
017	740	4	1 210	29	1 710
016	770	5	1 230	30	1 730
015	800	6	1 250	31	1 750
012½	875	7	1 270	32	1 770
010	950	8	1 290	33	1 790
09	970	9	1 310	34	1 810
08	990	10	1 330	35	1 830
07	1 010	11	1 350	36	1 850
06	1 030	12	1 370	37	1 870
05	1 050	13	1 390	38	1 890
04	1 070	14	1 410	39	1 910
03	1 090	15	1 430		
		16	1 450		
		17	1 470		
		18	1 490		

*Slender triangular cones about 7.5 cm high with a 2 cm base. They are placed upright in the kiln and the temperature taken when the tip begins to droop or when it touches the base.
†Temperatures given are those of incipient melting.

From H. Ries, *Clays, Their Occurrence, Properties and Use.* © 1927 John Wiley & Sons, N.Y. Reprinted by permission of John Wiley & Sons, Inc.

Physical and chemical changes take place when a clay is fired or burned, and the temperature at which the changes occur, the behavior of the clay during firing, and the properties of the finished product are all of importance. As clay is heated it first dehydrates, then organic matter, sulfur, and carbon dioxide are driven off and any iron compounds oxidize to the ferric state. Finally *vitrification* occurs over a more or less extended temperature range. The vitrification temperature of a particular clay may be determined by the use of pyrometric or Seger cones. These are thin cones especially fabricated to droop at particular temperatures thus giving a direct visual estimate of kiln temperature. Melting points of Seger cones are given in table 17.3.

Clays in a natural or refined state are also widely used for nonceramic purposes in which their very fine grain size, chemical stability, adsorptive capacity, and low unit cost are important factors. Porous substances such as paper, cloth, linoleum, or rubber may be sized, filled, or coated with clay; and paints, medicines, cleansers, or insecticides extended. The high adsorptive capacity of clay is exploited in the filtering and decolorizing of various oils and in general degreasing purposes. Other applications include the use of clay as a sealant in engineering works such as dams or ponds, as an additive in plaster and cement products; and its use in drilling muds, chemicals, or mild abrasives.

CONSTRUCTION AGGREGATES

Aggregate in engineering usage is that material which provides the bulk of solid body in such bonded materials as portland cement or macadam and in such unbonded applications as railway ballast, road and foundation base, topping for gravel roads, or french drains. It may be distinguished from fill on the basis of its uniform sizing. Aggregate is made from a wide variety of igneous, metamorphic, and sedimentary rocks and unlithified sediments. The following classification, simplified from the Kentucky Highway Department specifications, indicates the range of materials, criteria, and uses which should be considered:

1.0 Fine aggregate. Sand or finer-sized material graded to ± 20% of size specification.

1.1 *Natural sand.* Clean, sound, hard, durable particles containing < 1.25% deleterious substances such as coal, shale, mica, etc. Minimum gradation is as follows:

Sieve size	% passing
3/8 inch	100
No. 4	85–100
No. 16	40–80
No. 50	5–25
No. 100	0–5

1.2 *Crushed sand produced by crushing stone, slag, or gravel.* Limits for deleterious substances and gradation the same as for natural sand.

1.2.1 *Mortar sand suitable for use in portland cement.* Clean, hard, durable uncoated particles free from injurious amounts of organic or other deleterious substances; minimum gradation is as follows:

Sieve size	% passing
No. 8	100
No. 50	10–40
No. 100	0–10

1.3 *Mineral filler comprised of various inert rock dusts.* Must be thoroughly dry, free from clay, loam, lumps consisting of aggregations of fine particles and meet the following minimum gradation requirements:

Sieve size	% passing
No. 30	100
No. 80	95
No. 200	65

2.0 Coarse aggregates. Hard durable particles of crushed rock or slag.

2.1 *Crushed limestone.* Angular fragments of uniform quality.

2.2 *Crushed slag.* Angular fragments of air-cooled blast furnace slag or phosphate slag with a unit weight not less than 70 pounds per cubic foot ($16 kg/m^3$).

2.3 *Gravel.* Durable or natural or crushed fragments of rock reasonably uniform in density and quality. For use in bituminous surfacing at least 65% of the particles shall have one or more fractured faces. For use in portland cement concrete aggregate the maximum percent by weight of deleterious substances shall be

Friable particles	0.25%
Finer than 200 mesh	2.0
Chert with specific gravity below 2.35	1.5
Coal	0.5.

Depending upon its planned use, aggregate will typically be subjected to a number of the following tests:

Compacted weight Bulk specific gravity in closest packing

Wear Loss of weight when tumbled in a standard machine with steel balls

Soundness Immersion in sodium sulfate followed by drying and sieving

Coal and chert content Separation of coal or chert by heavy liquid flotation

Particle shape and texture of fine aggregates
Determination of porosity at several screen sizes:

$$\text{percent voids} = \frac{100\,(1 - W)}{VG}$$

where W = weight of material, V = volume of material, and G = bulk density, oven dry.

Specific gravity and absorption of aggregates
Pycnometer (see chapter 1) for fine aggregates and various measures of specific gravity for coarse aggregates. Let
A = weight of oven-dry sample in air,
B = weight of saturated sample air, and
C = weight of saturated sample in water:

Bulk oven-dry specific gravity $= \dfrac{A}{B - C}$

Saturated specific gravity $= \dfrac{B}{B - C}$

Apparent specific gravity $= \dfrac{A}{A - C}$

Absorption (water), percent
$= \dfrac{B - A}{A} \times 100$

Sieve analysis   (See chapter 10)

Fineness modulus of fine aggregate   Obtained by adding the cumulative percentages of a sample retained on each of a graded series of sieves and dividing by the sum of 100. (Figure 10.2 provides an example of the procedure.)

Percent shale, etc., in aggregate   Separated by handpicking or heavy liquid flotation:

$$\% \text{ shale} = \frac{W_s}{W_s + W_g} \times 100$$

where W_s = weight shale, etc., and W_g = weight of good aggregate.

Percent clay (plastic fines) in graded aggregate
Settling (elutriation) in calcium chloride solution, measurement of levels of top of clay suspension and sand in standard settling tube after standard settling time. Reported as sand equivalent, SE.

$$SE = \frac{\text{Sand reading}}{\text{Clay reading}} \times 100$$

FOSSIL FUELS

The economic health of the nation is rooted in the availability of an adequate supply of energy to serve for lighting, heat, transportation, and manufacture. Fuels are sometimes consumed in large stationary plants and the energy that is generated is transmitted via electrical systems to the point of use. Other needs such as space heating and vehicular power require local consumption. Many different energy sources are known and utilized such as water, solar, wind and nuclear power, but fossil fuels are dominant. These fuels are accumulations of the products of the decay of organic materials that are preserved as sedimentary rocks or as reservoirs of fluids trapped in the void spaces of rocks. Lignite and coal are plant-derived representatives of the former and petroleum and natural gas are animal-derived examples.

Petroleum
On death, organic matter may be preserved in an oxygen-poor environment and transformed to other hydrocarbon compounds. In the case of animals these compounds are liquids (oils) and gases, which accompanied by saline groundwater can migrate from source beds outward and upward through permeable zones. They may reach the surface as oil seeps or asphalt deposits or be trapped underground by a decrease in permeability or by a change in the configuration of the rocks (see fig. 14.6).

Underground accumulations of petroliferous fluids—pools or reservoirs—occur in porous and permeable rocks from which they may be recovered through drill holes. The three fluids present (gas, oil, water) may exist as separate phases arranged by density (fig. 17.6a). The gas phase, however, may be dissolved in either the oil or water. It is more readily dissolved in oil but at great depths may be dissolved in water as "geopressurized" gas. Release of pressure allows the exsolution of gas, and its expansion provides the driving force for primary production (flow) of an oil well. Figure 17.6a indicates the reservoir mechanics for such production; it should be noted that the well placement is critical.

Drill rig, Prudhoe Bay, Alaska

When the gas pressure becomes insufficient to lift the oil secondary recovery methods must be employed. Repressurization may be done through injection wells delivering gas to the cap area or by a *water drive* initiated as shown in figure 17.6b and c.

The complex flow of liquids into and out of a reservoir is a function of both the acting pressures and the permeability of the rocks; *reservoir mechanics* is the discipline that deals with the control and modification of these parameters. Pressure control is fundamentally a matter of balancing pumping pressures and output flow rates. Well spacing can help improve flow rates within the reservoir and natural permeabilities can be modified by deliberate fracturing of reservoir rocks. A common means is *hydrofracking* whereby a slurry of sand and water is injected into the reservoir rock at pressures in excess of lithostatic. The rock is wedged open by the high-pressure water and the crack supported by the sand grains.

Figure 17.6   Reservoir mechanics

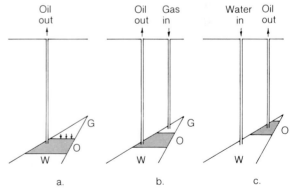

KEY

G   Gas-saturated rock
O   Oil-saturated rock
W   Water-saturated rock

Tar Sand and Oil Shale

Two subsidiary but potentially very important sources of liquid hydrocarbons are tar sand and oil shale. Tar and asphalt are heavy residual portions of natural oils that have been fractionated and whose lighter fraction has escaped. Tar and asphalt-impregnated rocks may be recovered by mining or the tars liquified by steam and recovered through boreholes. Tars and asphalts may be cracked, hydrogenated, and reconstituted into useful liquid and gaseous fuels.

Oil shale is misnamed; it contains no oil and is not always a shale. Rather, it is a fine-grained sedimentary rock that contains a significant amount of a solid organic substance called *kerogen*. On heating (pyrolysis), kerogen decomposes into tars, petroleum liquids, and methane-rich gases. Oil shales typically yield 10–50 gallons of oil and 1–3 thousand cubic feet (MCF) of gas per ton of rock. Retorting of mined rock may be done in surface plants or in-place pyrolysis can be accomplished by igniting and oxygenating broken rock underground. Extensive deposits of oil shale are found worldwide including both the eastern and western United States. The Green River Shales of Colorado and adjacent states are limy lake sediments and the eastern Chattanooga Shale and its equivalents are siliceous marine deposits.

Coal

Coal is a compact, stratified sedimentary rock composed largely of coalified plant remains mixed with a variable but subordinate amount of inorganic mineral matter. The fossilized plant fragments such as wood, bark, leaves, and spores are termed *macerals* and further categorized as vitrinite (plant matter), exinite (waxy and resinous components), and inertinite (charcoal). The coalification process takes place under reducing (oxygen-poor) conditions and over geologic time transforms plant matter in sequence to peat, lignite, brown coal, subbituminous coal, bituminous coal, and anthracite coal. This process is accompanied by a general increase in heating value and fixed carbon content and a decrease in volatile matter. The particular stage of coalification is termed *rank* and is the classification of coals ". . . according to their chemical and physical properties,

especially heating value, moisture content, carbon content, ash characteristics, and presence of associated minerals". (Schmitt, 1979).

The setting for coal formation is one of prolific plant growth coupled with a means for the preservation of dead material. Land plants were not abundant before the Carboniferous Period (see fig. 3.2) and all commercial coals are of this age or younger. Preservation of dead vegetation is readily accomplished in nature by covering it with oxygen-poor water such as is found in swamps. The distribution of coal bodies is thus coextensive with former swampy areas. Large areas of swamp today and in the past are related to low-lying coasts and associated stream systems. A useful means of assessing the quality of coal, the continuity of seams, and the nature of the enclosing rocks is to locate the coal deposit with respect to its position on a deltaic or alluvial plain. Small changes in water level in such environment caused extensive areas to be inundated or exposed with consequent changes in the nature of the sediments deposited at a given point. Oscillation of this kind led to repeated sequences of coal and other sedimentary rocks that in West Virginia, for example, gave rise to 117 coal beds of economic interest.

Coal, being a sedimentary rock, occurs in tabular beds or seams whose quality, thickness, areal extent, and location determine their economic viability. Coal quality is based upon heating value, the sulfur and ash content, and special properties (e.g., ability to coke). The standard means of measure is by *proximate analysis* whereby volatile, ash, moisture, fixed carbon, sulfur, and heating value contents of a sample are found by destructive distillation. *Ultimate analysis* is sometimes used in which C, H, O, N, S, and ash are determined.

Heating value is usually stated in British thermal units (Btu) per pound and is generally in the range of 9 000–16 000. Sulfides, sulfates, and organically-bound sulfur are normal components of coal, and sulfur oxides are released as gases on coal burning. Sulfur is a significant environmental pollutant and a probable cause of acid rain. Environmental regulations limit sulfur emissions and thus dictate the use of either low-sulfur coal or the installation of expensive emission controls when coal

Open-pit coal mine, Wyoming

is used. Unfortunately, only about one-third of the potential recoverable coal reserves of the United States are low sulfur ($\leq 0.7\%S$), and these are mostly to be found in the western states. Ash *per se* is generally not deleterious in coal burning, but it dilutes the energy content of the coal and requires arrangements for its disposal.

The thickness of a coal seam is important in the extraction and economics of coal production in both open pit and underground mining (as discussed in chapter 18). The ratio of the thickness of the overburden, which must be stripped away, to the thickness of the coal to be won (the stripping ratio) controls the economic depth of open pit operations. The minimum space, usually 22 inches, that allows miners and machines access to the seam establishes the minimum seam thickness that can be extracted in underground (deep) mining.

The location of a coal mine (or any production facility) with respect to the market for its product is an important criterion because of the high transportation costs involved. These may be a significant fraction of the overall production costs to the point of sale since few coal-using facilities are located adjacent to mines. The usual means of transport are by barge, the least expensive method, or by rail, truck, or slurry pipeline. The cost of transport may be thought of as a decrease in value of the coal as a function of distance from the minehead—the rate of decrease depending upon the transport means and transhipment representing a value loss with no increase in distance.

The coal business is big business and seems likely to remain so. Nearly all coal mined today is used directly as a solid fuel, after appropriate cleaning, or as coke produced by burning coal in a

Table 17.4 Recoverable Coal Reserves in the United States (Values in 10^6 Tons)

Coal Rank	Strippable	Deep	Total
Anthracite	68	2 166	2 334
Bituminous	13 597	54 596	63 340
Subbituminous	24 318	32 320	56 227
Lignite	8 895	—	8 895
Totals	46 876	89 082	135 696

Source: U.S. Bureau of Mines Information Circular 8531, 1971

reducing atmosphere. Coal, however, can serve as a feedstock for synthetic liquid and gaseous fuels (*synfuels*) made in surface plants or it can be gasified underground. Two-thirds of the coal presently produced in the United States is consumed by the electric utility industry (amounting to about 400 million tons in 1974). Even at this rate, adequate reserves remain; in 1971 the U.S. Bureau of Mines estimated the total recoverable coal reserves in the United States to be 135 696 in 10^6 tons (table 17.4). These reserves, in practice, must be reduced by an average of 50% to account for nonrecoverability and must be increased by unknown amounts in consequence of new discoveries and improved technology.

SUMMARY

The body of the earth is the source of nonrenewable raw materials that underpin our technology and economy. Because the distribution of raw materials is erratic, some regions and countries serve as suppliers and others as markets, thus providing the basis for trade and often political maneuvering.

Only a few of the more widely distributed and used mineral commodities have been discussed in this chapter; a complete listing of mineral materials and their uses would fill a large book. The following classification provides an indication of the range of uses for the earth's mineral wealth:

FUELS

Fossil—petroleum, natural gas, oil shale, tar sand, coal, lignite
Nuclear—ores of uranium, thorium

NONMETALLICS

Petrochemicals—petroleum, natural gas, asphalt, coal
Construction materials (fill, aggregate, dressed stone)—most rocks
Mortars and cements—limestone, dolostone, gypsum
Land plasters and fertilizers—limestone, dolostone, phosphate, gypsum
Glass and ceramics—glass sand, clay, feldspars
Abrasives—quartz, corundum, garnet, diamond
Pigments—ores of titanium, iron, cobalt, cadmium
Refractories—ores of magnesia, asbestos, alumina, chromite
Fluxes—quartz, fluorite, borates, limestone
Nuclear materials—beryllium, graphite
Lubricants—graphite, molybdenite
Absorbents—fullers earth, clay
Mordants—chromite, hematite
Basic chemicals—halite, sulfur
Gems—diamond, corundum, garnet, quartz, beryl
Pyrotechnics—strontium
Miscellaneous—barite (drilling muds), rare earths (sparkers), molding sands, soil stabilization

METALLICS

Ferrous metals—ores of iron, cobalt, chromium, vanadium, tungsten, manganese

Base metals—ores of copper, lead, zinc, tin, molybdenum

Light metals—ores of aluminum, magnesium, titanium, beryllium

Precious metals—ores of gold, silver, platinum group elements

Miscellaneous metals (minor metals)—ores of gallium, indium, cadmium, germanium, mercury, antimony, bismuth

---

ADDITIONAL READING

ASTM 1966. Tentative specifications for classification of coals by rank, ASTM 388–66. ASTM Standards, Part 19, Gaseous fuels, Coal and coke.

Boynton, R. S. 1980. *Chemistry and technology of lime and limestone.* 2d ed. New York: John Wiley & Sons.

Bureau of Mines, U.S. Department of Interior. Minerals Yearbook. V. I, *Metals, minerals and fuels;* V. II, *Area reports, domestic;* V. III, *Area reports, international.* Published Annually.

Cassidy, S. M., ed. 1973. *Elements of practical coal mining.* Soc. Min. Engineers, AIME.

Dapples, E. C., and Hopkins, M. E., eds. 1969. *Environments of coal deposition.* Geol. Soc. America Spec. Pub. 114.

Groves, A. W. 1958. *Gypsum and anhydrite.* London: H. M. Stationery Office.

Industrial Minerals. Published monthly by Metal Bulletin, PLC, London.

Ladoo, R. B., and Myers, W. M. 1951. *Nonmetallic minerals.* New York: McGraw-Hill.

Lamey, C. A. 1966. *Metallic and industrial mineral deposits.* New York: McGraw-Hill.

Schmitt, R. A. 1979. *Coal in America: An encyclopedia of reserves, production, and use.* New York: McGraw-Hill.

18

Mining

INTRODUCTION

Originally the term *engineering* was applied to works intended for military purposes; later *civil engineering* designated nonmilitary engineering activities. Mining was among the earlier fields in which civil engineering was practiced because of the special needs of excavation, haulage, construction, drainage, and ventilation; and *mining engineering* grew from civil engineering in response to this need. The close relationship of mining engineering and civil engineering exists today, and few large mining operations are conducted without the services of one or both kinds of engineers. Because of this it is pertinent to describe some of the principles of the recovery of natural resources by mining.

The basis for any mining venture is the presence of some natural material, ore, which may be recovered at a profit. *Profitability* is the key, and its prediction is a complex process that establishes the in-place value of the material and then debits this value by an assessment of all costs that will be incurred to the time and point of sale. Mining will properly be initiated only if this assessment indicates that the material will repay all discovery and exploitation costs and return a sufficient profit to underwrite the development and production costs of a new deposit.

The basic information concerning in-place value will usually be established by geologists after intensive mapping and sampling. Their goal is to delimit the ore body and establish the average concentration (grade, tenor) of the material of interest. The volume of the body is usually converted to tonnes of ore, and this times the average concentration yields total in-place marketable material. Total marketable material times value, of course, is total dollar value. Thus (volume) (bulk density) (ore grade) (value per unit) = total value

The geologist in the process of the field exploration is also expected to identify any potential advantages or disadvantages related to exploitation of the ore. Of particular importance are the *shape* of the body; the location of the *water table;* the chemical and physical *natures* of the ore, particularly as these might affect excavation, milling, or smelting; and potential environmental *hazards* and impacts.

The assessment of exploitation costs is a complex process since neither the unit value of an ore nor the costs of its recovery are constant in time. This assessment must thus include not only present-day estimates of costs for such things as plant construction, labor, taxes, transportation, and debt service on money but also a prediction of the probable cost changes over the life of the mine. The problem of estimating operational costs obviously requires information from many sources, which necessarily include engineers. Engineers must provide data on such things as mine design, mining rates, plant construction, transportation, and power costs. It is probably correct to say that accountants and engineers have found or lost as many mines as geologists.

All in all the discovery, the development, and the profitable operation of mines, quarries, and related means for recovering natural wealth within a framework of national needs for raw materials and environmental concerns represents an unparalleled challenge.

Mining is the general term used for the separation of an ore from the ground and *milling* refers to the processing of the ore into one or more marketable products. The combination of methods selected for the recovery of a given ore should, of course, be the optimum one to generate maximum profitability. Determining this combination is a particularly difficult task since it must include choices of mine geometry, machinery and operational plans, milling methods and equipment, and marketing strategy within a predicted economic framework; it will almost certainly involve a number of trial combinations in the planning stages.

The special province of a mining engineer is the separation of ore from the ground by mining, quarrying, dredging, or through bored holes. The separation of the ore minerals from **gangue** (waste rock) or the separation of impurities from bulk commodities is the responsibility of the mineral dresser. They cannot work independently; they must interact with both geologists and executives in making business decisions throughout the life of a mine.

Knowledge of quantity, geometry, physical properties, grade, and potential market of the ore in a deposit is fundamental to decisions relating to the optimum manner and rate of ore production to be employed. From these economic considerations a planned annual production rate is established. This rate may be simply the total recoverable tonnage divided by a predetermined lifetime (time to exhaustion) of the deposit or will more likely be dictated by the choice of mining method and the mill capacity vis-à-vis a desired level of return on investment.

It should always be kept in mind that the determination that a deposit apparently contains ore of sufficient quality and quantity to warrant its exploration does not automatically lead to a decision to bring it into production; *future* costs and markets must be considered and the balance sheet may not always be favorable.

Mining may be carried out by tunneling underground or excavating pits from the surface (open-pit mining). The latter, being cheaper and somewhat safer, will always be used when feasible. It is the common method for recovering materials that occur in large tonnages but have a low unit value (e.g., sand and gravel, iron ore, limestone, and disseminated copper ore). Underground mining tends to be restricted to materials whose high unit value allows this more costly extraction method to be employed at a profit.

EXTRACTION METHODS

General Principles

Open-pit mining is carried forward by the removal of unwanted cover material (overburden) in a mining step termed *stripping*. Excavation of the body to a depth controlled by the limits of ore, bank stability and pumping costs is then undertaken. The stripping and mining are usually done in sequential steps and, for large bodies, may involve a number of benches, or terraces, that provide both multiple working faces and haulageways. As work proceeds, stripping spoil will often be used to backfill the excavation as both a convenient method of waste disposal and an acceptable means of conforming to environmental statutes.

The design of an underground mining operation is largely controlled by the physical nature of the ore body and its geometry. Flat-lying tabular bodies must be mined differently than steeply inclined tabular bodies; methods for large irregular masses are not appropriate for small veins; and thick-bedded deposits present a different mining problem than thin beds. Indeed, each mineral deposit will require a custom-tailored design for its optimum recovery at lowest cost.

Regardless of the nature and shape of an ore body, however, certain principles will always apply. Advantage should always be taken of gravity; mine haulage slopes should incline slightly upward from the foot of the shaft allowing ore trains and water to travel downhill to a central collecting point. *Roof support* to protect working areas is critical and should be provided by unmined ore or rock (pillars); timber, steel or masonry sets, or by backfilling abandoned workings with broken rock or pumped-in rock slurry or slush. A general rule is to maintain only the size of opening needed for the work. *Ventilation* must be provided throughout the working area by means of fans circulating the flow in such a way that all working faces receive an adequate supply of fresh air. **Blasting** is done on a regular schedule known to all workers.

The planning and conduct of a mining program entails the three-dimensional visualization of a continuously changing workplace that must be safe, well-drained and ventilated, have appropriate power and transportation systems, and be able to generate ore at the predetermined rate and quality. Planning in such a way as to provide these needs and obviate potential and costly hazards is at least as important as the actual production activity. A well-planned and functioning mine is an accomplishment equivalent to, or surpassing, more visible products of engineering science.

Because of their modes of formation mineral deposits tend to be two dimensional sheetlike or tabular bodies, often more or less crumpled; linear blades or rods; or irregular three-dimensional masses. The mining of such bodies will require different approaches according to their thickness, attitude, amount of cover, and strength of the ore and its enclosing rocks. The strength of the enclosing wall rock determines the size of openings, the nature and amount of support, and is important in determining the mining plan. (Weak rocks, for example, may be deliberately caved.)

Mine openings must be large enough to provide easy access for men and machines. If the "mining width" is greater than the ore-body thickness, however, some wall rock will be mined with the ore thus reducing its grade. If the ore-body thickness is greater than the amount that can be safely removed in a single operation, several steps or lifts must be made. Flat-lying ore bodies may be mined on a single level and steeply dipping bodies can take advantage of gravity to bring ore down to transportation levels.

Basic to mine planning are the nature of the ore and the shape, size, and cover of the ore body since these dictate whether the ore may be extracted through boreholes, from open pits, or by underground methods. Borehole recovery is limited to underground fluids or fluidizable materials and has the advantages of low cost and minimum surface disturbance, but these are offset by low percent recovery. For solid ores, open-pit methods are often preferred because of their relatively lower cost, greater percent recovery, and safer operating conditions when compared with underground mining. Open-pit methods do, however, cause much greater disturbance of the surface and require extensive postmining reclamation.

Copper mine, Bingham, Utah

The range and variability of mineral extraction methods needed to handle the plethora of mining situations may be appreciated by an examination of the following categories of the more important means of recovery. In the selection of the method of mining for a given ore body, usually several methods are adapted to accommodate the size, shape, attitude, and strength of the ore body and walls. The resultant plan is seldom ideal from all points of view but should represent the best compromise between conflicting factors.

    I. Surface mining
        A. Open-pit (open cast, strip) mining (fig. 18.3)
        B. Quarrying
        C. Hydraulic mining   Excavation and movement of unconsolidated material using a water jet
        D. Dredging   Excavation by continuous bucket line or suction from a floating dredge. Ore processing done aboard
   II. Combined surface and underground mining
        A. Glory hole   Open-pit operations transferring ore through underground passages beneath the pit
  III. Underground stoping   Loosening and removal of ore from underground excavations called stopes.
        A. Naturally supported stopes
            1. Open stopes
            2. Open stopes with pillars

Cerro Bolívar iron mine, Venezuela

B. Artificially supported stopes
   1. Shrinkage stopes   Broken ore serves as support (fig. 18.9)
   2. Cut-and-fill stopes   Ore is removed and replaced by waste (fig. 18.10)
   3. Stulled stopes   Timber posts provide support
   4. Square set stopes   An open boxwork of timber provides support
C. Caved stopes
   1. Block caving   Support is removed from the sides and base of a large block of ore. Broken ore is drawn off through protected openings below (fig. 18.12)
   2. Sublevel caving   Ore body progressively caved from the top down (fig. 18.11)
   3. Top slicing   Caving method working from the top down and sides in
D. Combined methods

IV. Borehole recovery
   A. Auguring   Large diameter horizontal holes drilled with an Archimedes screw, usually for the recovery of coal
   B. Coaxial holes   Injection and recovery in a single hole; used to inject hot water and recover molten sulfur in the Frasch process
   C. Well field   Regularly arrayed injection and recovery holes

Open-Pit Mining and Quarrying

If there is any choice, an ore body will, with rare exceptions, be worked open to the sky. Under these circumstances standard civil engineering earth- and rock-moving techniques and equipment may be adapted to extraction of the ore, and higher productivity and recovery together with lower costs and greater safety may be achieved than in underground mining. *Quarrying* is the term used for an open-pit mining operation whose product is usually rock material exploited for itself alone (e.g., granite, slate, or marble). The principal distinction between quarrying and open-pit mining is the deliberate removal in quarrying of blocks or dimension stone that requires the use of special excavation techniques.

Overburden must usually be stripped away in order to expose an ore body for open-pit mining. The removal and replacement of this overlying soil or rock may represent a considerable fraction of the mining cost, and the economic viability of a deposit is affected by how much barren material must be included in calculations of ore grade. For flat-lying bedded deposits such as coal or limestone, the thickness of overburden that is to be removed per unit thickness of ore to be mined is called the **overburden ratio.** For example, the economic limit for stripping flat-lying coal seams in Kentucky is presently an overburden ratio of 12:1, and so a coal seam 1 m thick covered by 10 m of overburden may be worked by open-pit mining whereas the same seam under 15 m of cover is uneconomic to mine by this method.

The walls of an open-pit mine must be maintained at a safe angle, which for most operations is close to 45° but for unconsolidated material might be 30° or less. At an average wall angle of 45° each unit in depth requires the excavation of two units in pit breadth (fig. 18.1). Obviously, this is a losing proposition if a thick overburden must be stripped, and in many instances open-pit mines have been transformed into underground operations extending downward from the pit floor in order to exploit the deeper ore. The development of underground shafts and tunnels beneath an open-pit, which facilitates the removal of the ore, shifts the focus of activities below ground and transforms the open-pit operation into a *glory hole mine.* Such a mine receives ore from open-pit workings above and provides for ore-handling by draw holes, tunnels, and shafts beneath the pit.

The limiting depth for open-pit mining may be found from the geometry and grade of the ore body. For example (fig. 18.2), an ore body might be won from a successively deepened pit. Assuming the ore grade to be constant it can be seen that the average mined grade must continuously decrease because of dilution by wall rock. The quality of mineral matter mined may be held constant by selective mining, but the average grade of the deposit on which economic judgments are based decreases.

Figure 18.1 Breadth-depth relations in an open pit

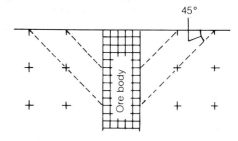

Figure 18.2 Volumes of ore and waste in a deepening pit

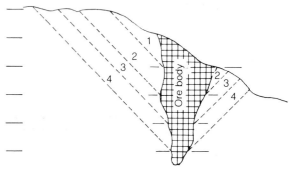

KEY

	Ore Volume	Waste Volume	Waste-Ore Ratio*
1	25	21	0.84
2	40	73	1.83
3	30	80	2.67
4	20	127	6.35

*Equals stripping ratio if the density of waste and ore is the same

The ratio of tonnage of barren overburden and wall rock removed to ore mined is the stripping ratio. For example, if the tonnages of ore, gangue, and overburden plus wall rock are respectively O, G and W tonnes, then $O/(O + G) \times 100 =$ ore grade, $W/O + G =$ stripping ratio, and $O/(O + G + W) \times 100 =$ mined grade.

Pits are usually developed as a series of steps or benches on the order of 15 m high, which spiral around the pit to serve as working and transportation platforms. Several bench faces of differing grade will probably be worked simultaneously to allow blending of the run-of-mine ore to a mill feed of constant quality. A plan of a typical open-pit mine is shown in figure 18.3.

Figure 18.3 Open-pit mine

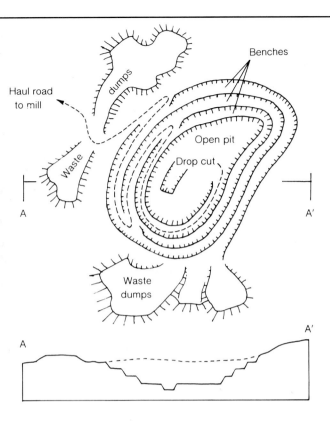

Underground Mining

The layout of an underground mine is necessarily more complex than that of an open-pit mine, and being designed for a particular ore body, it will vary from mine to mine. For all the great diversity of layout, some principles are common to underground operations, however, and the following basic mining techniques have been used for many years:

Undisturbed ore or rock is left in place as pillars or ribs to safeguard the integrity of mine roofs and openings. Gravity is used whenever possible to reduce handling costs. Ore is removed from below by stoping or caving; transportation and drainage tunnels slope downward from working faces, and ventilation is from below upwards.

Headroom is usually kept small to reduce the danger of rock falls from the back or roof.

Artificial support by timbers, roof bolts, packwalls, or backfills is provided whenever there is danger of roof collapse or falling rock.

The guiding factors in the choice of an underground mining method are the grade of the ore, the shape and attitude of the ore body, and the mechanical strength of the ore and the wall rocks (table 18.1). The initial stage of mine development, shared by all methods, is the provision of an *access system* to the ore body. A horizontal tunnel (adit) or an inclined or vertical shaft from the surface is the primary access opening from which a network of openings (analogous to a scaffolding used by builders) is developed to provide haulageways and access to working faces. Federal mining law requires the construction and maintenance of at least two primary access openings for reasons of safety. The second opening is often used for ventilation. Additional drives may be required to provide for underground ore transfer, drainage, ventilation, and backfilling of mined-out spaces. Terminology differs for these multitudinous openings but some of the more commonly used terms are defined in table 18.2.

Table 18.1 Selection of Underground Mining Methods

Type and Attitude of Ore Body	Strength of Ore and Walls (S, strong; W, weak)		Appropriate Mining Method
	Ore	Walls	
Thin tabular bodies with shallow dip	S	S	Room and pillar Casual pillar Open stopes
	W or S	W	Top slicing Longwall
Thick tabular bodies with shallow dip	S	S	Sublevel caving Room and pillar Cut-and-fill stoping
	W or S	W	Top slicing Sublevel caving
	W	S	Square set Cut-and-fill stoping Sublevel stoping
Narrow veins	W or S	W or S	Open stopes Stulled stopes
Thick veins with steep dip	S	S	Open stopes Sublevel stoping Shrinkage stoping Cut-and-fill stoping
	S	W	Cut-and-fill stoping Square set Top slicing Sublevel caving
	W	S	Open casual pillar Square set Top slicing Block caving Sublevel caving
	W	W	Square set Top slicing Sublevel caving
Massive bodies	S	S	Shrinkage stoping Sublevel stoping Cut-and-fill stoping
	W	W or S	Square set Top slicing Sublevel caving Block caving

Modified from L. H. Thomas, *An Introduction to Mining.* Copyright
© 1973 Methuen Australia Pty. Ltd. Reprinted by permission.

Table 18.2 Nomenclature for Mine Openings

Terms Peculiar to Metal Mining

Adit	A horizontal gallery driven from the surface and giving access to an ore body that is worked through it; used sometimes solely for drainage or ventilation or both. The term *tunnel* is frequently used in place of adit and has the same significance.
Crosscut	A horizontal gallery driven at right angles to the strike of a vein. When driven at an angle to the vein and across it the same term is applied.
Drift	A horizontal gallery driven along the course of a vein. The terms *footwall drift* and *hangingwall drift* are used when galleries are driven in the footwall and hangingwall respectively.
Incline	An excavation of the same nature as a shaft and used for the same purposes but driven at an angle from the vertical.
Level	All of the horizontal workings tributary to a given shaft station are collectively called a *level*. Levels are designated 100 ft., 200 ft., etc., the vertical depth from the surface determining.
Main level	Where the ore mined from several levels is hauled upon one level to the shaft, this level is designated as the ''main level.'' The term *main haulage* is used in the same sense.
Manway	Passages either vertical or inclined for the accommodation of ladders, pipes, etc.
Ore pass or chute	Vertical or inclined passageways for the downward movement of ore.
Ore pocket	Opening below a level for the temporary storage of ore at, and connected to, the shaft through which the ore is hoisted.
Raise	An excavation of restricted cross section, driven upward from a drift and in the ore body. It is used as a manway, timber chute, waste chute, ore chute, or for ventilation.
Shaft	A vertical excavation of restricted cross section and of relatively great depth used for access and working. Often compartmented.
Station	Junction of a level with a shaft.
Stope	An excavation underground, other than development workings, made for the purpose of removing ore.
Sublevel or intermediate level	A level driven from a raise or manway and not connected directly with the working shaft.
Underground shaft or incline	A shaft or incline that is driven from underground workings and not in the vein or ore body.
Winze	An excavation of restricted cross section driven downward from a drift and in the ore body. It is used for the same purposes as a raise.

Terms Peculiar to Coal Mines

Butt entry	A horizontal gallery driven parallel with the main cleat (jointing) of a coal seam.
Cross entry	A horizontal gallery driven at an angle or at right angles to the main entry.
Face entry	A horizontal gallery driven at right angles to the main coal cleat of the seam.
Gate	A horizontal gallery giving access to a working face in the long-wall method of mining.
Level, lift	Terms used to designate the working entries in the case of a coal seam that is dipping.
Main entry	The principal horizontal gallery giving access to a coal seam and used for ventilation, haulage, etc. Where two or more entries are driven in parallel the terms *double-entry, triple-entry*, etc. are used.
Mother gate	A horizontal gallery to which the gates are tributary.
Shaft, air shaft, working shaft	The significance of these terms is the same as the equivalent terms used in metal mining.
Slope	The main working gallery or entry of a coal seam that dips at an angle and in which trains of mine cars are hauled.
Sump	A pocket in which drainage water is collected.

From G. J. Young, *Elements of Mining.* © 1932 McGraw-Hill Book Co., N.Y. Reprinted by permission of the McGraw-Hill Book Co.

Figure 18.4 Mine plan: Room-and-pillar mining, retreating

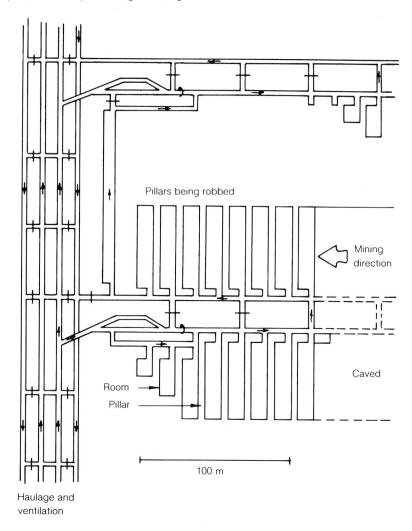

Pillars being robbed

Mining direction

Caved

Room

Pillar

100 m

Haulage and ventilation

KEY

Door

Stopping

Air flow

Mining may *advance outward* as a wave from the point of access to the ore body leaving pillars for support. When the periphery of the deposit or property line is reached, the mining reverses direction and *retreats* toward the mine entry point removing or "robbing" pillars and either backfilling or allowing the openings to cave. Alternately, mine development openings may be driven outward to the ore body perimeter and complete mining of the ore accomplished in a single retreating operation. Each procedure has certain advantages and disadvantages. Mining in advance can generate an immediate cash flow from the sale of the product to offset the heavy capital costs of mine development. Full-retreat mining generally allows the use of a simpler mining plan, a more efficient use of machinery, and

Figure 18.5 Mine plan: Longwall advancing

Section A–A'

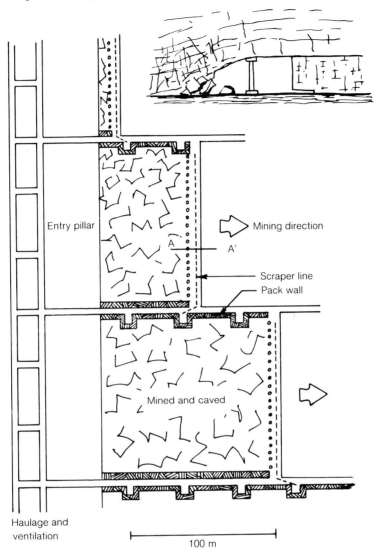

Entry pillar

⇨ Mining direction

A · A'

Scraper line

Pack wall

Mined and caved

Haulage and
ventilation

100 m

confers certain advantages of scale; but it requires extensive development work before any financial return is realized.

In flat-lying tabular bodies such as coal seams or stratiform lead–zinc deposits, the access and development openings are within the seams or ore bodies. A common mining plan for coal is a *room-and-pillar* system (fig. 18.4). Rooms extend outward like leaves from the haulageway stem and each stalk of rooms either begins at, or is eventually advanced to, the limit of the deposit. The pillars or ribs between the rooms may be broken through at regular intervals to increase coal recovery and improve ventilation.

Greater simplicity of plan and operation may be achieved if it is possible to mine long faces rather than a number of smaller faces. This is the basic concept of the *long-wall method* in which mining is either initiated at the perimeter and worked inward to the shaft or adit or worked outward from the shaft. Figure 18.5 shows the plan of a long-wall mine progressing outward. In long-wall mining, roof collapse

Figure 18.6 Access to the ore body

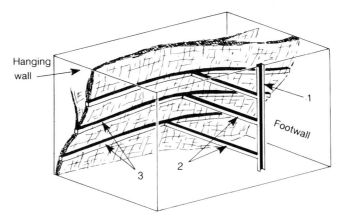

Hanging wall

Footwall

1

2

3

or caving is a preplanned part of the mining method, and the pressures generated are used to assist in breaking the coal at the working face.

Tabular bodies inclined more steeply than the angle of repose of broken rock are often mined by methods that are analogous to room-and-pillar mining wherein the working areas are called **stopes.** Stoping usually attacks the seam or vein from below to take advantage of gravity in moving ore from the working face. Occasionally, however, stoping may follow the ore body downward in an underhand, as compared with the more usual overhand, stope. Because of access, transport, and ventilation problems the layout of a stope mine also differs from the room-and-pillar plan by having more or less regularly spaced haulageways, or **levels,** leading from the shaft. Ore is usually stored temporarily within the stope where it provides a working platform for the miners, and access to the working face is by a timbered raise that is carried upward as the face advances. The development openings in this kind of mine are usually placed beneath the ore body in the footwall where they will not be affected by breakage and settlement of overlying rocks as mining proceeds. Figure 18.6 identifies the more important access openings.

Extraction of ore within the body takes place in rooms or stopes connected more or less directly to the **drifts** (fig. 18.7). This general plan is very flexible and can accommodate bodies with different

Figure 18.7 Arrangement of drift and stope

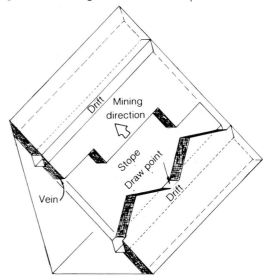

Drift

Mining direction

Stope

Draw point

Vein

Drift

width and dip, becoming a room-and-pillar arrangement when horizontal. The interconnection of stope and haulage drift is not always simple as, for example, in the Mount Isa copper mine, Australia (fig. 18.8).

As mining proceeds to greater depth or as the strengths of rock and ore decrease, the size of openings that may be maintained also decreases. Internal support in stopes and various other openings may be provided by artificial means; undisturbed ore

Figure 18.8 Drift-stope interconnection in a sublevel stoping operation

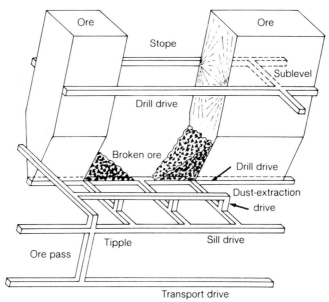

Figure 18.9 Shrinkage stoping

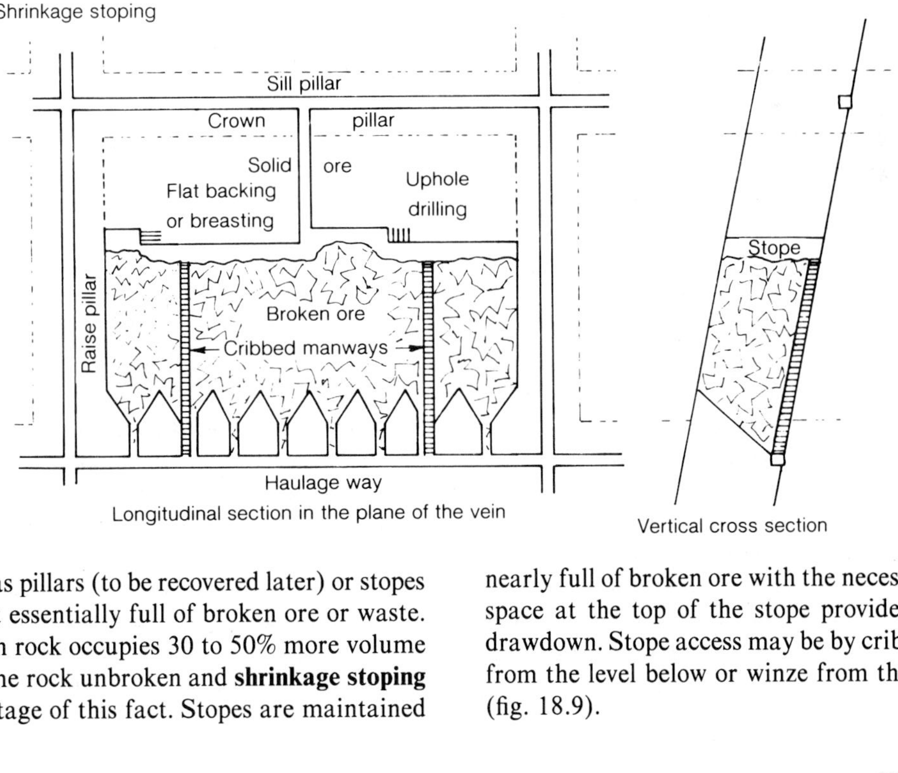

Longitudinal section in the plane of the vein

Vertical cross section

may be left as pillars (to be recovered later) or stopes may be kept essentially full of broken ore or waste.

Broken rock occupies 30 to 50% more volume than the same rock unbroken and **shrinkage stoping** takes advantage of this fact. Stopes are maintained nearly full of broken ore with the necessary working space at the top of the stope provided by careful drawdown. Stope access may be by cribbed manway from the level below or winze from the level above (fig. 18.9).

Figure 18.10 Working area at the top of a cut-and-fill stope

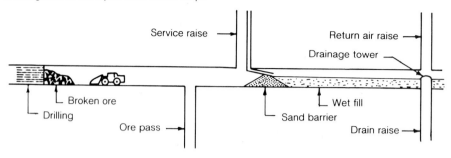

Figure 18.11 Sublevel caving

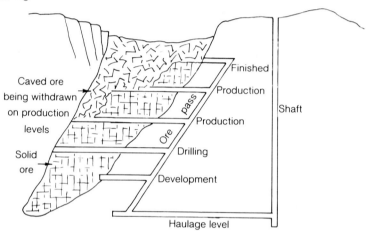

Stopes may also be maintained in a nearly full condition using waste rock from mining or milling. This is **cut-and-fill stoping** (fig. 18.10) and in modern mines tends to be the automatic first choice for a feasibility study since it is the most versatile of underground mining methods: (1) development costs are relatively low, (2) massive as well as tabular bodies can be stoped, (3) pillars and ore stringers into country rock can be recovered, (4) the design adapts well to mechanization, and (5) underground waste disposal obviates environmental problems.

For large, low-grade ore bodies or for mining in mechanically weak ground it may be advantageous to initiate and maintain a deliberate caving of the ore. Essentially, support is removed to the point that failure of the mass occurs and the self-crushed ore is drawn off through protected openings below

the body. Support may be removed from one layer at a time as in **sublevel caving** (fig 18.11) or from a large block of ore as in **block caving** (fig. 18.12). Advantages are low mining costs coupled with high production rates, though these are offset by high initial development costs, dilution of the ore by wall rock and capping, and disturbance of the surface.

Miscellaneous Extraction Methods

A few commodities exist in the ground in gaseous, liquid, or liquifiable form and may be extracted through **boreholes.** Included in these commodities are, of course, petroleum and natural gas and also easily melted sulfur, soluble salt, and several metals (notably copper and uranium) that may be dissolved by appropriate reagents. Recovery may be through single coaxial holes by the use of a field injection and recovery wells (see fig. 14.9).

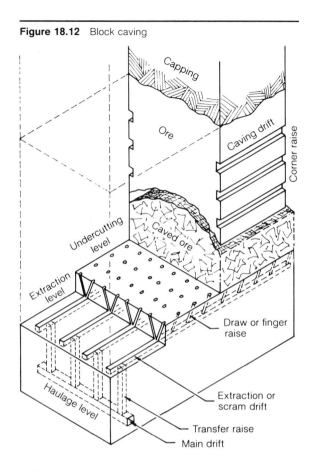

Figure 18.12 Block caving

Labels on figure: Capping · Ore · Caving drift · Corner raise · Undercutting level · Caved ore · Extraction level · Draw or finger raise · Extraction or scram drift · Haulage level · Transfer raise · Main drift

SUMMARY

The profitable separation of useful raw materials from the earth as well as excavation and tunneling for other purposes represents a major challenge to the geologist and engineer. A thorough understanding of the geometry and physical properties of rock bodies is obviously essential, detailed planning must be done to provide safe working conditions and a predetermined product flow, and consideration must be given to the environmental effects of the work.

Generally speaking, borehole extraction is the safest, the cheapest, and the least damaging to the environment. It is, unfortunately, applicable to only a relatively few mineral commodities and recovery efficiencies are low. Open-pit mining is environmentally damaging not only in terms of aesthetics but often by the exposure of waste materials whose leaching may contaminate water supplies. Where possible, open-pit extraction is the method of choice, however, because of the efficiency and hence cost effectiveness with which ore can be won. Underground mining causes some environmental damage in the way of subsidence (preventable) and in the production of mine wastes, some of which must be stored above ground. Because of high costs and relatively dangerous working conditions, underground mining will generally be selected for commodities of high unit value such as the metals.

ADDITIONAL READING

Biron, C., and Arioglu, E. 1983. *Design and supports in mines.* New York: John Wiley & Sons.

Cummins, A. B., and Givens, I. A. 1973. *SME mining engineering handbook.* American Institute Met. and Proj. Engineers, Inc.

Engineering and Mining Journal operating handbooks. Vol. 2, *Operating handbook of surface mining and exploration,* 1978; Vol. 3, *Operating handbook of underground mining,* 1978. New York: McGraw-Hill.

Engineering and Mining Journal. 1980. *Mining methods and equipment.* New York: McGraw-Hill.

Erickson, A. J., Jr., ed. 1984. *Applied mining technology.* Littleton, Colo.: Soc. of Mining Engineers of AIME.

Hustrulid, W. A., ed. 1982. *Underground mining methods handbook.* Littleton, Colo.: Soc. of Mining Engineers of AIME.

Lacey, W. C. 1982. *Mining geology.* Florence, Ky.: Van Nostrand Reinhold.

Peele, R., and Church, J. A. 1941. *Mining engineers handbook.* 3d ed. New York: John Wiley & Sons.

Phleider, E. P., ed. 1968. *Surface mining.* Littleton, Colo.: Soc. of Mining Engineers of AIME.

Rensburg, P. W. J., ed. 1971. *Planning open-pit mines.* Rotterdam: A. A. Balkema.

Thomas, L. J. 1973. *An introduction to mining.* Sydney: Hicks Smith and Sons.

Young, G. J. 1932. *Elements of mining.* 3d ed. New York: McGraw-Hill.

APPENDIXES

APPENDIX ONE

Dimensional Analysis[1]

Geometric and kinematic relationships between a model and the original may be given in terms of the *model ratio,* herein designated as ϕ, based upon ratios of the fundamental units of length and time. Using lowercase letters to designate the model and capital letters for the original, these ratios are as follows: length, l/L; area, l^2/L^2; volume, l^3/L^3; time, t/T; velocity lt/LT; and acceleration, lt^2/LT^2. Thus for a modeled length of 1:100, $\phi = .01 = l/L$ or .01 $L = l$, and for modeled acceleration of 1:100, .01 $LT^2 = lt^2$.

Dynamic modeling must consider both the body forces, which are proportional to mass and therefore to volume, L^3, and the surface forces, which are proportional to area, L^2. Consequently, as size decreases the body forces diminish more rapidly than the surface forces, and for sufficiently small volume elements the body forces may often be neglected. Mass relations between model and original are usually handled in terms of density, D, and volume, L^3,

since mass is awkward to handle in modeling experiments:

$$\frac{m}{M} = \frac{dl^3}{DL^3}.$$

The forces acting upon an element of mass occupying an element of volume are gravity, F_g, and inertial, F_i, body forces. $F_g = M_g$ and $F_i = Ma$ where g and a are respectively acceleration due to gravity and inertial forces; when a model is on the earth's surface, $F_a = F_g$ ($\phi = 1$) and size and time are no longer independent variables.

The surface forces on an infinitesimal volume may be decomposed into those acting perpendicular, σ_n, and parallel, τ, to a particular surface. These forces are termed *normal* and *shearing* stress and are proportional to the area acted upon.

1. Following M. K. Hubbert, The strength of the earth. *Bulletin of the American Association of Petroleum Geologists*, vol. 29, 1945.

The strain that results from the application of stress to an infinitesimal volume is describable as (1) volume strain measured as the change in volume divided by the original volume, $V + \Delta V/V$, and called the *volume modulus;* (2) length strain, $L + \Delta L/L$, or *Young's modulus;* (3) shear strain measured as shear displacement/thickness and called the *shear* or *rigidity modulus;* and (4) stress/strain = constant (Hooke's law) or the *modulus of elasticity.*

Since there is no acceleration in the usual geologic situations, the resultant of all forces acting (gravity, F_g, pressure, F_p, viscous resistance, F_r, elasticity, F_e, and inertia, F_i) on an element of volume or surface must be zero; and for dynamic similarity each vector force in the model must correspond in orientation and model magnitude to the forces in the original. The model ratio of two moduli is

$$\phi = \frac{(\text{stress/strain}) \text{ model}}{(\text{stress/strain}) \text{ original}}$$

$$= \frac{(\sigma n\tau /l) \text{ model}}{(\sigma n\tau /L) \text{ original}}$$

Strain is always in terms of L, and stress (force per unit area) in terms of F and L. Thus for length strain

$$\phi = \frac{f}{l^2}/l \div \frac{F}{L^2}/L \text{ or } fL^3 = Fl^3$$

APPENDIX TWO

Maps of the Conterminous United States Showing Features of Geologic and Engineering Interest

Map No.

1. Physiographic provinces
2. Coal fields
3. Igneous rocks and gneisses
4. High-calcium limestone and chalk
5. Loess
6. Glacial drift deposits
7. Swelling clay in near-surface rocks
8. Mean annual rainfall
9. Average annual runoff
10. Annual evaporation from a free water surface

11. Water surplus and deficiency
12. Groundwater provinces
13. Water-use regions
14. Maximum rainfall in 15-minute period
15. Average annual potential evapotranspiration
16. Seismic risk
17. Seismic-shaking probability
18. Landslide potential
19. Possible disposal sites for nuclear wastes
20. Geologic features important for deep waste-injection well evaluation
21. Distribution of tornadoes

Map 1 Physiographic provinces

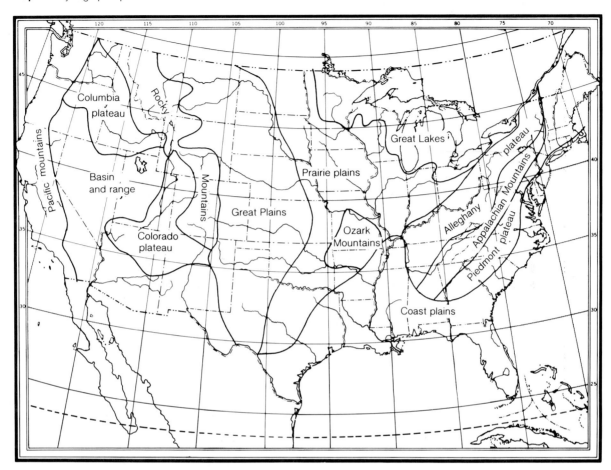

After Ransome, in *Engineering Geology,* 5th ed., by H. Ries
and T. L. Watson. © 1936 by John Wiley & Sons, N.Y.
Reprinted by permission of John Wiley & Sons, Inc.

Map 2 Coal fields

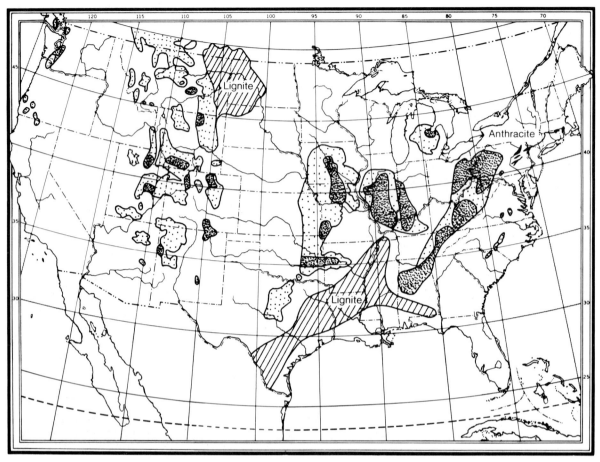

KEY

Important bituminous

Secondary bituminous

Source: U.S. Geological Survey.

Map 3 Igneous rocks and gneisses

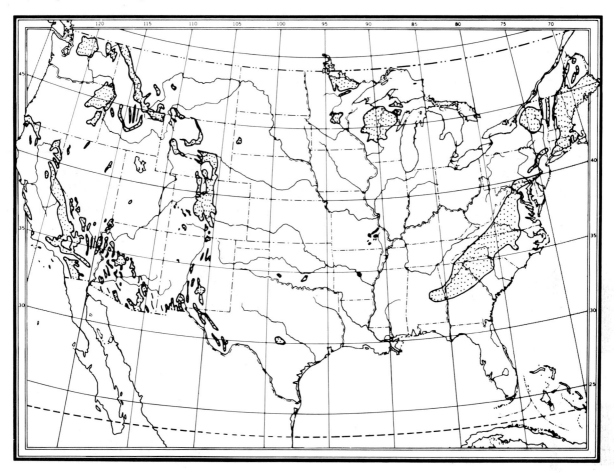

After G. P. Merrill, *Stones for Building and Decoration,* John
Wiley & Sons, N.Y., 1903.

Map 4 High-calcium limestone and chalk

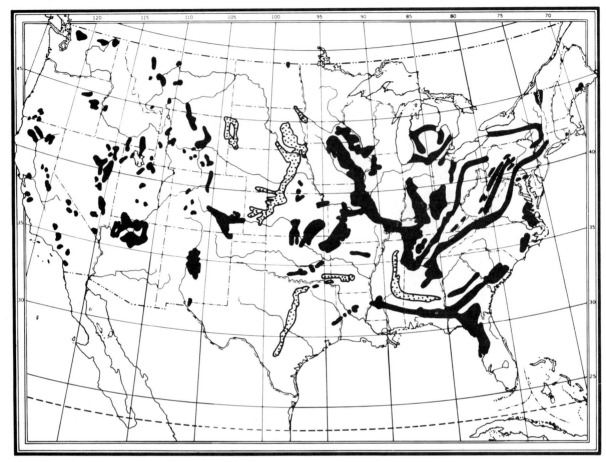

KEY

Limestone
Chalk

Source: U.S. Bureau of Mines.

Map 5 Loess

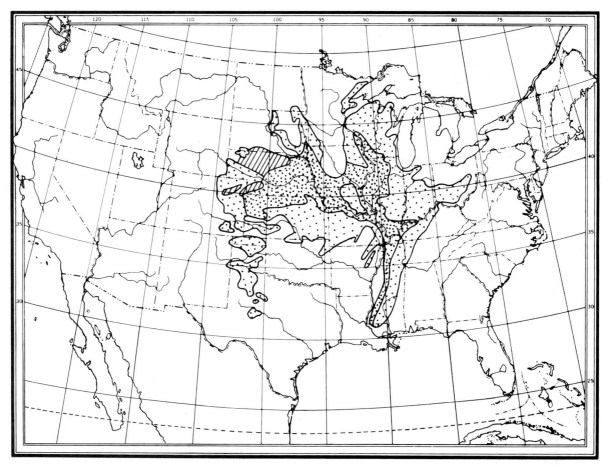

KEY

  > 8 feet (2.4 m)
  < 8 feet
  Sandhills

From map of Pleistocene Eolian Deposits of the United
States, Geological Society of America, 1952.

Map 6 Glacial drift deposits

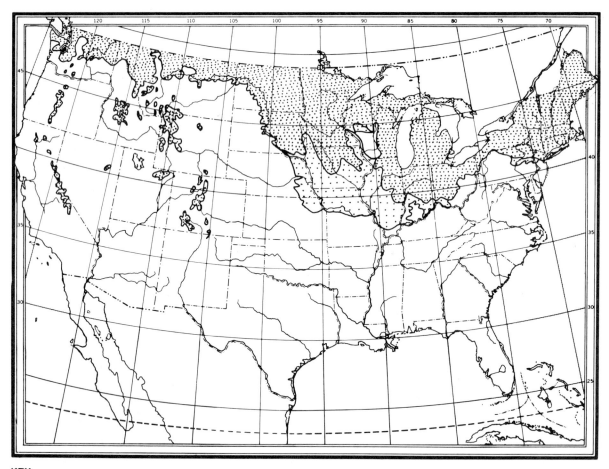

KEY

 Wisconsin drift

Pre-Wisconsin drift

From the Glacial Map of North America, R. F. Flint *chm.*,
Geological Society of America Special Paper 60, 1945.

Map 7 Swelling clay in near-surface rocks

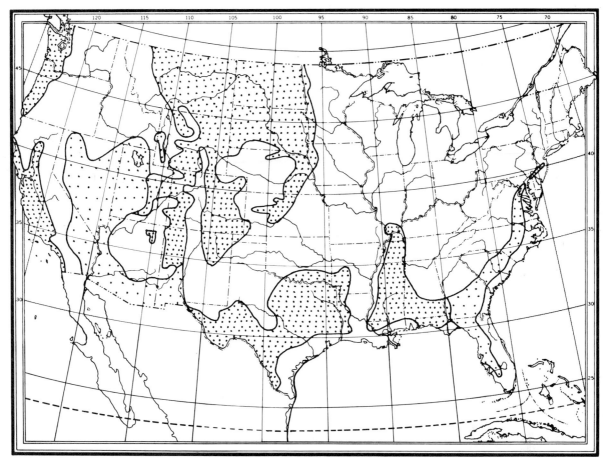

KEY

Regionally abundant
Locally abundant
Locally present

After G. D. Robinson and A. M. Spiekes *eds.* U.S. Geological
Survey Professional Paper 950, 1978.

Map 8 Mean annual rainfall (contours in inches)

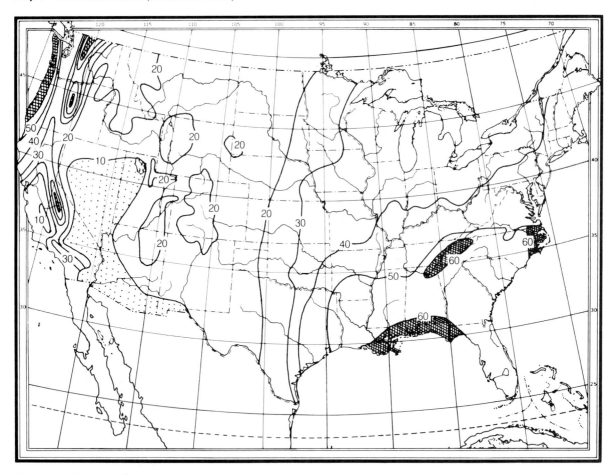

Source: U.S. Geological Survey.

Map 9 Average annual runoff

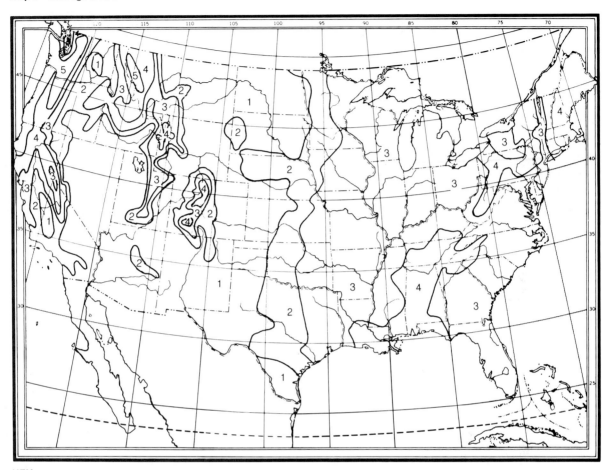

KEY

1 0–1 in.
2 1–5 in.
3 5–20 in.
4 20–40 in.
5 > 40 in.

From *River of LIfe*, U.S. Department of the Interior
Conservation Yearbook Series, Vol. 6.

Map 10 Annual evaporation from a free water surface
(contours in cm)

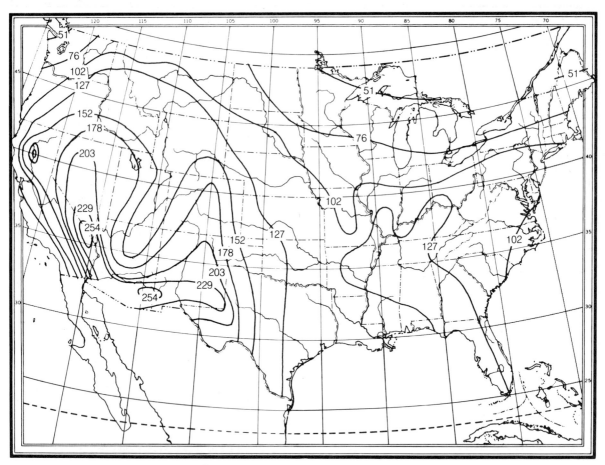

From Strahler and Strahler, *Environmental Geoscience*.
Reprinted by permission of John Wiley & Sons, Inc.

Map 11 Water surplus and deficiency

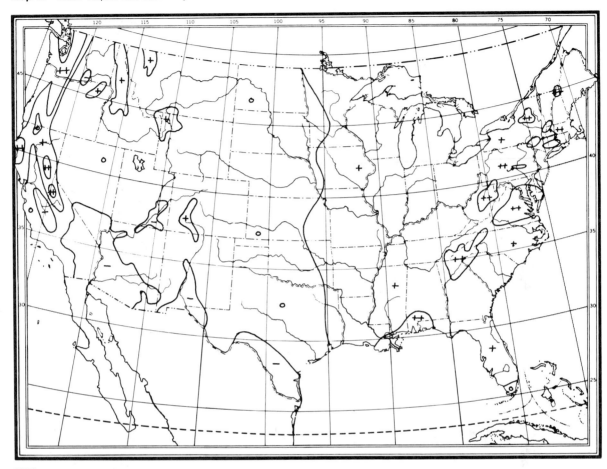

KEY

++ 20–80 in. annual surplus
+ 0–20 in.
o 0– –20 in.
– –20–<–40 in.

From *River of Life,* U.S. Department of the Interior
Conservation Yearbook Series, Vol. 6.

Map 12   Groundwater provinces

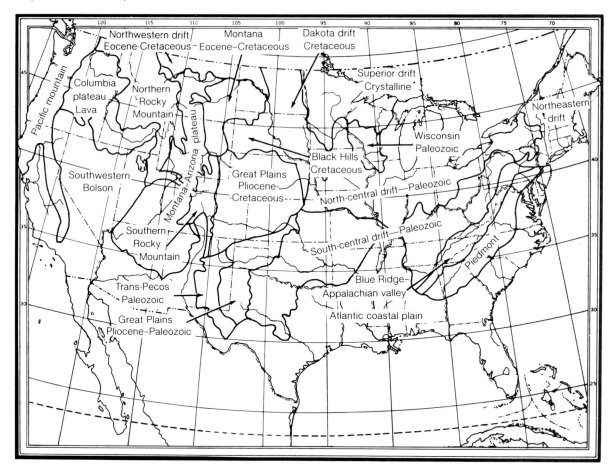

After O. E. Meinzer, U.S. Geological Survey Water Supply
Paper 489, 1923.

Map 13   Water-use regions

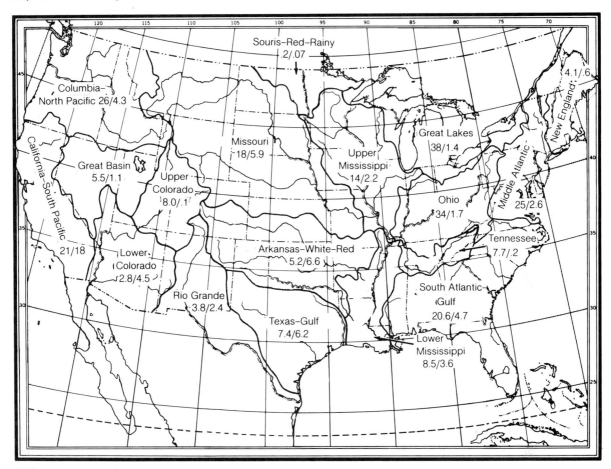

Souris–Red–Rainy
.2/.07

Columbia–
North Pacific 26/4.3

California–South Pacific

Great Basin
5.5/1.1

Upper
Colorado
8.0/.1

Missouri
18/5.9

Upper
Mississippi
14/2.2

Great Lakes
38/1.4

New England .4.1/.6

Middle Atlantic
25/2.6

Ohio
34/1.7

Tennessee
7.7/.2

21/18

Lower
Colorado
2.8/4.5

Arkansas–White–Red
5.2/6.6

South Atlantic–
Gulf
20.6/4.7

Rio Grande
3.8/2.4

Texas–Gulf
7.4/6.2

Lower
Mississippi
8.5/3.6

KEY

Surface water withdrawal/groundwater withdrawal: for 1970,
10^9 gal/day

After C. R. Murray and E. B. Reeves, U.S. Geological Survey
Circular 676, 1972.

Map 14 Maximum rainfall in mm during a 15-minute period
to be expected once in 100 years

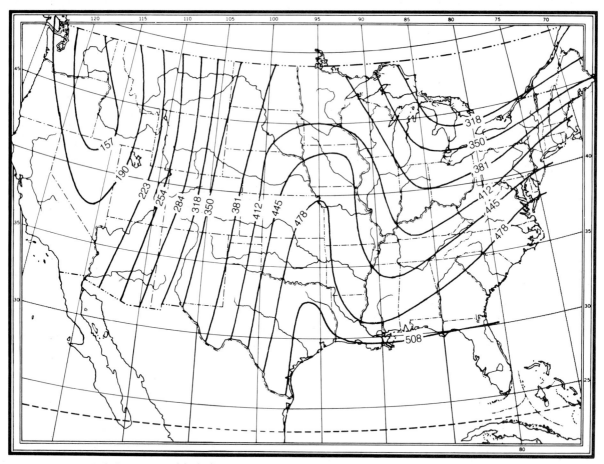

After D. L. Yarnell, U.S. Department of Agriculture
Miscellaneous Publication 204, 1935.

Map 15 Average annual potential evapotranspiration
(contours in cm)

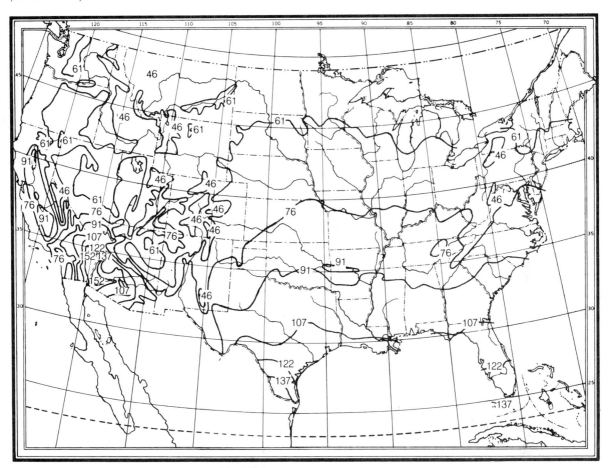

After Thornewaite, *Geographical Review,* Vol. 38, 1948, by
courtesy of the American Geographical Society.

Map 16 Seismic risk

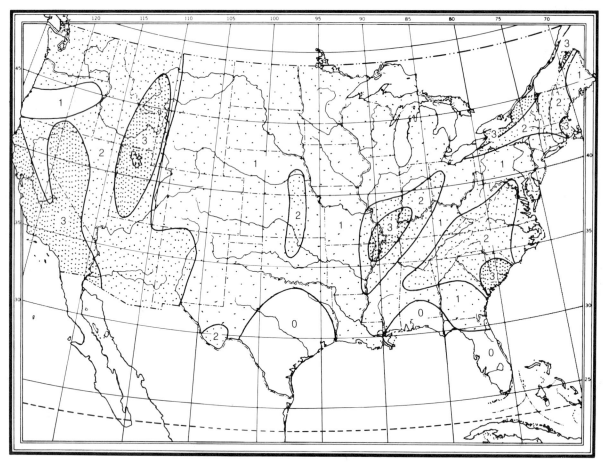

KEY

0	No damage expected
1	Minor damage, intensity V–VI
2	Moderate damage, intensity VII
3	Major damage, intensity VII+

Source: U.S. Geological Survey.

Map 17 Seismic-shaking probability. Areas in which shaking by an earthquake can be expected to exceed 0.10 g at least once in 50 years

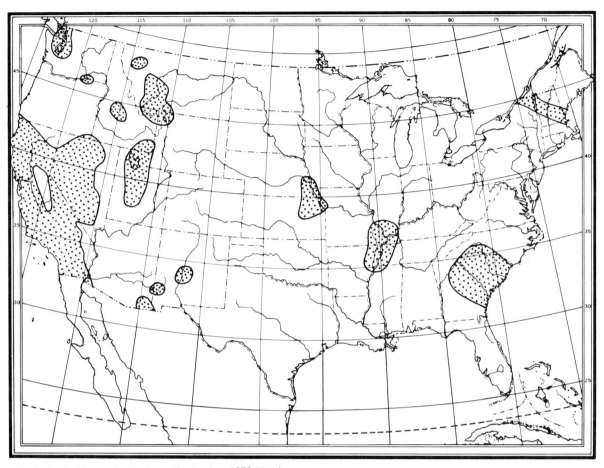

U.S. Geological Survey in *Geotimes,* September, 1976. Used with permission of *Geotimes.*

Map 18 Landslide potential

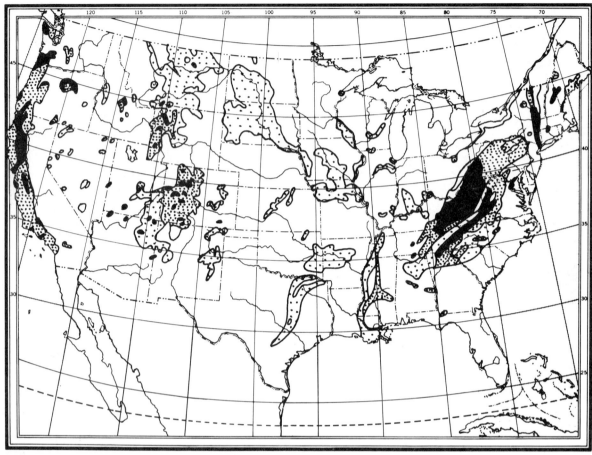

KEY

■ High risk
▨ Moderate risk
▢ May slide if disturbed
□ Low risk

After G. D. Robinson and A. M. Spiekes *eds.* U.S. Geological
Survey Professional Paper 950, 1978.

Map 19 Possible disposal sites for nuclear wastes

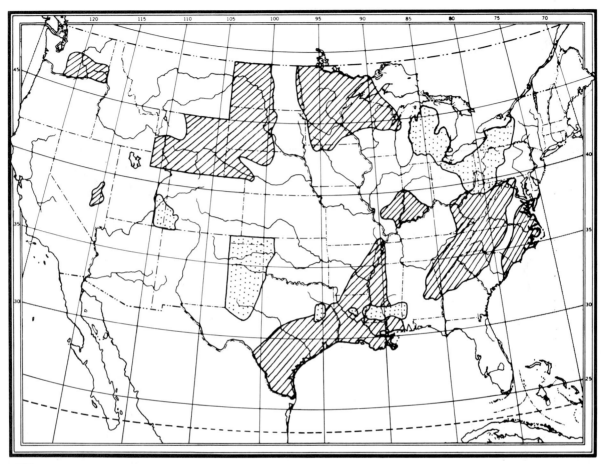

KEY

Rock salt deposits
Clay, shale, limestone, granite

Source: General Accounting Office Report, 1977.

Map 20 Geologic features important to deep waste-injection
well evaluation

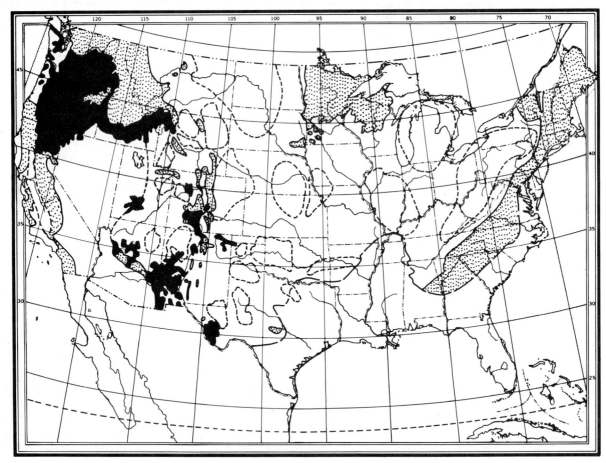

KEY

- - - - - Basin outline
- - - Coastal plain limit
▬▬ Volcanic rocks
░░░ Crystalline rocks

After U.S. Bureau of Mines in D. L. Warner, AAPG Memoir 10,
J. E. Galley *ed.* © 1968 American Association of Petroleum
Geologists. Used with permission of the Association.

Map 21 Distribution of tornadoes 1955–67

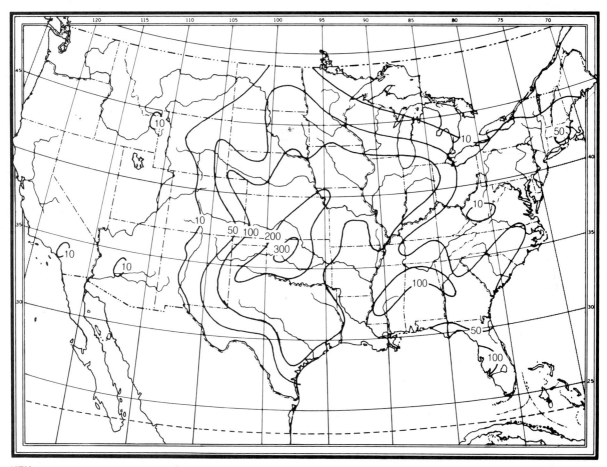

KEY

Contour Interval	Annual Frequency
10	1
50	4
200	17
300	25

After M. E. Pautz, ESSA Technical Memo WBTM FCST 12,
Office of Meterological Operations, 1969.

APPENDIX THREE

Table A3.1 Greek Alphabet

alpha	A	α	nu	N	ν	
beta	B	β	xi	Ξ	ξ	
gamma	Γ	γ	omicron	O	o	
delta	Δ	δ	pi	Π	π	
epsilon	E	ϵ	rho	P	ρ	
zeta	Z	ζ	sigma	Σ	σ	
eta	H	η	tau	T	τ	
theta	Θ	θ	upsilon	Υ	υ	
iota	I	ι	phi	Φ	ϕ	
kappa	K	κ	chi	X	χ	
lambda	Λ	λ	psi	Ψ	ψ	
mu	M	μ	omega	Ω	ω	

Table A3.2 Conversions

Length		
To convert	*Into*	*Multiply by*
chain, surveyor's (ch)	links	100
	rods	4
	feet	66
inches (in)	centimeters	2.540
kilometers (km)	feet	3 281
	meters	1 000
	miles	.621 4
meters (m)	centimeters	100
	feet	3.281
	kilometers	.001
microns (μ)	meters	10^{-6}
miles (mi)	kilometers	1.609
mils (mil)	inches	.001
rod (rd)	chain, surveyors	.25
	feet	16.5
	meters	5.029
	link	25

Area		
To convert	*Into*	*Multiply by*
acres (A)	square chains	100
	square feet	43 560
	hectares	.404 7
	square meters	4 046.9
hectares (ha)	acres	2.471
	square feet	1.076×10^5
	square meters	10^{-4}
square kilometer (km²)	square miles	.386 1
square meters (m²)	square yards	1.196
square miles (mi²)	square kilometers	2.590
square yard (yd²)	square meters	.836 1

Table A3.2 Conversions (continued)

Volume

To convert	Into	Multiply by
cubic centimeters (cm³, cc)	cubic inches	.061 0
	cubic meters	10^{-6}
	liters	.001
	milliliters	.999 9
cubic feet (ft³)	cubic inches	1 728
	cubic meters	.028 3
	liters	28.32
	quarts (US liquid)	29.92
cubic meters (m³)	cubic centimeters	10^6
	cubic feet	35.31
	cubic yards	1.307 9
	gallons (US liquid)	264.2
	liters	1 000
cubic yards (yd³)	cubic feet	27
	cubic meters	.764 55
gallons, US (gal)	cubic meters	3.785×10^{-3}
	liters	3.785
	gallons, imperial	.832 67
liters (l)	cubic centimeters	1 000
	cubic meters	.001
	quarts (US liquid)	1.057

Weight

To convert	Into	Multiply by
carats, metric (c)	milligrams	200
	grams	.2
gallons of water	pounds of water	8.345 3
grams (g)	kilograms	.001
	milligrams	1 000
	ounces, troy	.032 15
kilograms (kg)	grams	1 000
	long tons	$9 842 \times 10^{-4}$
	short tons	$1 102 \times 10^{-3}$
micrograms (μ, γ)	grams	10^{-6}
milligrams (mg)	grams	.001
ounces, avoirdupois (oz. av)	ounces, troy	.911 5
ounces, troy (oz. t)	grams	31.103 5
	ounces, avoir	1.097 1
	pennyweights, troy	20
	pounds, troy	.083 3
pennyweights, troy (dwt)	ounces, troy	.05
	grams	1.555 2
pounds, avoirdupois (lb. av)	grams	453.592
	pounds, troy	1.215 3
	ounces, troy	14.583 3
	short ton	.000 5
pounds, troy (lb. t)	grams	373.242
	ounces, troy	12
	pennyweights, troy	240
	pounds, avoir.	.822 9

Table A3.2 Conversions (continued)

Weight (continued)

To convert	Into	Multiply by
tons, long (tn.l)	kilograms	1 016
	pounds, avoir.	2 240
	tons, short	1.120
	tonnes	1.016
tonnes, metric tons (t)	kilograms	1 000
	pounds	2 204.62
	tons, short	1.102 3
tons, short (tn.s)	kilograms	907.185
	pounds, avoir.	2 000
	tonnes	.907 2

Density

To convert	Into	Multiply by
grams per cubic centimeter (g/cm^3)	pounds per cubic foot	62.43
	pounds per cubic inch	.036 13
kilograms per cubic meter (kg/m^3)	grams per cubic centimeter	.001
	pounds per cubic inch	3.613×10^{-5}
	pounds per cubic foot	.062 4
pounds per cubic foot (lb. av/ft^3)	pounds per cubic inch	5.787×10^{-4}
	grams per cubic centimeter	.016 02
	kilograms per cubic meter	16.018
pounds per cubic inch (lb. av/in^3)	grams per cubic centimeter	27.680
	kilograms per cubic meter	2.768×10^4

Concentration

To convert	Into	Multiply by
grams per tonne (g/t)	parts per million	1
	percent	.001
	milligrams per kilogram	1
karat (k)—one of gold to 24 of mixture	milligrams per gram	41.667
	parts per million	41 666.7
	percent	4.166 7
milligrams per assay ton	troy ounces per short ton	1
milligrams per liter (mg/l)	parts per million	1
ounce, troy per ton, short (oz. t/tn.s)	parts per million	34.286
parts per million (ppm)	grams per short ton	.907 2
	grams per tonne	1
	troy ounces per short ton	.029 17
	percent	10^{-4}
percent (%)	parts per million	10^4

REFERENCES

Bowen, N. L. 1928. *The evolution of the igneous rocks.* Princeton, N.J.: Princeton University Press.

Bretschneider, C. L. 1952. *The generation and decay of wind waves in deep water.* Transactions American Geophysical Union, v. 33.

Busk, H. G. 1929. *Earth flexures, their geometry and analysis in geological section, with special reference to the problem of oil finding.* Cambridge: Cambridge University Press.

Caldwell, J. M. 1956. *Wave action and sand movement near Anaheim Bay, California.* U.S. Army Beach Erosion Board Tech. Memo No. 68.

Carroll, D. 1959. Cation exchange capacities in clay minerals and soils. *Bull. Geol. Society of America ,* v. 70.

Casagrande, A. 1932. Research on the Atterberg limits of soils. *Public Roads,* v. 13.

Department of the Navy, Naval Facilities Engineering Command. 1982. Soil Mechanics Design Manual 7.1.

Dolan, J., and Ferm, J. C. 1969. Crescentic landforms along the Atlantic Coast of the United States. *Science* v. 159, no. 3815.

Duncan, N. 1969. *Engineering geology and rock mechanics.* 2 vols. Glasgow: Blackie & Son.

Dyer, K. R. 1973. *Estuaries: A physical introduction.* New York: John Wiley & Sons.

Flinn, D. 1962. On folding during three-dimensional progressive deformations. *Quart. Jour. Geol. Soc. London,* v. 118.

Johnson, J. E.; O'Brien, M. P.; and Isaacs, J. D. 1948. *Graphical construction of wave refraction diagrams.* U.S. Hydrographic Office Pub. 605.

Keller, E. A. 1982. *Environmental geology.* 3d ed. Columbus, Ohio: Charles E. Merrill.

King, C. A. M. 1959. *Beaches and coasts.* London: Edward Arnold.

Leopold, L. B.; Clarke, F. E.; Hanshaw, B. B.; and Balsley, J. R. 1971. *A procedure for evaluating environmental impact.* U.S. Geological Survey Circular 645.

Mead, W. J. 1936. *Engineering geology of damsites.* International Committee on Large Dams of the World. Power Conference, 2d, Congress on Large Dams, Washington, D.C., Transactions v. 4.

Moore, B. R. 1970. Scour and fill processes in a deep river hole, Louisville, Kentucky. *Journal of Sedimentary Petrology.*

Munk, W. H., and Taylor, M. A. 1947. Refraction of ocean waves, a process of linking underwater topography to beach erosion. *Journal of Geology,* v. 55.

Parker, G. G.; Shown, L. M.; and Ratzlaff, K. W. 1964. Officers Cave, a pseudo-karst feature in altered tuff and volcanic ash of the John Day Formation in eastern Oregon. *Geological Society of America, Bulletin,* v. 75.

Piper, A. M. 1970. *Disposal of liquid wastes by injection underground: Neither myth nor millenium.* U.S. Geological Survey, Circular 631.

Prandtl, L. 1920. Über die Härte plastischer Körper. Gottingen: Nach. Ges. Wiss.

Putnam, A. L. 1972. *Effect of urban development on floods in the Piedmont Province of North Carolina.* U.S. Geological Survey Open File Report 72–304.

Royster, D. L. 1978. Excavation characteristic designations for materials identified for field investigations. *Bulletin Ass'n. Engineering Geologists,* vol. XV.

Schmitt, R. A. 1979. *Coal in America.* New York: McGraw-Hill Inc.

Shephard, F. P. 1973. *Submarine geology.* 3d ed. New York: Harper & Row.

Terzaghi, K. 1925. Principles of soil mechanics: IV, Settlement and consolidation of clay. *Engineering News-Record,* pp. 874–78.

———. 1926. Simplified soil tests for subgrades and their significance. *Public Roads,* v. 7.

Valentin, D. J. P., and Palmer, H. D. 1978. *Coastal sedimentation.* Stroudsburg, Pa.: Hutchinson Ross.

Vanoni, V. A.; Brooks, N. H.; and Kennedy, J. F. 1960. *Lecture notes on sediment transport and channel stability.* California Institute of Technology, Keck Laboratory, Report no. KH-RI.

GLOSSARY

Definitions of geological terms can also be found in the *Dictionary of Geological Terms* published by the American Geological Institute.

abrasion grinding

active zone zone of annual freeze and thaw in permafrost regions

adsorption adhesion of ions to particle surfaces

aftershock earthquake that follows a larger earthquake

aggregates mechanical mixtures of solid particles

aggregation formation of extended coherent solids from individual particles

"A" line divides soils of different properties with respect to their plastic and liquid limits

angle of repose maximum stable slope for granular materials

anisotropy condition of having different properties in different directions

anticline convex upward fold

aphanitic fine-grained igneous rock

aquiclude rock having a low hydraulic conductivity

aquifer rock having a high hydraulic conductivity

argillaceous term applied to all rocks or substances composed of clay minerals or having a notable proportion of clay in their composition

artesian flow groundwater under sufficient head to rise above the aquifer

aspect ratio ratio of length to width of a particle

assay determination of chemical composition

Atterberg limits collective term relating the mechanical properties of clay aggregates to their water content

augering means of drilling shallow holes

avalanche large rapid downslope movement of material

axial line line of intersection of an axial plane of a fold and a bedding surface

axial plane imaginary plane of symmetry between the limbs of a fold

bank storage temporary storage of flood waters in river banks

base levels theoretical downward limits of stream erosion

batholith very large body of deep-seated igneous rock

beach gently sloping shore of a body of water

bearing direction of a line

bedding layering in a sequence of sediments or sedimentary rocks

block caving in mining, a method involving the undercutting of large ore volumes

block diagram three-dimensional perspective representation

body waves seismic waves transmitted within rocks

boundary layer lower contact of a body of moving water

Boussinesq formula relates pressure at a buried displaced point to surface load in rigid solids

breakwater fixed structure for dispersing wave energy

breccia rock composed of angular particles cemented by a finer groundmass

bulk density weight-per-unit volume of sediment or rock

Busk construction means of reconstructing concentric folds from surface outcrop data

calcareous rocks sedimentary rocks dominantly made of carbonate minerals

capillary fringe soil zone above the water table in which soil pores are partially water filled

carbonation weathering process generating carbonate minerals

cataclasis rock fracture resulting in broken and rotated grains

caverns large underground openings (caves), usually formed by the dissolution of carbonate rocks

cementation process of precipitating a binding material around rock-forming grains

cinder cones accumulation of tephra in conical piles around volcanic vents

clapotis standing water wave formed by reflection from an obstacle

clastic rock fragmental sedimentary rock composed of detrital material

clay term variously used for a group of minerals, a very fine particle size, or a class of mechanical properties

clay minerals hydrated aluminosilicates of very fine grain

cleavage in minerals, the tendency to split in a fixed direction; in rocks, closely-spaced parallel surfaces of rupture

coefficient of consolidation measure of the change in thickness of a loaded soil with time

coefficient of runoff ratio of surface water runoff to the original rainfall

cohesion capacity of sticking or adhering

columnar basalt basalt jointed into long, parallel, polygonal rods

compaction decrease in volume of sediments or soil due to compressive stress

compaction shale rock composed of clay particles held together with water bonds

compression index change in void ratio as a function of applied pressure

cone of depression dewatered volume around a pumping well

confining pressure equal, all-sided pressure from surrounding rock and liquids on buried material

conformity geologic contact in sedimentary rocks in which contact surfaces are parallel and may be either horizontal or inclined

connate fluids fluids entrapped in sedimentary rock at the time of its deposition

consolidation any process whereby loose, soft, or liquid material becomes firm and coherent

contact the surface where two different rocks come together

Coriolis forces forces related to the rotation of Earth that affect the movement of matter, deflecting it in a clockwise motion (Northern Hemisphere) or in a counter-clockwise motion (Southern Hemisphere)

corrasion mechanical abrasion performed by moving erosive agents

corrosion chemical erosion

creep imperceptibly slow, downslope flowage of soil or slow deformation of solids under stress

crest line line connecting the highest points on the same bed in an anticline

cross section a vertical profile

crust outer layer of the earth

current bedding irregular sediment layering produced by water current flow

cut-and-fill bedding channel erosion of bedding followed by deposition

Darcy relation describes the flow parameters of subsurface water

delta alluvial deposit at the mouth of a river

delta tract farthest downstream portion of a river course

density mass per unit volume

denudation removal of surface materials by erosive agents

detrital particles in sedimentary rock derived from preexisting rocks

diamond drilling use of a diamond-studded bit to recover a cylindrical rock core

diatomaceous friable earthy deposit composed of nearly pure silica made by the microscopic plants called diatoms

dike cross-cutting tabular igneous rock body

dilatancy change in volume in a granular aggregate with a change in packing

dip maximum angle between an inclined and a horizontal plane

discharge volume of water per unit time passing through a given cross section

disconformity irregular geologic contact in sedimentary rocks bounded above and below by parallel bedding

drag folds minor folds formed by the shearing motion of adjacent rocks

drumlins whaleback gravel hills formed by the readvance of glacial ice over ground moraine

duricrusts case-hardened layers in or on soil

earthflow slow downslope flow of earth lubricated by water

earthquake sudden motion in the earth caused by an abrupt release of accumulated stress

elements substances each composed of the same kind of atoms

epicenter point on the earth's surface directly above an earthquake focus

epochs divisions of geologic time

eras major divisions of geologic time

erosion sum of processes whereby earth materials are loosened and removed

esker serpentine ridge of sediment deposited by a stream originally having ice banks

eutectic point lowest melting temperature of a mixture of given components

eutectic point lowest melting temperature of a mixture of given components

exchange capacities relative abilities of solids, especially clay minerals, to adsorb ions

expansive soil soil whose volume increases when wet

exsolution unmixing; separation of solid phases in the solid state

fabric size, shape, and arrangement of constituent particles in a rock

faults fractures in rock showing significant displacement of the sides parallel to the fracture

fetch continuous area of water over which the wind blows in a constant direction

field capacity amount of water held in soil above the water table

filter pressing straining out of liquid from a partially crystalline igneous rock

flocculation clumping of clay particles

flood the overtopping of stream banks

flood plain portion of a river valley built of sediments and covered with water at flood stage

fluid injection pumping of a fluid into an underground reservoir

fluidization process whereby gas mixes with solid fragments and causes them to flow like a fluid

focus source of seismic waves

folds bending of strata

foliation laminated structure in rocks

footwall rock mass beneath an inclined fault or vein

formation primary unit of formal rock description or geologic mapping

fossil remains or traces of animals or plants

fracture capacity of rocks to part along closely spaced parallel surfaces

frost heaving lifting of the surface by freezing within the soil

gangue nonvaluable minerals accompanying ore

geotechnics engineering discipline dealing with the mechanics of soil masses

geothermal gradient change of temperature with depth within the earth

gilgai ridged microtopography developed in permafrost regions

gossan reddish surface residuum remaining when sulfide minerals have been leached away

gouge claylike material formed by maceration in a fault

graben down-faulted block

graded bedding sedimentary layering within which a regular change in grain size occurs

grading distribution of grain sizes in sediments and soil

groins shore protection walls extending seaward

groundmass finer-grained material between phenocrysts in a porphyritic igneous rock

groundwater subsurface water in the zone of saturation beneath the water table

habit usually assumed external form of a mineral crystal or group of crystals

half-life time period in which one-half of the initial number of radioactive elements disintegrate

hanging wall rock mass above an inclined fault or vein

hardpans impermeable layers in the soil composed chiefly of clay

head fluid pressure due to differences in height in two water surfaces

heaving lifting of the surface due to intrusion of water or freezing within a soil mass

horst up-faulted block

hydration chemical combination of water with a substance

hydraulic conductivity ratio of flow velocity to driving force for water flow in porous media

hydrologic cycle cycle of phenomena through which water passes above, on, and under the earth's surface

hydrolysis dissociation of water molecules

hypabyssal rock igneous rock emplaced at shallow depths, usually as sill or dike

ice wedges vertical, wedge-shaped veins of ground ice found in permafrost regions

igneous rock rock formed by the solidification of magma or lava

intensity, seismic effects of an earthquake on humans, their structures and the earth's surface; described by the Modified Mercalli scale

isoclinal fold fold whose limbs are parallel

isotopes elements differing in the number of their neutrons

isotropy condition of having the same properties in all directions

joints rock fractures along which no appreciable motion has taken place

kame sand or gravel hills deposited by streams flowing over a glacial front

kaolinite a common clay mineral

karst topography developed on calcareous rocks and characterized by sink holes, disappearing streams, and caverns

kettle hole closed depression in glacial drift caused by the wasting away of a detached ice mass

Landsat satellite dedicated to earth surface photography

laterite reddish soil leached of soluble elements and enriched in iron and aluminum hydrated oxides

lava fluid igneous rock at the surface

levee artificial or natural mound along the bank of a stream

liquid limit water content of soil at which its consistency changes from plastic to liquid

liquidus line representing the maximum solubility of a solid phase in a liquid on a temperature-composition diagram

lithification conversion of sediment into sedimentary rock

loess deposit of wind-transported dust

longshore drift material transport parallel to the shoreline by waves and currents

Love waves transverse seismic waves propagated along the surface of the earth

magma molten rock material at depth in the earth

magnitude, seismic *see* Richter scale

mantle portion of the earth between its outer crust and inner core

meanders a series of looplike bends in the course of a stream

Mercalli Intensity scale *see* intensity, seismic

metamorphism process by which solid rocks are transformed in mineralogy and fabric by heat and pressure

mineral deposit any valuable concentration of minerals

minerals naturally occurring inorganic compounds, usually crystalline, having fixed limits of chemical composition

Modified Mercalli scale *see* intensity, seismic

Mohr diagram graphical representation of the stress-strain relations in a solid

Mohs scale of hardness ten minerals arranged in order of their ability to scratch one another

montmorillonite a group of clay minerals characterized by swelling in water and extreme colloidal behavior

moraines rock debris deposited directly from a glacier

mountain tract upstream portion of a river course

mudcracks dessication cracks in mud usually outlining polygonal blocks

muskegs poorly drained swampy areas in northern latitudes

nonclastic rock sedimentary rock of chemical or biochemical origin formed by precipitation in a body of water

nonconformity stratified rocks resting upon igneous or metamorphic rocks

normal fault fault in which the hanging wall has moved down with respect to the footwall

nuée ardente highly heated cloud of gas-supported volcanic ash

open-pit mining ore extraction from a pit open to the sky

ore an aggregate of minerals than can be worked at a profit

ore bodies solid and continuous masses of ore

ore deposit ore of sufficient quality and quantity to warrant extraction

overburden ratio ratio of waste removed to ore recovered in open-pit mining

packing arrangement of solid particles in a rock

peat partially decomposed and disintegrated vegetable matter typical of swamps

pegmatitic very coarse-grained igneous rock texture

percolation test used to establish the hydraulic conductivity (permeability) of soil or rock

periods major worldwide standard geologic time units

permafrost permanently frozen ground

permeability *see* hydraulic conductivity

phaneritic medium-grained igneous rock texture

phreatic zone *see* vadose zone

pieziometric surface *see* potentiometric surface

pillar column or rib of rock left in place during mining to provide support

pingo large ice-cored mound found in permafrost regions

plain tract lower reaches of a stream course

plasticity index range of water content over which a soil exhibits plastic behavior

plastic limit lowest moisture content of soil at which it behaves plastically

plate tectonics global-scale dynamics involving the movement of rigid plates comprising the earth's crust and upper mantle

plunge inclination of an inclined line, especially the axial line of a fold

plutonic rock igneous rock formed at great depth

polymorphs different minerals having the same composition

pore pressure pressure of contained liquid on surrounding particles and buoyant with respect to the particles; equals the weight of the connected water column ± polar attractive forces

porosity ratio of volume of interstices in a rock to its total volume

porphyritic igneous rock texture in which larger crystals (phenocrysts) are set in a finer-grained groundmass

potentiometric surface surface to which water in an aquifer will rise under hydrostatic pressure

P waves compressional-extensional seismic body waves

pyroclastic fragmental volcanic material formed by explosive ejection

quick clays clay–water mixtures responding mechanically as fluids

quicksand sand–water suspensoid responding mechanically as a fluid

radioactivity spontaneous transformation of one nuclide to another

Rayleigh waves surface seismic wave causing ground roll

recharge addition of water to the groundwater zone of saturation

remote sensing acquisition of information from a distance, usually employing cameras or radar operated from aircraft or satellites

resistivity resistance to flow of electrical current in a three-dimensional medium

reverse fault fault in which the hanging wall has moved up with respect to the footwall

Richter scale numerical values representing powers of ten and indicating the relative magnitude of earthquakes

rippability relative ease of excavation by ripping

salinity measure of the amount of dissolved solids in water

saltation jumping

sampling selection of a small amount representative of a greater amount

seismic intensity *see* intensity, seismic

seismic magnitude *see* Richter scale

seismic sea waves *see* tsunamis

shear stress acting parallel to a real or imaginary surface

sheeting jointing approximately parallel to the earth's surface

sill tabular, comformable, intrusive, igneous rock body

sinkhole funnel-shaped depression in the surface leading downward to a cavern

slickensides striated and polished surfaces resulting from friction in a fault plane

slide downslope movement of a cohesive rock or soil mass

slump movement of a cohesive soil mass on a spoon-shaped failure surface

soil mechanics *see* geotechnics

solid solution solution of one solid in another, usually by ionic substitution in minerals

solidus boundary on a temperature-composition diagram above which liquid and solid are in equilibrium and below which solids only can exist

solifluction slow downslope flowage of soil in permafrost regions

sorting degree of size similarity of fragments

specific gravity weight of a substance with respect to an equal volume of water

stock intrusive, discordant, igneous body of moderate size

stoping loosening and removal of ore in an underground mine

strain deformation as the result of the application of stress

strain ellipsoid representation of principal strains by the length of ellipsoid axes

strath low grassland along river

stratum single sedimentary layer

stress force per unit area to which each part of a body is subjected

striae fine parallel lines or grooves

strike line of intersection of an inclined and horizontal surface

subduction descent of one tectonic plate beneath another

subsidence downward movement of the earth's surface

supergene enrichment downward enrichment of ore by descending groundwater

superposition order in which rocks are deposited above one another

surface waves seismic waves propagated along the surface

suspensoid fluid containing suspended solid particles

S wave transverse seismic body wave

synclines folds in rocks in which the strata dip inward from both sides

talus fallen disintegrated rock fragments collected at the foot of a slope

tephra solid ejecta of a volcano

ternary diagram triangular plot representing the interaction of three components

thixotropy regain of strength in dispersed clay aggregates when allowed to stand

thrust fault *see* reverse fault

tsunamis earthquake-related, very long period ocean waves

turbulence flow in which the streamlines are nonparallel

unconformity erosional surface separating younger from older rocks

vadose zone unsaturated soil zone above the water table

valley tract central portion of a stream course

vertisol *see* expansive soil

vesiculation exsolution and expansion of gas bubbles in lava

void ratio ratio of void space to solids in a given volume of rock or soil

void space space between rock particles; interstices, pores

water table upper surface of the zone of groundwater saturation

weathering collective processes that disaggregate and decompose rock

Westergaard formula relates pressure at a buried displaced point to surface load in plastic soils

worst surface most likely surface on which failure by slump will occur

xenolith foreign rock fragment incorporated in an igneous rock body

zone of aeration *see* vadose zone

CREDITS

Photos

Chapter 1
Page 14, top: Drexel Wells, **bottom, left:** Wendell Wilson, **bottom, right:** USGS.

Chapter 2
Page 23, top: W. H. Dennen, **bottom:** USGS; **pages 24, 26:** W. R. Brown; **page 30:** J. C. Ferm.

Chapter 3
Pages 48, 49: W. H. Dennen; **page 53:** USGS; **page 54:** W. H. Dennen; **page 55:** W. R. Brown.

Chapter 4
Page 74, top: V. Erickson: **bottom:** W. H. Dennen.

Chapter 5
Page 83: W. H. Dennen; **page 86:** USGS; **pages 91, 92:** W. H. Dennen.

Chapter 6
Page 109: University of Kentucky, Department of Geology; **pages 110, 113:** USGS; **pages 116, 118:** W. H. Dennen.

Chapter 7
Pages 136, 137, 138: W. L. Dennen; **page 141:** C. D. Dennen; **page 142:** USGS.

Chapter 8
Page 154: University of Kentucky, Department of Geology.

Chapter 9
Pages 174, 175: USGS; **page 182:** W. L. Dennen; **page 184:** USGS; **page 185, top:** W. L. Dennen, **bottom:** USGS; **page 187:** C. D. Dennen.

Chapter 10
Page 198: R. Bayan.

Chapter 11
Figure 11.6: W. H. Dennen; **page 217:** W. H. Dennen; **page 226:** C. D. Dennen.

Chapter 12
Pages 246, 248, 251: W. H. Dennen; **pages 256, 257:** USGS; **page 259:** W. L. Dennen.

Chapter 13
Pages 267, top: University of Kentucky, Department of Geology, **bottom:** C. D. Dennen; **page 268, top:** W. H. Dennen, **bottom:** USGS; **pages 269, 273, 287:** USGS.

Chapter 14
Pages 299, top: USGS; **bottom:** W. H. Dennen; **page 301, bottom:** C. D. Dennen, **top:** W. H. Dennen.

Chapter 16
Page 327: USGS; **pages 328, 331, 336, 339:** W. H. Dennen.

Chapter 17
Pages 357, 359: USGS.

Chapter 19
Pages 366, 367: W. L. Dennen.

Illustrations

Chapter 1

Figure 1.1, Illustration in Table 1.4, Illustration in Table 1.5, Figure 1.3, Illustration in Table 1.8, and Figure 1.5: From W. H. Dennen, *Principles of Mineralogy.* © 1960 The Ronald Press Co., N.Y. Reprinted by permission of John Wiley & Sons, Inc.

Illustration in Table 1.2: Redrawn by permission from M. J. Buerger, *Elementary Crystallography,* John Wiley & Sons, Inc., N.Y. Copyright 1956 by M. J. Buerger.

Figure 1.6: Redrawn from A. N. Winchell, *Elements of Mineralogy.* © 1942 Prentice-Hall, Inc., Englewood Cliffs, N.J. Used with permission of Prentice-Hall, Inc.

Figure 1.8: Redrawn from W. A. White and E. Pichler, Illinois State Geological Survey Circular 266, 1959. Used with permission of the Illinois State Geological Survey.

Chapter 2

Figure 2.4 and Figure 2.5: From W. H. Dennen, *Principles of Mineralogy.*
© 1960 The Ronald Press Co., N.Y. Reprinted by permission of John Wiley & Sons, Inc.

Excerpt, page 39: From N. Duncan, *Engineering Geology and Rock Mechanics,* Volume 1. © Blackie and Sons, Ltd. Reprinted with permission of Blackie and Sons, Ltd.

Chapter 3

Figure 3.10: From W. H. Dennen, *Principles of Mineralogy.* © 1960 The Ronald Press Co., N.Y. Reprinted by permission of John Wiley & Sons, Inc.

Chapter 4

Figure 4.1: After M. W. Senstius, *American Scientist,* vol. 46. Used with permission of Sigma Xi, The Scientific Research Society.

Figure 4.3: Modified from H. D. Foth and L. M. Turk, *Fundamentals of Soil Science,* 5th ed. © 1972 John Wiley & Sons, N.Y. Reprinted by permission of John Wiley & Sons, Inc.

Figure 4.5: Modified from A. N. Strahler and A. H. Strahler, *Environmental Geoscience.* © 1973 Hamilton Publishing Co., Santa Barbara, Calif. Reprinted by permission of John Wiley & Sons, Inc.

Figure 4.7: Modified from A. A. Meyerhoff in *Problems of Plate Tectonics.* © 1978 Canadian Association of Petroleum Geology. Used with permission of the Society of Petroleum Geology.

Figure 4.9: Modified from F. W. Thwaites, *Outline of Glacial Geology,* 1956.

Figure 4.12: Redrawn from V. N. Vijayvergiya and R. A. Sullivan, *Bull. Assoc. of Engineering Geologist,* vol. 11. © 1974. Used with permission of the Association of Engineering Geologists.

Chapter 5

Figure 5.3: Modified from W. B. Langbein and S. A. Shumm, *Trans. Am. Geophysical Union,* vol. 39. © 1958 American Geophysical Union. Used with permission of the American Geophysical Union.

Figure 5.8: Modified from F. Hjulstrom, in *Recent Marine Sediments,* P. D. Trask, ed. © 1939 Society of Economic Paleontologists and Mineralogists. Used with permission of the Society.

Figure 5.10: Redrawn from V. A. Vanoni, *W. M. Keck Laboratory Report No. KHR–1,* California Institute of Technology. © 1960. Used with permission of the author.

Chapter 6

Figure 6.2: Drawn from data in B. K. Hough, *Basic Soil Engineering,* 2d. ed. © 1969 The Ronald Press Co., N.Y. Used with permission of John Wiley & Sons, Inc.

Figure 6.12: Redrawn from C. F. Tolman, *Ground Water.* © 1937 McGraw-Hill Book Company, N.Y. Used with permission of the McGraw-Hill Book Company.

Figure 6.18, Source: Boston Soc. C. E. Boston, Mass. Reprinted by permission.

Figure 6.19: After R. F. Black in *Applied Sedimentation,* P. D. Trask, (ed.), 1950. Reprinted by permission of National Academy Press.

Figure 6.20: Drawn from data in S. W. Muller, *Permafrost or Permanently Frozen Ground and Related Engineering Problems.* © 1947 J. W. Edwards, Ann Arbor, Mich. Used with permission of J. W. Edwards Publisher, Inc.

Figure 6.21: Modified after C. F. S. Sharpe, Landslides and Related Phenomena, Columbia University Press, 1938. Used with permission of the author.

Figure 6.23: Redrawn from K. A. Linnel and C. W. Kaplar in *Proc. Permafrost International Conference,* Building Advisory Board, K. B. Woods *chm.*
© 1973 National Academy of Science. Used with permission of the National Academy of Science Press.

Chapter 7

Figure 7.1: Simplified from H. J. McLellan, *Elements of Physical Oceanography.* © 1965 Pergamon Press. Used with permission of Pergamon Press.

Figure 7.5: Redrawn from B. H. Ketchum, *Journal of the Water Pollution Control Federation,* Washington, D.C. © 1952 Water Pollution Control Federation. Used with permission of the Water Pollution Control Federation.

Figure 7.6a, Figure 7.17, and Extract on page 147. © 1972 C. A. M. King. Reprinted by permission of St. Martin's Press, New York. Also by permission of Edward Arnold Publishers, Ltd., London.

Figure 7.9: After R. C. H. Russel and D. H. McMillan in *Beaches and Coasts,* C. A. M. King *ed.* © 1959 Edward Arnold, London. Used with permission of Edward Arnold Publishers, Ltd.

Figure 7.10: From C. L. Bretschneider, *Shore Protection Manual* vol. 1, U.S. Army Corps of Engineers Coastal Engineering Research Center, 1973. Used with permission of C. L. Bretschneider.

Figure 7.11: Modified from C. L. Bretschneider in *Estuary and Coastline Hydrodynamics,* A. T. Ippen ed., McGraw-Hill Book Co., N.Y. © 1963 A. T. Ippen. Used with permission of the author's estate.

Figure 7.26: Redrawn after J. Dolan and J. C. Ferm, *Science,* Vol. 159, No. 3815, 1969. Used with permission of J. Dolan.

Excerpt, page 147: After H. Valentin, Die Küste der Erde, Peterman's Geog. Mitt. Ergänsungheft, vol. 246, 1952.

Excerpt, page 147: Following F. P. Shephard, *Submarine Geology,* 3d ed. © 1973 by Francis P. Shephard. Used with permission of Harper & Row, Publishers, Inc.

Extract on page 129. From K. R. Dyer, *Estuaries, a Physical Introduction,* 1973. Used with permission of John Wiley & Sons, Inc.

Chapter 8

Figure 8.9 and Figure 8.11: Redrawn from R. G. Park, *Foundations of Structural Geology.* © 1983 Blackie and Sons, Ltd., Glasgow. Used with permission of Blackie and Sons, Ltd.

Chapter 10

Figure 10.6: Redrawn from A. H. Strahler, *Environmental Geoscience.* © 1973 Hamilton Publishing Co., Santa Barbara, CA. Used with permission of John Wiley & Sons, Inc.

Chapter 11

Figure 11.5: After Seed and Idriss, 1979.
Figure 11.6: Simplified from MOSAIC, National Science Foundation, July/August 1979. After work by H. B. Seed, I. M. Idriss and F. I. Makdisi.
Figure 11.7: Redrawn after R. E. Grim, *Applied Clay Minerology.* © 1962 McGraw-Hill Book Co., N.Y. Used with the permission of the McGraw-Hill Book Co.
Figure 11.11: From A. W. Skempton and R. D. Northey, *Geotechnique,* vol. III, 1952. Used with the permission of the Institution of Civil Engineers, London.
Figure 11.23 and Figure 11.28: Redrawn from P. D. Krynine and W. R. Judd, *Principles of Engineering Geology and Geotechnics.* © 1957 McGraw-Hill Book Co., N.Y. Used with permission of the McGraw-Hill Book Co.
Figure 11.24: After L. Prandtl (1920) in P. D. Krynine and W. R. Judd, *Principles of Engineering Geology and Geotechnics.* © 1957 McGraw-Hill Book Co., N.Y. Used with permission of the McGraw-Hill Book Co.
Figure 11.29: Modified from K. Terzaghi in *Economic Geology 50th Anniversary Volume.* © 1955 Economic Geology. Used with the permission of Economic Geology.

Chapter 12

Figure 12.10: After J. R. Murphy and L. J. O'Brien, *Bulletin of Seismological Society of America,* vol. 67, no. 3. © Seismological Society of America. Used with permission of the Society.
Figure 12.11a and Figure 12.11b: Modified from Sawkins, et al, *The Evolving Earth,* 1978, MacMillan Publishing Co., Inc., New York.
Figure 12.15: U.S. Geological Survey. Modified after map in *Geotimes,* September, 1976. Used with permission of *Geotimes.*
Figure 12.19: Based on diagram from Soiltest, Inc. Used with permission of the Kurt Lesker Co., Clairton, PA.

Figure 12.22: Redrawn from A. N. Strahler, *The Earth Sciences,* 2d. ed. © 1971 by Arthur N. Strahler. Reprinted by permission of Harper and Row, Publishers, Inc.
Figure 12.24 and Figure 12.25: Redrawn from R. W. Goranson, *American Journal of Science,* Vol. 21, ser. 5. © 1931. Used with permission of the American Journal of Science.

Chapter 13

Figure 13.22: From Terstriep, et al, Illinois State Water Survey Publication, 1976. Used with permission of the Illinois State Water Survey.

Chapter 14

Figure 14.7: After G. Dickenson in AAPG Bulletin, vol. 67, no. 2. © American Association of Petroleum Geologists. Used with permission of the Association.
Figure 14.8: Modified after M. Muskat, *Physical Principles of Oil Production.* © 1949 McGraw-Hill Book Co., N.Y. Used with the permission of the McGraw-Hill Book Co.

Chapter 15

Figure 15.2: From E. A. Keller, *Environmental Geology.* © 1976 Charles E. Merrill Publishing Company. Reprinted by permission of the Charles E. Merrill Publishing Co.
Figure 15.4: After D. L. Royster in *Bulletin of the Association of Engineering Geologists,* vol. 15. © 1974 Association of Engineering Geologists. Used with permission of the Association.
Excerpt, page 315: Adapted from Edward A. Keller, *Environmental Geology,* 3d ed. © 1982 Charles E. Merrill Publishing Company. Reprinted by permission of Charles E. Merrill Publishing Company.

Chapter 17

Figure 17.2 Drawn from data in R. S. Boynton, *Chemistry and Technology of Lime and Limestone.* © 1967 John Wiley & Sons, N.Y. Reprinted by permission of John Wiley & Sons, Inc.
Figure 17.3 After R. S. Boynton, *Chemistry and Technology of Lime and Limestone.* © 1967 John Wiley & Sons, N.Y. Reprinted by permission of John Wiley & Sons, Inc.
Figure 17.4 Redrawn from J. L. Eades and R. E. Grim, *U.S. Highway Research Board Meeting, January 1960.* © 1960 National Academy of Science. Used with permission of the National Academy of Science.

INDEX